YALE AGRARIAN STUDIES

THE INSTITUTION FOR SOCIAL AND POLICY STUDIES AT YALE UNIVERSITY

The Yale ISPS Series

Agrarian Studies

SYNTHETIC WORK AT THE CUTTING EDGE

Edited by James C. Scott and Nina Bhatt

YALE UNIVERSITY PRESS

NEW HAVEN AND LONDON

This is a book in the Yale Agrarian Studies Series,
James C. Scott, series editor.

Designed by Nancy Ovedovitz and set in Times Roman type by Tseng Information Systems, Inc. Printed in the United States of America by Vail-Ballou Press, Binghamton, N.Y.

Library of Congress Cataloging-in-Publication Data
Agrarian studies : synthetic work at the cutting edge /
edited by James C. Scott and Nina Bhatt.
p. cm. — (Yale agrarian studies series)
(The Yale ISPS series)
Includes bibliographical references and index.
ISBN 0-300-08500-1 (hc) — ISBN 0-300-08502-8 (pkb.)
1. Rural conditions. 2. Agriculture and state—History.
3. Peasantry—History. 4. Rural development—History.
I. Scott, James C. II. Bhatt, Nina. III. Series.
IV. Yale agrarian studies.
HN8 .A34 2001
307.72—dc21 00-012915

A catalogue record for this book is available from the British Library.

The paper in this book meets the guidelines for permanence and durability of the Committee on Production Guidelines for Book Longevity of the Council on Library Resources.

10 9 8 7 6 5 4 3 2 1

For Marvel Kay Mansfield, the heart and soul of the program in Agrarian Studies

Contents

Acknowledgments

The chapters in this book are a distinguished but very small sample of the some 260 papers thus far presented at the Program in Agrarian Studies Colloquia at Yale University from 1991 to the present. They are selected for their quality and for both coherence and range. On grounds of quality alone, many other papers might as easily have been included. Whereas we thank the contributors represented here for offering us their work, we are doubly conscious of the rich intellectual stew provided by all our visitors, out of which we can pick only a few tasty morsels.

A handful of people were instrumental in the conceiving of this volume and bringing it to fruition. Carol Pollard first suggested the idea, helped set up the machinery to make it a reality, and shepherded it through to publication by Yale University Press. We are grateful to Donald Green, director of the Institution for Social and Policy Studies, for encouraging us and underwriting some of the publication costs. John S. Covell, our editor at Yale University Press, guided the volume through the administrative shoals with his customary professionalism and quiet authority.

This seems, finally, an appropriate place to recognize the central contribution that the manager of the Program in Agrarian Studies, Marvel Kay Mansfield, has made to all our endeavors. The first name is both real and descriptive. Kay is usually referred to as the heart and soul of the program, and this designation, too, is no hyperbole. Her generosity of spirit, her competence, and her attention to all the intangibles that make a satisfying "home" out of a mere program continue to be indispensable to everything we have achieved. We gratefully dedicate this volume to her.

Introduction

JAMES C. SCOTT

How to prepare the reader for the sheer intellectual excitement and originality to be found in this, the first collection of papers selected from more than two hundred presented to the Program in Agrarian Studies Colloquia at Yale University over the past ten years? In one way or another, each chapter opens a door to new and bracing views of the terrain. Having no paper of my own represented in the collection, I feel no embarrassment in being an aggressive tout for the work represented here. I can virtually guarantee that no social scientists or historians who read this book will fail to acknowledge that this work recasts much of what they knew, or what they thought they knew, about rural society and the processes of state-making, collective violence, collective representations, and understandings of nature, class, development, and market integration. These contributions, each excellent in its originality and synthesis, are not only an account of an intellectual breakthrough but a point of departure for work that will be reconceived in its wake. This work represents, collectively, something of an intellectual boatyard that will launch many ships.

The Program in Agrarian Studies at Yale has, in the course of the past decade, become an intellectual crossroads for much of the most innovative and daring work on rural society. Dissatisfied with the thin (if elegant) reasoning of formal, economistic theories and with the generally ahistorical tone of most work on "development" and "modernization" in the academy and development agencies, we sought, from whatever quarter, work that transcended these limitations. Although the Yale faculty incubating the program were drawn largely from the social sciences (principally history, anthropology, political science, economics, sociology, law, and forestry and environmental studies), the weekly colloquia, from which the papers collected here were chosen, cast a far broader net. On the agnostic principle that excellent and original work on rural society might be found in virtually any field, we have invited farmers, agronomists, art historians, philosophers, archaeologists, novelists, folklorists, environmental activists, poets, and World Bank officials to contribute papers to our weekly colloquium.

The major intellectual premise of the Program in Agrarian Studies is that the most valuable work on rural society—whether on Han Dynasty China, contemporary Mozambique, or nineteenth-century North America—draws on elements of popular reasoning of rural people about their environment, about economic and political justice, about markets, about agricultural practices, about household structure and family life, about violence, about customary law. Our perspective is not the radical populist belief that folk wisdom is a complete and self-sufficient alternative to economic social theory. Nor do we imagine that popular beliefs are unitary and not riven with conflicts arising from gender, class, religious, and ethnic differences, among others. The "folk" never speak with a single voice, except when states and social movements attempt to ventriloquize on their behalf, and seldom do they speak in unambiguous terms. What has, however, underwritten our perspective is the conviction that theoretical and conceptual work touching on agrarian societies cannot be profitably undertaken without encountering the lived experience of the peoples making up those societies. The Program in Agrarian Studies was established to foster just such encounters between practical reason and theory.

Such encounters are necessarily disrespectful of disciplinary and area studies turf. In this spirit, the second intellectual premise of Agrarian Studies was that virtually every inquiry into rural life, whenever and wherever situated, could be illuminated by work addressing similar problems in quite different contexts. Two brief examples will help illustrate the point. In the 1960s R. C. Cobb, the distinguished English historian of France, was completing a book on subsistence crises and food supply problems in France between the Revolution and 1820, based entirely on research in the departmental archives within the affected areas. By chance, he found himself lecturing in India at a time when food shortages and riots were common. The very processes whose dynamics he was attempting to uncover in early nineteenth-century France were there, unfolding before his very eyes. Not only that, there was a large and distinguished literature on the subject in India. There was hardly a one-to-one relationship between, say, a food riot in Marseille in 1803 and one in Orissa in 1965, but there were enough commonalities and diagnostic differences to allow Cobb to understand more fully what was distinctive and what was generic about French subsistence crises. A second example may be drawn from my own work. In the course of my fieldwork in a Malay Moslem village, I uncovered a high level of resistance to paying the *zakat,* the Moslem tithe, consisting, for farmers, of one-tenth of the harvest. After mapping the contours of this resistance, the moral reasoning behind it, and the techniques of resistance, it dawned on me that the zakat and the Catholic tithe in France were essentially the same tax (both the Christian tithe and the Moslem zakat deriving from the Hebrew tax system) and that there might be a literature on the subject. As it turned out, Le Roy Ladurie and other exponents of *la longue durée* in historiography had written extensively about the tithe. I thus had arrayed before me the statistical profile of tithe (*la dime*) collection and resistance for nearly five centuries. Furthermore, I had what Le Roy Ladurie did not have: firsthand accounts of scores of peasants explaining why they resented the zakat and how they largely managed to circumvent it successfully. Once again, the cases were different in crucial ways. But those differences as well as the parallels, swirling around a history of collective tax evasion, permitted me to make educated guesses and to pose issues that would not have arisen out of a single case.

None of the chapters in this book are explicitly comparative case studies. Nevertheless, they have all been deeply touched by comparative reading across regions, across historical periods, and across disciplines. They are, in other words, analyses conducted in the shadow of a broad interdisciplinary encounter. The imperial past, and present, of Western scholarship has generally ensured that scholars from Latin America and the former colonies would be conversant with much of the Western literature touching on their fields of interest by virtue of their training and the structure of their disciplines. Western scholars, by contrast, have typically been more "provincial," if only because it was easy to assume that their own work was central by definition. Thus it is that, speaking broadly, Western scholarship can benefit even more in its breadth and imagination from the encounter with non-Western scholarship than the other way around.

The first two chapters, by Peter Taylor and Hermann Rebel, respectively, go a long way toward showing how states and legal systems in Central Europe have shaped the peasant household and its strategies. Whereas the household was once taken as the "natural" point of departure by social analysts of rural life, these chapters show that the state's fiscal strategy, and its interest in creating viable taxpaying units and subjects for conscription, have greatly inflected the productive and reproductive strategies of every household. Just as historians have shown us that the Russian or the Javanese village was, in large part, an artifact of the state fiscal apparatus, Taylor and Rebel show us that the "peasant family" is no more a natural unit than the "peasant village." Taylor's analysis of eighteenth-century Hesse is a mesmerizing blend of demography, fiscal history, law, and peasant production brought finally into vivid relief by a striking reading of the fairy tales collected by the Brothers Grimm during that period. Rebel's analysis of Austria is more than a "denaturalization" of the peasant family; it is a chilling account of how state policy and peasant strategies combined to systematically marginalize, to the point of negligence and murder, that segment of the rural population which threatened the dynastic strategies of the tribute-paying household. Rebel suggests that this "disorder-necessary-for-order," abetted by state institutions flying false flags of "welfare institutions," prepared the way, organizationally and psychologically, for the lethal marginalizations of the Jews and Gypsies under National Socialism.

Each of the following two chapters represents an important synthesis of agrarian change in Western Europe from the seventeenth to the nineteenth centuries. Peter Jones lays out the revolutionary legacy for property and class relations in France. In particular, the question of whether the Revolution in rural France represented a defense of common property or a true bourgeois revolution, establishing untrammeled private freehold property of land, is assessed with great care as well as a daring attempt at an overall view of the peasantry's role in the Revolution. Here a whole series of facile conclusions is replaced by evidence showing the difference between the state legal codes, on the one hand, and agrarian practices at the local level, on the other; the revolutionary legacy cannot be read directly off the decrees and legal codes emanating from Paris. Nor can the defense of common property by smallholders, tenants, and laborers be automatically assumed; once the stampede to divide the commons began, the small fry wanted their patch, too. Liana Vardi's quarry is nothing less than the representation of the peasant in French, Flemish, and English painting and poetry from the sixteenth to the nineteenth centuries. Here the issue is not the facts of production, class, and labor, but the social images of the countryside and rural life. She shows, with powerful images of the harvest, the transition from representations of exertion and communal labor, to classical images from Virgil and Ovid in which physical labor, when it appears at all, is more domesticated and more celebratory, to broad vistas of landscapes in which only a few docile, virtuous peasants (if any) make an appearance. Above all, Vardi connects these shifts in the social imaginary to the changing fear of rural unrest and the changing view of the peasantry: a view in which they were less feared as brutal, insurrectional savages and more cast as domesticated, folkloric guardians of an Edenic past.

The theme of common property figures again in Paul Greenough's "*Naturae Ferae,*" the first of three chapters examining agrarian themes in South Asia, for some time the site of some of the most imaginative work on class, environment, and demography. Greenough's chapter, along with the subsequent chapters by

David Arnold and David Ludden, sets out to overturn one of the reigning narratives of the subcontinent's agrarian history. In this case, the orthodoxy he wishes to unhorse is nothing less than the standard postcolonial narrative of environmental history: of a peasantry whose practices were harmonious with sustainable resource use until disturbed by colonialism and market forces. It is, he aptly notes, a secular, nationalist version of the Fall in Christian mythology. This narrative is now under attack on several fronts, but Greenough's assault is particularly well documented. He shows that the peasantry, with good reason, feared tigers and undertook to trap them, kill them, destroy their habitat, or drive them away lest they menace cultivators and livestock. Failing that, cultivation and grazing were abandoned. Preservation of tigers and their habitat was the farthest thing from their minds.

The idea of a "wild frontier" so prevalent in colonial historiography is, in turn, the target of David Arnold's paper on disease and resistance. Arnold proposes that the frontier in colonial India was not some imaginary line between civilization and wilderness but rather a real epidemiological barrier defined by disease vectors, particularly malaria. Valley and hill were ecologically and biotically distinct, and the susceptibility of valley populations to diseases to which hill populations were relatively immune formed a physical barrier to the colonization of the hills. Only clearance and deforestation, or the use of DDT by the mid-twentieth century, could break this frontier decisively. Arnold's concept of frontiers defined by disease vectors might, if further developed, reshape our views of civilizational discourses, not only in India but elsewhere. It would, for example, comport well with the literature on the advancing and retreating frontiers defined by the tsetse fly in Africa.

David Ludden launches a broadside critique of the intellectual movement called Subaltern Studies. Inasmuch as Subaltern Studies—itself a critique of much colonial and nationalist historiography—has become influential in African and Latin American studies, the intellectual stakes here are substantial. Ludden argues forcefully that the stark dichotomies characteristic of Subaltern Studies work—dichotomies between subaltern and elite, official and vernacular, modern and traditional, landlords and peasants—cannot hold up, as they ignore most of the subtle variation introduced by kinship, caste, and patron-client relations. Above all, he argues, such dichotomies miss the great differences in production regimes between the wet, densely populated, rice-growing coastal areas of India, on the one hand, and the drier, interior, millet-and-wheat areas on the other: differences that foster greatly divergent class structures and property regimes. Whereas Ludden may overlook the signal contribution of Subaltern Studies to reading colonial documents "against the grain" and the radical revisions its exponents have forced to nationalist-triumphalist histories, his critique is a bracing attack against views of the Indian village as a stable, isolated, pre-commercial, timeless sort of place.

The last section of the book brings together considerations of broad theoretical import, seen through the lenses, respectively, of classical political economy (Ronald J. Herring), a reconfigured neoclassical economics (Herman E. Daly), and long-duration micro-history (Hans Medick).

Just as no analysis of agricultural economics is plausible without reference to the household labor analysis of A. V. Chayanov, no rural political economy is plausible without reference to Karl Polanyi's *The Great Transformation.* Ronald Herring examines Polanyi's analysis in the course of a subtle and theoretically astute account of struggles over the commodification of land and labor in Kerala, India. Rejecting Polanyi's reification of society and his romanticization of premarket society, Herring nonetheless shows the great value of Polanyi's account of political efforts to restrict injurious market outcomes. Farmers, tenants, and laborers each strive to invoke moral economies or market logic when it suits their interests. Each struggle over how markets are to be regulated is, in turn, vitally affected more by the "historical origins and institutional structure of any particular economy" than by some abstract logic of market relations per se.

The relationship between the economy (the market economy, that is) and the larger society surrounding it is the subject of Herman Daly's powerful critique of neoclassical methods of measuring growth, income, and welfare. It is a critique both from within neoclassical parameters and, then, from without. A more convincing and brief case for sustainable resource policies would be hard to find. The distinction between "cultivated natural capital" (such as shrimp farms carved from mangrove swamps, or eucalyptus plantations), on the one hand, and "natural capital" (such as "wild" shrimp or unmanaged, old-growth, mixed forests), on the other, marks, I think, a radical shift in perspective.

As Daly remarked in an earlier version of his chapter, "At the limit, all other species become cultivated natural capital, bred and managed at the smaller population size to make room for humans and their furniture. Instrumental values, such as redundancy, resilience, stability, and sustainability, will be sacrificed, along with the intrinsic value of life enjoyment by sentient subhuman species, in the interests of efficiency defined as anything that increases the human scale. One imagines, in this vision, a vast zoo in which the only species whose habitat and breeding are not 'managed' is Homo sapiens."

Ending this volume with Hans Medick's spirited defense of micro-history is entirely fitting. So much of the great and lasting work on rural society—work with theoretical bite—has taken, as its point of departure, the holistic, detailed, precise, and often deep historical study of particular places. Medick, in his work on Laichingen, Germany, as well as in his earlier work on proto-industrialization, shows that the analysis of closely observed microcosms, far from being a flight from theory, can be designed to answer large questions in theoretically powerful terms.

Writing on behalf of the graduate students and faculty associated with the Program in Agrarian Studies, I wish to acknowledge our large intellectual debt to the scholars represented here for allowing us to demonstrate the imagination and theoretical reach of contemporary rural studies.

PART I *State Formation and Peasant Histories*

CHAPTER ONE

Some Ideological Aspects of the Articulation between Kin and Tribute: State Formation, Military System, and Social Life in Hesse-Cassel, 1688–1815

PETER TAYLOR

In 1962 Otto Büsch wrote about "social militarism" in order to describe how deeply eighteenth-century Prussian military institutions became entwined with German structures of authority at every level of society.[1] Fifteen years later Michel Foucault connected the same developments in the entire Western world to a new technology of power he called "discipline."[2] Much of the later work of Gerd Oestreich has elaborated on this theme for early modern territorial states in Germany.[3] Despite the fruitfulness of this work, much of it remains limited to the dimension of what power holders intended and what they accomplished. Particularly true of Foucault's account (where even the forms and substance of "resistance" are controlled by the forces of discipline), this one-sidedness obscures the extent to which subaltern peoples adapted and adjusted their own ways of living to better fit and take advantage of what authorities were doing. The work of this chapter is to begin to rebalance the perspectives of social militarism.

An example of the adaptation and adjustment to which I refer is noted by Hermann Rebel and appears in Theodor Adorno's comments on Eugen Kogon's observation that the cruelest guards in the German death camps could be identified as the dispossessed sons of peasants from Eastern and Central Europe.[4] This Adorno attributed to the great cultural gap between town and country, and he called for a "debarbarization of the countryside." For him the barbarism of rural life was not the result of any failure to "modernize," rather it stemmed from the need of a capitalist society to provide an image and goal of "shelteredness" for workers in urban society. The ideological re-creation of the peasant family as a "stable" and "sheltered" world of *gemütlichkeit,* that is, an imagined escape from *anomie* in modern life, required, according to Adorno, that "surplus" children be cast out and cut off, that they be tragically sacrificed to the economic viability of rural households squeezed between state domination and economic exploitation. Thus, a peasant tendency to disadvantage some children as they ordered work and property through kinship was intensified by articulation with urban capitalism. According to Rebel, the historical experience of these sacrificed children resulted in "their exercise of pathologically debased and yet officially authorized forms of power."[5] In this chapter, by telling the story of the emerging German territorial state of Hesse-Cassel between 1688 and 1815, I discuss a similar ideological articulation between groups that organize work differently. More generally, this account represents an exploration of the process of state formation as a process of articulation and rearticulation between a kin-ordered peasant society and a tribute-taking state.

The concept of "articulation" comes from theoretical work in Marxist social theory. In shifting the attention of Marxist social analysis back to social formations and modes of production, Eric Wolf made one of the most important theoretical contributions of the 1980s.[6] What made this focus so interesting was the uncoupling of modes of production from set historical progressions (that is, feudalism to capitalism to socialism) so that it became possible to see historically existing social formations as encompassing different groups of people practicing different means of mobilizing, organizing, and distributing the benefits of social labor. This "articulation" of different modes of production within a single social formation enables a highly complex form of class analysis. The analyst may now see class formation as occurring simultaneously along as many axes as there are identifiable modes of production. Further, we now have the theoretical language to speak about the relation between classes practicing different organizations of social labor and to speak about individuals experiencing multiple class positions simultaneously. These class positions may either reinforce or contradict one another. Finally, these insights allow a far more sophisticated understanding of the relation between ideology and social formation—an understanding that escapes strictly determinist base-superstructure analogies and concepts of hegemony, which see dominant ideologies as products of single dominating classes.

Social Formation and State Formation in Hesse-Cassel, 1688–1815

By 1688 the process of state formation had been under way for more than a century in Hesse-Cassel. Although the rulers, or Landgraves, of the territory still acknowledged the nominal overlordship of the Holy Roman Empire and some of its legal and military institutions, they had established their rights to collect military taxes, had successfully claimed a monopoly on the military services of their subjects, and had created an officialdom who owed primary loyalty to them. Within their territory they still shared the aspects of sovereignty with a landed aristocracy that occasionally practiced demesne farming (*Gutsherrschaft*) but primarily lived from collected tributes as judicial overlords (*Gerichtsherrn*), as landlords (*Grundherrn*), and as important creditors of the Hessian peasantry.[7] The aristocracy possessed a legal claim to control the amounts and means of taxation which it expressed through the Hessian Diet (*Landtag*). Many landed aristocrats added to their income, influence, and power by serving as officials in the Landgrave's government or as officers in his army, thus complicating the loyalties of his officialdom.

After 1688, Landgrave Karl accelerated the pace of state formation, thereby decisively eroding the corporate powers of the aristocracy. This gradual shift in balance resulted from the Landgrave's intensified participation in a developing market for trained military units. On the basis of "subsidy treaties," many princes of small and medium-sized territories (including both Austria and Prussia) received payment from Europe's great dynasties to raise, to equip, to train, and to officer many of the military units deployed in eighteenth-century warfare. The year 1688 saw Landgrave Karl begin a "subsidy" relationship with the English crown, which lasted until England withdrew from continental fighting in 1815. Here I shall not address the shifting articulation between the English and Hessian states but instead will focus on the changing relationship between the Hessian state and the peasant society that provided both the men and the wealth that made this system work. It is sufficient to say now that the subsidy markets drove those changes in social formation and ideology that are the subject of my account.

The wealth that the Hessian state acquired from the market in military units provided not only a motivation for further state formation but also financial means to do it. First, Landgraves used subsidy wealth as a patronage tool to forge a political consensus among Hessian elites in support of this peculiar set of institutions. The Hessian landed aristocracy continued their nearly unquestioned support of subsidy institutions well into the 1770s even though such practices eroded the Diet's control of the Landgrave's budget.[8] They did so because the military system provided positions for their sons, credit for themselves, and general tax relief for the aristocracy. They cooperated by putting their own estate administrative apparatus at the service of the state for the purposes of recruitment and tax collection.

Second, subsidy wealth provided the means for the frequent administrative elaborations that recruitment and taxation among the Hessian peasantry required. In some measure, the political and financial independence gained from subsidy wealth made such reforms possible. Thus, the Landgrave's participation in the international subsidy market substantively influenced the three crucial dimensions of state formation: administrative elaboration, intensification of taxation, and the growth of military institutions.

Hesse-Cassel's peculiar path to state formation provoked little overt peasant resistance. Only during the War of the Spanish Succession and the Seven Years' War did officials report any substantial disobedience to recruitment and taxation measures. Despite participation in wars in every decade after 1730, only occasional reports surfaced of a slow hemorrhage of poor men fleeing to avoid military service. A new efficient tax system after the 1740s and 1750s and the formal introduction of conscription in 1762 seemed to ease the most obvious tensions by reducing favoritism among officials. Only in 1773, when the state intervened to forestall adjustment of peasant dynastic strategies concerning conscription, did the system come unglued enough to precipitate a crisis among both aristocrats and peasants. William IX resolved the crisis caused by his father only by completely overhauling the military system in the 1780s and 1790s.

The social impact of articulation between kin-ordered producers and tributary groups was most visible in the way Hessian peasants adjusted their dynastic strategies to the military system. Hessian practices of administration, recruitment, and taxation all seemed to fit well enough into the dynastic strategies of important villagers—peasant officials, holders of large tenures, church elders, and wealthy "retired" peasants—so that these members of the peasant elite continued to work with judicial overlords to deliver recruits to the militia and the standing army, and to enforce taxation. Precisely how some peasant dynasties found advantages in the earlier stages of subsidy state formation remained shrouded in the oblique explanations of reforms made by Frederick II (1760–1785) and his predecessor, William VIII (1730–1760). These laws implied that past administrators, their subalterns, and peasant village officials did not function without "partiality . . . or from other moving causes."[9] Rather, they turned administrative activity into a hierarchy of patron-client relationships that shifted the burden of taxes and recruitment onto poorer and less influential peasant dynasties. Thus, as in the system of domination (*Herrschaft*) that David Sabean has described for Württemburg, authorities exchanged protection (often from their own administrative actions) for obedience.[10] Reform carried out under the auspices of "rationalizing" Enlightenment language sought to attack this mode of domination and replace it with another less partial and more efficient one. Throughout the eighteenth century such reforms foundered on the shoals of an administrative paradox—that is, in order to carry out the reforms, reformers needed to invoke the authority of the very system of patronage they hoped to neutralize.[11] Despite the contradictions, in 1762 administrators created a wealth-based system of military service exemptions, which partly neutralized the operation of past patronage networks. In instituting administrative reforms, authorities posed as protectors of the needy, who, they claimed, bore an unfair burden of taxation and recruitment.[12]

In response to the recruitment reform of 1762, rural heads of households married their children to different kinds of people and practiced partible property devolution more frequently than in the past.[13] Their adjustments to the institution of selective conscription varied according to the classes of wealth established by the law itself. Peasants with three hectares or more were themselves exempted from service. Those who held eighteen or more hectares could exempt a single son whom they designated as their sole heir. Members of the peasant elite, whose heirs the law exempted, allowed their other sons to be drafted, whereas before they had sought marriages for them. Peasants without sufficient wealth to win exemptions for heirs began matching their sons earlier than in the past. In order that these young grooms acquire military exemptions, farmers divided their estates and bought the necessary plow teams to work them. These farmers needed to keep their sons close to home because they could not afford to hire farm labor as frequently as wealthier villagers could. Thus, the lesser peasant dynasties offered some manipulation of, and resistance to, state strategies of recruitment.

The patterns of adjustment to the 1762 law contributed to a crisis of intimate authority in the households and villages of Hesse-Cassel. Among lesser households, parents accomplished division of estates only by reducing the dowries of daughters, sacrificing the economic viability of their farms, and burdening their estates with debt. New couples on the mini-estates became increasingly dependent on their own wage labor and that of their children and so adopted reproductive strategies that maximized the number of births. The elder generation had lost an effective sanction that had kept land-household ratios in relative balance for more than fifty years, and as a result, population grew for more than a century.[14]

Moreover, past marital practice had always meant that some brides from less wealthy dynasties married above their stations to the cadet sons of the peasant elite. This marriage planners accomplished as they concentrated credit and other resources to provide attractive dowries for their daughters. Such plans could no longer be easily enacted, as sons commanded a greater share of dynastic resources to acquire draft exemptions. Additionally, wealthy parents with cadet sons no longer provided these men with sufficient marriage portions to purchase an exemption-giving farm. Marriage alliances between such children had been a way of extending the influence of wealthy families over more property, while poorer families acquired protection and work in association with the wealthier family. With the institution of administratively "neutral" taxation and recruitment processes, the influence of the wealthy diminished in value, but such marriages became more risky for poorer parents as well. Cadet sons from wealthy households became specific targets of recruiters, and once a man entered military service, his regimental commander froze his assets, holding them as a surety against desertion. To wed a soldier or a potential soldier always involved some risk that the wealth of the bride's family could be frozen. Wealthier parents found that letting the dispossessed brothers of their heir enter the army delayed the need to provide large marriage portions for them, provided a potential source of income, and removed disgruntled presences from households.[15] In pursuing this line of adjustments, peasants diminished the number of affinal ties between the wealthy and the poor and thus threatened the networks of authority on which they had previously based labor, marriage, and credit relationships. This prepared the way for sharpened class divisions within peasant communities and dynasties.[16]

The state could not sustain these arrangements and still maintain its strong position in the subsidy market. In 1773 Frederick II, aware that many peasants used the selective service provisions of the draft law to avoid service in the subsidy army, sought to regulate the processes of peasant marriage and property devolution to forestall such adjustments. Crisis resulted when this naked aggression against the fundamental operation of kin-ordered production combined in 1776 with massive military recruitment to fill the terms of the subsidy treaty with the English. Nearly nineteen thousand Hessian boys went to help win the "hearts and minds" of rebellious colonists in America. Despite the surge of income from subsidies, aristocrats and their subaltern officials claimed that the new inheritance practice violated natural law, disrupted military recruitment, and damaged the common good. They pointed to deranged property relations and further recruitment inroads after 1777 as causes of farm bankruptcies, peasant flight to avoid the draft, labor shortages, credit shortages, and collapsing land values. Further, privileged peasant officials began to withdraw their cooperation from the Hessian administrative system by failing to show up for quarterly village court sessions and by showing increasing disrespect for their aristocratic patrons who could no longer protect them from the military system. Administration itself became increasingly difficult despite attempts to buy the loyalty of the rural population with tax remissions and that of the aristocracy with patronage loans and tax-farming concessions.[17]

The social formation never fully recovered after the crisis years of the American campaign (1776–83). Aristocratic criticism intensified in the remaining time the system existed. William IX had to rebuild his administrative apparatus without relying on the patrimonial officialdom of his father. Although peasant elites managed to recover some of their authority in families and villages, population continued to grow out of control—a problem mitigated only by three massive waves of emigration beginning in the mid-nineteenth century.

In the pages following this truncated summary of the material articulations of kin and tribute in Hesse-Cassel, I explore in much greater detail how both peasants and aristocratic officials talked about this social formation. In particular, I point out that even in proposing strategies of resistance or offering implicitly critical assessments of administrative practice, both peasants and bureaucrats wove a cultural hegemony that cast out as useless pariahs substantial groups in the Hessian population. It is my belief that in these ideological articulations between kin and tribute we see a systematically distorted language similar to that which attracted dispossessed peasant sons to the fascist enterprises of the twentieth century.

The Military System and the Ideology of Connectedness in Peasant Oral Tradition

Hessian landed aristocrats produced their own stories about how peasant heads of households aided the Hessian military system and, as we would say, articulated with a kin-ordered group. One aristocrat spoke of "the farmer who bears arms" who "tells the son his adventures," making "the lad eager to tread in the footsteps of the elder," and of the boy who "trains his feeble arms early" so that "he is quickly formed into a soldier."[18] Similarly, another aristocrat claimed that peasant sons were not scared by military duty and that "in most cases the soldier's oath makes their youth vigorous and cheerful."[19] This account of a militarized peasant culture hegemonically smoothing recruitment and training was surely an oversimplified fantasy. Other sources tell us that the recruitment process and soldiers' oaths depressed and terrified youths in the rural population, turning them into sullen and stubborn subjects.[20] Military training manuals, with their recommended petty brutalities and endless dull exercises, clearly indicated that officers who produced them believed that peasant children remained a refractory material, which they could reduce to military discipline only with great difficulty. They recommended the development of barracks to separate recruits from their social context more completely. Enlightenment bureaucrats even saw the army as a great pedagogical machine to reform stubborn natures and rehabilitate even hardened criminals.[21]

To provide a more critical account than the fantasies of some social romantics requires careful attention to what peasants themselves said about the military system. Yet few people recorded peasant speech. When they did, the transcripts of peasant discourse were systematically hidden (distorted) in a variety of ways.[22]

Peasants' words were mediated through officials who did not share their experience or who would not understand them unless they petitioned in the distorted language of deference. The most obvious hidden transcripts of Hessian peasant discourse are the fragments of Hessian oral tradition recorded and published by the Brothers Grimm beginning in 1813 as *Die Kinder und Hausmärchen der Brüder Grimm.* Even with this source we must face the possibility that the relations of domination endogenous to Hessian peasant society added a measure of ideological distortion to the translucent veils of symbolism that hid peasant meanings from their rulers. That the language of family and love is not free of elements that prepare the ground for historical victimization has long been recognized by anthropologists.

Claude Lévi-Strauss and Eric Wolf, among others, have observed that ideologies of groups practicing kin-ordered production center upon the fundamental distinction between groups whose members are connected by descent (consanguines, however defined) and members of groups who may become connected through marriage (potential and actual affines).[23] David Sabean has found these categories at the center of much peasant discourse in the villages of Württemburg in the three centuries before 1800. Sabean observed three crucial categories—enemies (*Feinde*), friends (*Freunde*), and relatives (*Verwandte*).[24] Enemies and friends were, respectively, potential affines and affines, that is, the group of marriageable people. Relatives or, in the nomenclature of Hessian peasants, blood friends (*Blutfreunde*) were covered by the incest taboo enforced by religious authorities.

Sabean's peasants said that they expected certain kinds of danger from nonrelatives, including intentional aggression and envy, which seemed appropriate to competition among equals. The danger represented by relatives and recognized in peasant discourse could only be unintentional—a kind of genetic pollution—because the opposite of hatred (friendship or love) was appropriate to relations between members of this group. Peasants saw marriage as a way to (only) partially diminish the dangers of enmity by converting enemies into friends from whom one could expect support. In discourse, as in the practice of kin-ordering production, marriage played a central role for German peasants.

Thinking in terms of enemy, friend, and relative laid the groundwork for the experience of historical victimization because of ideological limitations and blind spots that obscured tensions within the contexts in which peasants told stories. As Wolf has suggested, the language of affinity and consanguinity—the language of connectedness—has great difficulty with the unconnected. Those people we call orphans and strangers.[25] These limitations became crucial because the bureaucratic process of recruitment, the legal status of soldiers, and the concrete experiences of military life changed peasant boys into disconnected strangers and even functional orphans by denying them the normal courses of social interaction like credit, property, and marriage. Compounding the direct operations of the military bureaucracy, related state intervention in the process of peasant inheritance extended the difficulties to soldiers' sisters. Peasant tales provided both an exploration of the meaning of militarily related disconnectedness and some suggestions for relieving it. At the same time, they also revealed a pattern of thinking through which Hessian bureaucrats could expand and intensify an already present victimization.

The discourse of peasant oral tradition envisioned military service as a tragic condition from which only the sisters of soldiers could rescue them. The Brothers Grimm collected a number of folk tales from Marie Müller and Dorothea Viehmann, women of rural backgrounds who belonged to the generation that provided the bulk of soldiers for the campaigns of 1776–83.[26] Three relevant tales have received much attention.[27] "The Three Ravens," "The Twelve Brothers," and "The Six Swans" all share the same fundamental structure—young brothers magically transformed from human to animal form and then rescued by their sisters' performance of a magical ritual of renunciation.[28] The similarity of structure suggests that these are improvisations on a theme which juxtaposes two worlds: one of normal family life and one of a world transformed where "people become . . . birds, 'unstable male cosmogones,' whose destiny is to be socially disconnected."[29] Performers Müller and Viehmann connected this particular state of social death and disconnectedness with military status through the symbolism of the birds into which the boys are transformed.[30] In Müller's versions the boys become ravens or swans, and in Viehmann's they fly away as ravens. In Teutonic mythology, ravens accompany Odin, god of war, to the field of battle to feast off the dead, and swans are the Valkyrie, another soldier symbol.[31] With these similarities and connections in mind, we turn to the variations in the tales, which appear to be meditations on the reasons for disconnection, on the processes by which it occurs, and finally on the meaning of existing in the world of the socially dead.

In all three versions of the tale, magical transformation comes out of the context of family conflict, anger, and fear. In the oldest extant text, an angry mother curses her three sons for having played during Sunday church services, and as a result "they were turned into ravens and flew away."[32] It is difficult to tell from the language of the tale whether the transformation of her sons was a punishment of the woman for her anger, or whether it was punishment of the boys for violating sacred ritual. The use of passive construction in the quoted sentence enhances the belief that the woman did not intend to transform her sons and that they became functional orphans not because of the enmity of a relative but because of pollution of either kin relations or sacred ritual. The lack of harmful intent is even more explicit in the 1819 version of the tale, in which a father curses his sons for breaking a pitcher they were supposed to use to carry baptismal water for their newborn sister.[33] Although the father wishes out loud for the transformation in the moment of anger, the tale implies immediate and deep regret over the consequences of that expression. Further, the parents' own shame at what has occurred leads them to conceal the existence of the brothers from their sister. She discovers them only through the gossip of neighbors who blame her for what happened to them. But here again, the change is treated as a kind of misfortune that lacks any intent on the part of those responsible. These narrative constructions only barely stay within the rule that consanguines may not intentionally harm one another.

In a far more complex variant, "The Six Swans," old Marie told of the magical metamorphosis as the result of a premeditated act of a stepmother who sees the children of a past marriage as a threat. Although the wicked woman converts the six brothers to swans by throwing magic shirts upon them, their sister escapes notice, remains unharmed, and is left to cope with the consequences of the militarization of her brothers. The success of the stepmother represents the failure of the father's attempts to protect his descendants from her maliciousness. Through the result of an intentional act and conscious enmity, this orphanage remains within the boundaries observed by Sabean because a stepmother was seen as no real relative. In this view, her action is structurally determined from the stance of the children because she is neither an affine nor a potential affine (she is already married) and so her enmity cannot be converted to friendship. The aid expected from affines was the reason the father married the woman in the first place. He needed his present mother-in-law's aid to help him find his way out of a dark forest. She, in turn, extracted the promise that he marry her daughter, something which made him deeply uneasy.

Whereas Marie's stories stayed barely within the boundaries of the ideology of connectedness, Dorothea Viehmann's treatment did not, but finessed the issue in another way. In "The Twelve Brothers," a father actually expresses the intention to kill his twelve sons so that his newborn and long-awaited daughter may inherit his patrimony. However, he neither kills the boys nor transforms them into avian form. The boys' mother warns them, and they flee the conflicted family situation to hide in the forest. When their sister later discovers their existence and searches them out, her act of picking flowers out of their garden accomplishes their transmutation into ravens. Unlike the father, the girl in this circumstance has nothing but the best intentions. Although Frau Viehmann touched on the ideologically dangerous idea that disinheritance (orphanage) and military service were related to choices that parents made between children, she did not carry through an analysis in which blood kin could be fatally and intentionally divided by interest and emotion. The sister's unintentional act is made to bear the burden of tragedy.

In all versions of this story, boys are cast out of families, and, orphaned, they become strangers to their relatives. As Hermann Rebel and I have already pointed out, the position of these narrative characters replicates many aspects of the experience of Hessian draftees dispossessed by inheritance practice and later law and, as a result, subject to military service as "the household's most expendable people."[34] Although clearly parental decisions to advantage one child over others played crucial roles in the process of subjecting some to involuntary military servitude, none of the stories can finally confront this terrible fact directly. Story texts contain ambiguous language about intent, or ambiguous constructions of consanguinal and affinal relationships, in order to avoid the issue. Even when the harmful intent of choosing between children entered Viehmann's variant directly, she finessed it and gave it no large role because another blood relation intervened to deflect that intent. Through avoidance, the hegemony of connectedness reigned, giving few conceptual handles to transform social relations of the kin-ordered mode of production.

Although the tales fail to provide the conceptual apparatus to face and transform relationships between kin, they do provide insight into the meaning of disconnectedness for those who suffered it. The avian symbolism of the transformations, and the dark forests or glass mountains in which the transformed male children live, provide insights into the ways military orphanage might be perceived. In Germanic popular culture, birds were symbols not only of soldiers but also of outlaws. Medieval and early modern legal sentences proclaimed traitors and other serious criminals to be "free as birds" (*Vogelfrei*), meaning that they no longer enjoyed the protection of law and might be killed with impunity.[35] This "freedom" condemned the disconnected to live in the social forest of family trees (*Arbores Consanguinitas*) in "The Twelve Brothers" and "The Six Swans." As birds, they might fly among the trees, perch in them briefly, wander among their trunks, but they could never be of their substance or draw on their resources for protection.

The condition of being physically present in a society of families but without the connections to it or membership in it is most clearly symbolized in Marie's story "The Three Ravens." The transformed children are condemned to live not in the forest but locked in a glass mountain. Thus, imprisoned behind walls through which they could see and be seen but through which no human contact could be made, they exist in a state physically present but socially absent. No better representation of social victimization can be made, and no clearer picture of the meaning and experience of militarily caused disconnectedness may be drawn.

Although the discourse of peasant folktales could not directly confront the culpability of the directors of kin-ordered production in the victimization of peasant children, it did address issues related to rescuing the boys from the state of disconnectedness. In all three stories their sisters reconnect them to the world of families through manipulation of the rules of property devolution and marriage. Sisters perform symbolic transfers of property, undertake vows of silence, and marry to effect the salvation of their brothers from military servitude. Such acts seem to reflect some of the shifts seen in peasant marriage and devolutionary behavior in response to the extremes of the Frederician military system. Although these women become heroines through their acts of self-sacrifice and renunciation, so constructing their acts blurs the source of their own victimization and reveals a tension between the value of loyalty to parents and lineage and that of loyalty to siblings. However, the stories are clear that women are to bear (bare)—more fully than even their brothers—the ideological contradictions of the Hessian military system.

In "The Three Ravens," the self-sacrificing heroine traveled to the ends of the earth to find a cure for her brothers' removal from normal human existence. Before she literally found the key to her heart's desire, the girl confronted the sun and the moon, both of whom were consumers of human flesh and who offered no help. That sun and moon consumed the flesh of children seemed reminiscent of those rulers who took young boys to serve involuntarily in their armies. It seemed appropriate that the solution to the boys' retransformation did not lie in appeals to higher authorities but in the symbolic manipulation of devolutionary rules. The stars forced the moon to give the girl a chicken-leg bone (*Hinkelbeinchen*), which unlocked the glass mountain in which the boy-ravens were imprisoned. In the version of the tale discussed below, the chicken bone is replaced by the girl's finger bone. The joints of the human body had long been used to calculate devolutionary paths in Germany.[36] It seems reasonable to assume that this narrative suggested the manipulation of inheritance as a key to releasing the brothers from their condition. The sister simply went to the glass mountain and unlocked it with the bone key and left a ring on the ravens' table. The retransformation of the brothers occurred when the ravens recognized their sister and her love. In the ritual exchange of wedding ceremonies, the ring was a sign of what each member of the couple brought to the marriage and gave to the other. It is unclear whether the woman, in giving the ring to her brothers, renounced marriage or simply her dowry.

In the 1819 variant of the tale under discussion, "The Seven Ravens," Marie Müller's symbolism is much less opaque. First, the bone key is not the gift of "the stars" but of "the morning star," Venus. This clearly associates it with the

Germanic custom of "the morning gift" (*Morgengabe*), which, in the nomenclature of Germanic inheritance custom, was provided by the groom's family to the bride the morning after the marriage was consummated. Second, the sister loses the gift, and instead of unlocking the glass mountain with it, she must cut off her own little finger to perform the task. Losing the gift stands as a loss of the benefits of marriage itself, while cutting off the finger implies a renunciation of a marriage portion from her own family.

This renunciation is consistent with the altered patterns of marriage that were part of the peasant response to the draft law of 1762. Some parents diminished their daughters' opportunities for a good match in order to acquire exemptions for sons. In some cases the renunciation of dowry resources meant that the prospects for marriage became seriously restricted, if not lost. But once again, it appears that causal agency is blurred. Although heroic acts of self-sacrifice were not inconceivable, it seems more likely that such sacrifices were enforced by those who arranged the patterns of property devolution and, ultimately, through the inheritance law of 1773, by the Landgrave himself.

In the other versions of the tales under scrutiny, renunciation does not mean the end of marriage prospects but rather becomes the opportunity to meditate on another range of problems created by reduced inheritance portions. The heroic sisters of "The Twelve Brothers" and "The Six Swans" both renounce their portions by undertaking long vows of silence, which they begin while sitting in trees in the forest. In each case the sister is found by a hunting king who falls in love and marries her. In each case the king's mother disapproves of the marriage. In "The Six Swans" she complains, "This slut who can't talk. . . . Who knows where she comes from? She's not worthy of a king." In this manner women with little or no dowry are associated with the promiscuous or with prostitutes. The inheritance portion stands as a marker of personal worthiness, the sign of the strength of kin connections, of having a known past.

Ultimately, because the new bride makes no claims on her own kin through her silence, she does not possess the resources to convert the enmity of the king's mother to friendship. As a result she suffers a dangerous attack, which further elaborates the dangers of renunciation and the distortions of marital exchange wrought by military servitude upon the process of ordering production through kinship. In "The Twelve Brothers," the attack merely takes the form of constant slander by the mother-in-law: "Even if she's dumb and can't speak, she could laugh once in awhile. Anybody who doesn't laugh has a guilty conscience." In "The Six Swans" the attack is more elaborate and more revealing. The king's mother steals the children of the marriage, accuses the queen of eating the babies, and manufactures evidence for her charge by smearing the queen's mouth with blood as she sleeps. The mother-in-law's dishonesty points out that not bringing in a portion, not uniting the fortunes of two families in marriage, is tantamount to cannibalism of one's own children. This crime goes far beyond the polluting danger that kin usually represented to one another. The mother-in-law accuses the silent girl of enmity against her own descendants—of practicing witchcraft against them.[37] As in "The Twelve Brothers," where the crime remains unspecified, the new bride is sentenced to be burned as a witch for her (alleged) crimes.

The heroism of the sister is not punished; rather, at the last moment the term of the vow of silence expires, her brothers return to human form, she is able to tell her story, and the mother-in-law is punished in her stead. This resolution is not as straightforward as it would appear, however, for although the sham cannibalism is exposed, the girl's innocence is not established in any symbolic sense. Instead the conditions of her guilt may now be revealed and explained as she regains her voice. The problem of her descendants is simply ignored, as the stepmother is executed and the tale is resolved in favor of sibling love.

If viewed from the point of view of the executed stepmother, though, the sister may not be guilty of literal cannibalism, but she is guilty of sacrificing the welfare of her own children to that of her brothers. The common interest in descendants converts affines into friends as much as do contractually shared resources. But here the failure to contribute by the sister is a sign of lack of common interest—in fact, a sign of conflicts of interest between two different groups of consanguines (brothers and children). This conflict of interest is structural in the predominantly patrilineal and patrilocal inheritance and marriage practices of most Hessian regions. To increase portions of sons is to decrease that of daughters and their children, and vice versa. In medieval Germanic epics this structural conflict is ideologically and perhaps practically blurred by assigning brothers as protectors of their sisters' male children.[38] To some extent this pattern is repeated in ritual parenting relationships.[39] In these stories, the children, but mostly the sisters and mothers-in-law, bear the burden of such tensions—again a structural necessity because brothers in the militarized Hessian context needed protection more immediately than their potential nieces or nephews.

In Hesse-Cassel, the intensified need for protection of this kind grew out of the victimization of conscription and the inheritance reforms of 1773, which intentionally dispossessed and enslaved. But these practices dispossessed women, too, both directly and through peasant adjustments to conscription, and the tales fail to render that process fully transparent. As the contradictions of kin-ordered groups are blurred at the beginning of the tales by insufficiently specified causal agency, so also are they blurred at the end by too much agency wrongly attributed. The victimization of women is masked to the extent that tales picture suffering as

Table 1.1. *The Likelihood That a Soldier or Cohort Member Would Be an Heir or Do Better*

Group	% Soldiers (N=108)	% Cohort (N=115)	% Cohort Males (N=58)	% Cohort Females (N=57)
Heirs/better	27.8	38.9	44.6	29.1
Left parish	24.1	38.9	36.2	40.0

Note: Tables 1.1 and 1.2 compare a group of 108 soldiers recruited from a parish of 1,000 persons under the Kanton ordinance of 1762 with a cohort of 58 males and 57 females who were born between 1743 and 1748. The cohort members were contemporaries of the soldiers, and some members of the cohort were also part of the soldiers' group. The tables were created by combining a parish register, village tax records, and military master lists through normal record linkage.

the result of intentional acts of other women. The heroic self-sacrifice of sisters and the attempts of the stepmothers to maintain the value of the patriline cause the sister and her mother-in-law to suffer. However, it seems unlikely that women in the real living conditions to which the tales refer obliquely, experienced any dominant role in the arrangements of marriage and inheritance that caused their suffering.

The shift in devolutionary patterns between 1762 and 1773 showed that this conflict was no mere fantasy of storytelling women. Furthermore, marriage contracts from after 1783 reveal that some parents expected both brothers and sisters to take in siblings returning from the American war and thereby to cannibalize the inheritance portions of their own children. The connection between soldiers and their sisters extended to their objective life chances. Soldiers and women of the cohort were nearly equally unlikely to find a family situation in the parish that was as good as, or better than, the one they were born into (Table 1.1). This was a significantly poorer performance than all men of the cohort (which included some draftees). These differences remained even though soldiers came from all social strata in representative proportions, with the exception of the very wealthiest (Table 1.2). Despite their lesser chances for a good life, soldiers remained in the parish more frequently than contemporaries of either gender.[40]

The elements of Hessian oral tradition spoke in complicated ways, but not always lucidly, of conditions in which peasants and soldiers found themselves. At times the lack of clarity went beyond the veiling of language that may have served to hide the discourse from authorities. Rather, some ambiguities we may relate to a hegemony of a kin-ordered elite—a hegemony that, first, could not

Table 1.2. *Household of Origin for Soldiers and Their Contemporaries in Parish Oberweimar*

Group	Land Held (ha)	% Soldiers from (N=108)	% Cohort Members from (N=113)
Poor	<3	54.8	53.2
Middling	3–18	19.2	18.0
Wealthy	18–33	17.3	15.3
Very wealthy	>33	8.7	13.5

Note: See table 1.1 for an explanation of the data in this table.

face tensions between interests of parents and children or allow such tensions to remain fully exposed, and, second, one that vigorously affirmed a picture of society as a forest of family trees. Such a society had no room for those disconnected ones we call orphans and strangers. The very condition of disconnectedness was a curse that required undoing. Although a strategy of quiet resistance and rescue appears in the tales—one that was actually practiced—these rescue operations merely confirmed the influence of those whose position was maintained by the operation of the rules of peasant inheritance. Such a strategy was working within the system of kin-orderedness and thus helped support it. In so doing it promoted spreading the social pain of disconnectedness from peasant boys to their sisters, thus imagining and underwriting the creation of a whole new category of necessary victims of the military system. Finally, the stories suggest not that the military system disconnected people, but that military service was an environmental consequence of disconnection. As perceptive and as poignant as it was, this oral tradition provided only a critical beginning to transform either involuntary military servitude or the class structure of kin-ordered society that marked some for victimization and left others untouched. It still required some minimal therapeutic intervention.

The Bureaucratic Redefinition of Peasant Marginality

It is not usual to portray cultural materials of peasants that encode quiet strategies of resistance as contributing to a cultural hegemony. However, the strategies of resistance in Hessian oral tradition went beyond the simple failures of vision thus far marked out. The peasant vision of "disconnectedness" provided imagery and material for a tributary ideology of "marginality," which dovetailed nicely with theories of state management we now call cameralist.

Mack Walker has called cameralism a "baroque science . . . whose symmetry depended on its being seen from a distance, and whose rationality was seen in abstract outline from the top, subject to giddiness when attention moved from that [top] to concrete detail."[41] From the top, then, cameralism was the administrative science of taking tribute from the many corporate monads of German society, many of which were parts of kin-ordered enterprises of production.[42] In the general outlines of this theory, the state, in filling its fisc for military purposes, performed a superficial skimming operation that left towns, guilds, villages, and peasant households to order their own internal relations. For teachers of cameralist science like Christian Wolf, briefly professor at Phillip's University in Marburg, development meant enhancing revenues and could be seen only as a process of harmonizing the conflicts between, and encouraging the interdependence of, corporations so that they would form "rational" (read "obedient") tax-paying "communities."[43] The state encouragement of "community" formation had a strong moral element, which considered greed, laziness, and the ritual profligacy of popular culture as corrosive forms of undisciplined behavior that authorities saw within their purview to attack legally.[44]

Beyond this moralizing chord in the Protestant or Counter-Reformation key, cameralist ideology also grew out of a humanist literature of household management. An early modern vision of the "whole house" (*das Ganze Haus*) extended a Renaissance prescriptive social psychology for the new noble who was to serve under the new prince. A "new" noble was to fashion a self through neostoic self-conquest and to create a *virtù* based on work and learning that could be juxtaposed to the *fortuna* of accident, envy, and hatred.[45] Like Justus Lipsius, other reformers of the mid-seventeenth century advocated the diffusion of this vision beyond the nobility to subaltern officials and even peasants in an effort to encourage the just functioning of the state and its many elements. An entire prescriptive genre—"housefather" literature—grew to aid this diffusion. The roots of this discourse extended back to the genre of estate management literature of Xenophon and Aristotle.[46]

Housefather literature gave the crucial figure in the reimagined *oikos,* the head, not only technical knowledge to produce, but the moral imperative that had been given to the senses, intelligence, and reason of the knight in the earlier literature of noble neo-stoicism. This task of disciplining the conflicting desires of the household's children and servants served the state's purpose of eliciting a more orderly and just production of tribute. State officials and consistorial employees in eighteenth-century Hessian realms promoted this ideology of household "self-discipline" as the task of its *Brotherr* (loaf master or head of the house). These ideas appeared in both law and in the catechisms used to teach the children of peasants in preparation for their confirmation in the Christian community.[47] The "good" householder or loaf master directed the "self" of a monad suspended between the state and the individual, and performed the same harmonizing tasks for the individual members of the household that the state performed for corporations. Peasants also came to measure success of the household by amounts of tribute rendered, thus paralleling the fiscal measure the state held for itself.

Not surprisingly, the role of the whole house in the scheme of Hessian cameralism revealed itself in the structure of the documents that the state constructed to measure and to establish the tax liability of those units. Administratively, the most effective of these documents was the Kadaster (cadastre), which the state began creating in 1736.[48] Each Hessian cadastre actually attempted to harmonize the greedy selfishness of monads by reifying property rights and establishing a standard measure of value for them.[49] The Kadaster used the tax florin (*Steuergulden*) as the fundamental unit on this yardstick. In making this measure of taxpaying ability and liability, officials attempted to create not only a measure of absolute value but one that allowed them to separate good managers from bad.[50]

Although the Hessian Kadaster stemmed from and reflected impulses of cameralism and its vision of the peasant "community" as a harmonized collection of whole houses, the existing conditions of peasant society fit this vision poorly. The problems stemmed from the poor fit between the notions of "taxpayer" and "Brotherr" (household head) as well as the disjunction between the images of the village "community" as a divided and bounded parcel of land, as the corporate owner of common property, and as a collection of whole houses. The difficulties appeared in the discrepancy between the number of household files in a given village cadastre and in the number of actual functioning households.[51]

In the Kadaster for the village of Niederweimar, for example, there were fifty-four household files but only thirty-seven houses ("whole houses" in cameralist terms) when officials first inscribed the cadastral summary in 1747. The extra files fell into three separate categories, none of which was entirely consistent with cameralist ideology.[52] One category that caused difficulty were people who paid *Kontribution* on land in the village but lived in another village. A second difficult category consisted of people who obviously lived in the village but who held no house. The proportion of people entered under these conditions could become quite large, as in the village of Allna, with eleven such files out of forty-four.[53] The retirement arrangements of prosperous farmers could be a third way in which cadastral discourse conflicted with the cameralist vision of the whole house. Many times these retirement arrangements included either rooms in the house of the heir or a separate cottage that remained part of the estate. In cadastral discourse these separate houses frequently had their own separate files, with garden and crop lands attached in some instances and other times not. In most cases, the separation of these households in the Kadaster was an illusion. Census data show that these retirees still ate with the heir's family.[54] Moreover, marriage contracts suggest that they did not even pay the taxes on the land and buildings in their file, but that this obligation was taken on by the heir as well.[55]

Thus, the mechanical boundaries of the whole house did violence to a peasant society that conceived of itself in terms of an organic forest of family trees whose roots and branches extended in space and time. It was precisely these extensions that created difficulties for the cadastre. In the eyes of the state they became a group of people who were marginal and thus dangerous and unwanted, and state and village officials harassed them in innumerable ways. They were defined as a problem to be solved through police, welfare institutions, and ultimately military recruitment.

Yet even if social, economic, and geographic marginality created substantial problems of coherence for the discourse of the Hessian Kadaster, the same real and imagined difficulties provided an ideological escape for an interventionist practice of recruitment in the context of a management theory that emphasized corporate integrity. For Hessian officials, recruitment was never really simply a matter of skimming the surpluses of primary production in the way that collecting the Kontribution was. As Eric Wolf might put it, state recruiters set about the task of acquiring "rights in persons" that peasants had bound up in the symbolic language of kinship and inheritance.[56] "Unlocking" such rights required an intervention in peasant production that became ever more direct and ever less consistent with cameralist theory. Initially, officials cloaked this intervention in a discourse of "marginality," which allowed the systematic assignment of military service obligations to those defined officially as outside the corporate integrity of the "household" and "community" monads. Such persons remained in short supply, so that by 1762, with the military reforms of Frederick II, state policy actually extended the discourse of marginality to members of the whole house. Then in 1773, when even the supply of persons so defined did not meet the needs of military recruiters, the state created them through the legal manipulation of the symbolic grid of inheritance itself.

The military system created by Landgrave Karl and the Count of Lippe in the 1680s divided the army between a militia and a subsidy army. Only recruitment for the standing subsidy army presented problems of ideological coherence. Military service in the militia by members of peasant households was justified by a number of different arguments that remained consistent with the idea of the household as monad. Involuntary service in militia units, even when it included extensive time spent in training, involved the defense of "home, hearth, and faith," as Landgrave Moritz had put it.[57] Thus, the defense of the peasant household stood at the center of an obligation that could be read as a reimagining of feudal obligations to defend the territory against attack and help to maintain law and order within it (*Landfolge* and *Gerichtsfolge*).[58] The most important aspect of military service justified in this way was that it was limited to durations of emergency within territorial boundaries and did not amount to a personal obligation of particular individuals.

Early on, the basis of military service in the subsidy army caused little difficulty because it remained a private law and contractual relationship between the individual soldier and his military overlord. These contractual relationships stemmed from the voluntary contracts between mercenary captains and their soldiers of the previous two centuries. As past mercenaries had, soldiers symbolized and sealed the agreement to serve by the acceptance of a recruitment bounty called *Handgeld.* In accepting the bounty, the recruit had "sold his skin" and could be thus used for purposes other than the common defense.

This legal nicety meant that the Landgraves' subsidy business remained within acceptable legal practice, for it was a proprietarial rather than an imperial or territorial army that he leased to foreign powers. However, the practice of contractual military service could be squared with the ideology of the whole house only by assuming that soldiers who came to serve in the subsidy army had no obligations to, or place in, the kin-ordered households of peasants. Indeed, recruitment law certainly bowed deeply in this direction. As late as 1733, the Landgrave instructed recruitment commissioners to fill the ranks of the subsidy army with those "masterless loafers (*herrenlose Müssigänger*) whom you can find in the cities and domainal administrative divisions entrusted to you . . . [and] see to it with the greatest industry that except for such loafers that no poor subject necessary for agriculture is taken as a volunteer."[59] Even the very intrusive legislation of 1762 still required that recruiting officials check to make sure recruits did not have prior obligations as servants to a peasant Brotherr or to the household of a noble (*adelige Hintersasse*).[60]

The phrase "masterless servants and loafers" (*herrenlose Gesinde und Müssigänger*) began appearing in Hessian legislation immediately following the Thirty Years' War.[61] Frequently, in association with "dismissed soldiers" (*abgedankte Soldaten*), officials used it after wars to justify draconian policing and monitoring measures against mobile populations. The free movement of war refugees or migrant workers violated the cameralist assumption that labor should be kept at home and that laborers from outside should be encouraged to settle within.[62] Moreover, any substantial population living outside the authority of a Brotherr seemed an intolerable threat to order and proper distribution of welfare payments.

Finally, the same formulation appealed to the dis-ease expressed in popular culture for persons disconnected from kin-ordered enterprises.

To target "masterless servants and loafers" in villages implied that military recruitment swept the land clean of a dangerous and surplus population. It remained consistent with conscriptive recruitment practices of other princes at the time. However, there is some evidence that as early as 1702, Hessian recruiters for the subsidy army could find sufficient numbers of thus qualified recruits neither within nor outside the territorial borders. In an effort to bring regiments up to strength for the opportunities presented by the War of the Spanish Succession, recruiters received instructions to inquire about "parents who have different sons on hand whom they can best do without in their [the parents'] trade or subsistence production (*Nahrung*)."[63] Officials backed off this broader casting of the recruitment net after the wartime crisis, but the need to cast it so widely squares with the picture that demographers have given us—that early eighteenth-century Hesse-Cassel was still recovering from the terrible depredations and losses of the Thirty Years' War.[64]

Another indication that recruiters could not successfully restrict themselves to those that cameralist theory rendered as dangerous deadbeats came in early attempts to blur the clear legal distinction between militia service and standing-army service. Although militia service was involuntary and subsidy service was technically free-will enlistment, many of the Landgrave's recruiters used the militia as a pool of recruits. A commonplace for European armies of the time, such recruiting frequently occurred when militia and subsidy units mustered together for regular yearly exercises. Officials mystified the questionable legality of the practice by following militia enrollment procedures, which required that every recruit be paid a bounty.[65] Technically this meant that even militia soldiers taken involuntarily from households now had "sold their skins" or delivered the rights to their persons to the Landgrave should his recruiters choose to exercise them.

The Landgrave's officials recognized that they needed to balance the manpower needs of the army with the labor needs of the rural economy. As early as 1734, law more carefully defined marginality by ordering muster commissioners to take only those recruits who "in their current place of residence do not have so much to lose and can be used for military service without disadvantage to the public (*Publico*)."[66] A 1741 administrative clarification defined "disadvantage to the public" more precisely as "reduction of military taxes (*Kontribution*) and damage to agricultural production."[67] The discourse of marginality still did not explicitly admit that the Landgrave took his recruits from functioning peasant households, but those who fit poorly into cameralist categories were clearly targeted.[68]

Intrusions into intra-household relationships, however, grew gradually more explicit in the course of the 1730s and 1740s. Officials first formulated a language of direct official intervention into household relations in a nonmilitary connection. Following the theory of the Marburg cameralist Christian Wolf, that the master-servant relationship constituted an "authoritarian society from which advantages on both sides are derived," in 1736 Hessian officials instituted an ordinance that commanded village authorities to regulate contracts between Brotherrn and their employees.[69] This servants' ordinance, which went beyond regulating wages and establishing the master's moral obligation to discipline servants, forbade masters of the house to hire any servant without documentary evidence of good performance from the last employer.[70] Such a system of recommendations established in law the classic double bind of the young job seeker who cannot get a job without experience and can get no experience without a job. The same law added to the tension by ordering village mayors to "get rid of all foreign and masterless servants who have become a burden to the community, tolerating only those who perform their day labor well. Particular attention is to be paid to village inhabitants who would prefer not to serve, and in order that they become accustomed to work." Such legal language worked as much to create a group of masterless (and, we might add, easily recruitable) young men as to assure that all lived under the authority of a household head or master.

During the 1740s state intrusion into households went further still. A Brotherr lost the capacity to keep his male servants off village recruitment lists, and officials required them not only to prove that a contractual relationship existed between master and servant but that the servant was indispensable to the local agricultural economy.[71] In practice, this meant that day laborers, their sons, and the apprentices of village artisans became the primary focus of recruiters. W. B. Blome, a tax official, pointed out the contradictions of this policy: "If one needed recruits fast . . . these [marginal] sorts were the easiest to give up," but recruiting them interrupted their training and reduced their capacity to support themselves after military service.[72]

Recruitment thus tended to create the very marginals that the army was supposed to be cleaning up. Even worse, according to a 1747 administrative order, "marginals (*Unvermögende*) at the time of a muster hire themselves outside of the territory or simply flee," leaving military service to the wealthy who have more reason to stay.[73] The problem became so severe that in the years immediately preceding the Seven Years' War the Landgrave had to promise that "no-one will be taken into military service against their will or with force especially if they cannot go without the decline of their estates or otherwise damaging the subsistence of their household."[74] Marginality or superfluity was becoming ever more determined not by needs of village economies, or abstract public needs defined by the fisc, but rather in terms of the whole house itself. This conception of marginality reached back to the abandoned recruitment law of 1702, which sought out those sons that the household could spare.

After the Seven Years' War the military reforms of Frederick II represented a watershed and a fruition in the developing language of marginality in recruitment policy. Already substantially blurred by the practice of subsidy recruitment from militia units, the new army simply eliminated the legal distinctions between the two kinds of units and placed the obligation to serve on a more universal basis.[75] Keeping true to the cameralist principal of balance, this conscription law selectively exercised the newly asserted rights to subsidy service.[76] The Landgrave allowed recruits for service in regiments most likely to see subsidy duty (field regiments) to be taken "only if they are the household's most expendable people, or if agriculture and other necessary occupations will not be interrupted."[77] If he limited service in field regiments to "the household's most expendable people," then he reserved service in garrison regiments (former militia units) to those youngsters who "are not entirely expendable to parents who own estates."[78]

A person was not expendable if he owned a house, or an estate requiring a plow team (*Anspannig*), or if he were the heir of a well-provided peasant, a master craftsman, an apprentice to a necessary trade, or a journeyman serving a widow. Earlier law had defined a "well-provided peasant" (*stark begütherte Bauer*) as one who paid more than one Reichsthaler per month in Kontribution. The Landgrave instructed his officials to examine carefully the documentary evidence of contractual relationships between peasants and evaluate the relative necessity of particular laborers. They were also to "ensure that immature youths and especially those under twenty years of age neither marry nor inherit estates in order to escape military service."[79] Whereas at the beginning of the eighteenth century military recruiters swept villages clean of perceived superfluous people, now the state was reaching deeply into the kin-ordered household and explicitly vetoing the decisions of householders in the most crucial acts of assigning rights to the surpluses of primary accumulation to one party or another. Officials did so to acquire rights in persons for the state.

We have come a long way from the language of a household as a Leibnitzian monad and approach more closely Justi's police state (*Polizeistaat*) vision of householders as intermediate authorities of discipline. Although perhaps Justi's vision more accurately reflected past practice as well as practice after 1762, this new, more transparent language permitted state officials to think in terms of directly manipulating the rules of marriage, inheritance, and property devolution in such a way that more persons could be claimed for military service. This they proceeded to do in 1773 through an ordinance known as the *Hufen* edict, which regulated devolutionary practice.[80]

Hesse-Cassel was one of the territories in Germany where the rural subject population practiced both partible and impartible inheritance. Land held of a variety of feudal tenures (*Lehnland*) was restricted to impartibility in law and, for the most part, in practice. Allodial land (*Erbland*) was subject to no restrictions and passed as the holder saw fit. Regardless of partible or impartible practice, an ideology of equal division existed in law and made its appearance in practice. This meant that peasants practicing impartibility might pass their estates whole to a single heir, but those children who did not receive the land and buildings of the estate still had an equal claim to its resources. These claims took the form of debt or payments in cash and kind spread over several years and sometimes more than one generation. Additionally, should either one of the couple passing resources to the next generation still live, they too were entitled to claims on its resources to support their continued existence. As Hermann Rebel has shown, such claims became golden chains that bound labor and marriage prospects to heirs and their elders who controlled the inheritance and marriage networks.[81] Partible inheritance (*Realteilung*) required actual division of properties and was seldom entirely equal because few people were inclined to divide buildings and movables equally. This meant that parents usually advantaged one heir over others and put the heir in a privileged position for ordering the networks of marriage and labor. More frequently, peasants used partible land as part of the settlements of children on impartible estates. That is, such land tended to be marginal parcels that passed from impartible estate to impartible estate as part of marriage settlements of the siblings of heirs. Whether land was partible or impartible, devolutionary practice remained multi-staged and always centered around the marriage settlement of the principal heir. The symbolic grid of inheritance—the assignment of heir status and co-heir claims—was the discourse of kin-ordering of production by peasants.[82] The movable and immovable properties that passed from generation to generation constituted a fund created by "primitive accumulation."[83]

Most past authorities writing about such edicts as the *Hufen* edict discuss how they mandated impartibility and then focus on their fiscal "rationality."[84] For these accounts, the important achievements of such legislation were to prohibit the division (*Zerstücklung*) of peasant farms, establish a legal framework for impartibility, and harmonize the interests of state, heir, retiring farmer, and dispossessed siblings. Manifestly, state officials wished to keep peasant farms sufficiently large so that they provided both subsistence for those who held them and taxes and tribute for the state and landed aristocrats. The official line in the Hufen edict of 1773 expressed the belief that "a divided estate (*Erbgut*) only very rarely suffices to support an entire family (*Familie*) . . . and therefore such families are not able to render the dues (*Zinsen*), the labor burdens (*Dienste*), taxes, military taxes (*Kontribution*), and all other obligations" that fell upon them as tenure holders.[85]

The Hufen edict of 1773 went far beyond asserting the impartibility of closed estates. Expressing concern that when dispossessed "co-heirs demand the so-called true value of their portions from the possessor in cash, it frequently comes to pass that the farm is still further divided or so encumbered with debt that the owners do not hold on to large segments of it."[86] The new Hufen edict sought to limit co-heirs' portions and the benefits claimable by retiring elders to less than the total assessed value of estates.[87] Past law, in fact, required that estates passed impartibly still bear the burden of compensating (*Abfindungssummen* and *Herausgifte*) those heirs of the previous holder who did not receive the estate. Moreover, should a tenure pass while either one of the couple who passed it still lived, the survivor was entitled to a retirement portion (*Altenteil*). Further, the new law prohibited land and buildings from being separated from estates for retirement purposes and prohibited retirement funds except in direst need.

Thus, the Hufen edict directed a massive redistribution of surpluses toward heirs within the kin-ordered households of rural Hesse-Cassel. Using cadastral assessments, it set a limit on the total of all compensatory portions at 80 Reichsthaler on estates assessed at a value of 90 Reichsthaler and 100 Reichsthaler if that estate included a farmstead (*Haus und Hof*). The amount fell or climbed ten Reichsthaler for every Albus per month less or more owed in Kontribution. This meant that the proportion available to take care of co-heirs climbed with value of the estate and diminished as it diminished. If an estate paid 20 Albus in monthly Kontribution, 105 percent of its assessed value would be available for compensation, while an estate that paid 5 Albus would permit 66 percent of its assessed value to be distributed among the dispossessed co-heirs. The law put an 80 Reichsthaler limit on estates paying 10 Albus per month in Kontribution. The tax rate was 1 Heller per Steuergulden of assessed value per month. There were 27 Albus per Steuergulden and 36 Albus per Reichsthaler. A payment of 10 Albus per month means 120 Heller per month or 120 Gulden assessed value or 3,240 Albus. Divide that by 36 Gulden per Reichsthaler to yield a result of 90 Reichsthaler assessed value. This allotment seems very generous and certainly in line with the principles of equally divided value that lay behind *Anerbenrecht* devolutionary practice. However, when one recognizes that officials based Kontribution payments on an assessed value after about one-third had been deducted for tribute payments, the picture looks less rosy. Beyond this, Kersten Krüger and Klaus Greve have calculated that Steuergulden assessments probably amounted to about a quarter of market values to begin with. Finally, all debts had to be subtracted from the value of the estate as well. What looks like a very generous 89 percent of an estate's value allotted to compensation comes out to be less than 15 percent. This becomes more shocking when we learn that peasants at the time probably valued their estates up to five times the gross assessed tax value and more than twice the market value for the purposes of devolutionary settlements.[88]

According to Ingrao, this bold stroke had been intentionally designed "to effect a massive shake-out," particularly among the poorest peasants.[89] Apparently the hope was to save some households on these tenures by forcing children into more promising careers, the most important of which was the military.[90] In short, the Hufen edict consciously created marginals in both households and in villages to better fill the ranks of its subsidy army. Further, it attempted to define exactly who these designated failures would be by legally encoding the peasant practice of giving sons precedence over daughters and elder sons over younger.[91] Whether the policy makers had hoped simply to preserve the tax-paying capacity of peasant farms or whether they had also intended initially to create a new pool of conscripts is not entirely clear from the sources. As things worked out, the latter alternative occurred rather more frequently than the former.

To unlock the household thus required both a new level of ideological elaboration and new controls over the ways the heads of households distributed rights in persons within rural society. To justify recruitment practice, Hessian officials relied on an ideology of marginality. Officials associated those under no household authority or community authority with dismissed soldiers and postwar refugees to justify a draconian policy of monitoring and policing measures against mobile populations. Thus, officials could cast military recruitment, along with that to workhouses and orphanages, as a way of supporting the rule of "housefathers" and "community leaders" by sweeping society clean of those who remained under no corporate authority. In short, they tried to cast military recruitment as a kind of skimming of surplus people.

Both recruitment practice, with its open use of fraud, and the demographic history of the region suggest that no large numbers of such surplus people existed within Hessian villages until after the mid-eighteenth century.[92] Therefore, officials reached ever more explicitly into the internal relations of the "communities" and "households" by dividing households between "good" and "bad" and siblings and servants between the "needed" and the "extra." To do so was not consistent with cameralist vision of the households as externally bounded, internally homogeneous units, nor was it consistent with the forest of families pictured by peasant oral tradition, but such discourse is "immarginating" in Foucault's sense.[93] This is to suggest that in defining and attempting to harmonize such social corporations, the Hessian state (and others like it) had provided some of the fundamental tools required to discipline, punish, and control them rather than provide the grounds for their independence and shelteredness. It further defined some insiders as outsiders, thus releasing them from the authority of kin-ordered production to the authority of the state.

T. Jackson Lears has suggested that one of the most useful aspects of the idea of cultural hegemony is that we may view dominant sets of ideas as mélanges of symbols woven together from both ruling and ruled groups.[94] Thus, the power of hegemonic ideas is that they do, at least partially, speak to many experiences in social formations and thus can evoke some emotional assent to their appropriateness from both rulers and ruled. In short, like Eric Wolf's modes of production, ideologies also articulate with one another. This I wish to suggest has occurred between the Hessian peasant version of "disconnectedness" and its bureaucratic counterpart, "marginality." For peasants, being pruned from the forest of families was to no longer be able to draw on the power in the blood and even to be removed from the realm where enmity was converted to friendship through marriage in a society that knew no other way to distribute labor and its fruits. To enter such a state required rescue operations, which demanded from peasant women their own sacrifices to resolve contradictions within the system. State ideology took advantage of this experience by picturing military service as a means of reincorporation of the marginal despite their disconnection from whole houses. At the same time, by manipulating the categories of cameralism, bureaucrats widened the definition of marginality by circumscribing family trees to whole houses and membership in whole houses to the unexpendable (*unentbehrlich*). The manifold contradictions of both discourses point to class conflicts within peasant families and between those families and the tributary state. Such contradictions remained largely hidden and blurred in what Fredric Jameson has called the "political unconscious," to be experienced in daily life as those double binds faced by the sisters of draftees.[95] Similar articulations allowed the bureaucratic creation of marginals in the twentieth century who subsequently were reimagined as racial vermin to be easily exterminated as the necessary victims of a greater and purer Germany where everyone left drew from the power in the blood.

Notes

1. Otto Büsch, *Militärsystem und Sozial-leben im alten Preußen, 1713–1807: Die Anfänge der sozialen Militärisierung der Preußisch-deutschen Gesellschaft* (Frankfurt, 1981), pp. 167–75.

2. Michel Foucault, *Discipline and Punish: The Birth of the Prison* (New York, 1979), pp. 135–36.
3. Gerhard Oestreich, *Geist und Gestalt des frühmodernen Staates* (Berlin, 1978), p. 196.
4. Hermann Rebel, "Cultural Hegemony and Class Experience: A Critical Reading of Recent Ethnological-Historical Approaches (Parts 1 and 2)," *American Ethnologist* 16, no. 1 (1989): 119.
5. Ibid.
6. Eric R. Wolf, *Europe and the People without History* (Berkeley, 1982), pp. 75–100.
7. David Warren Sabean, *The Power in the Blood: Popular Culture and Village Discourse in Early Modern Germany* (Cambridge, 1984), pp. 20–27.
8. Charles W. Ingrao, *The Hessian Mercenary State: Ideas, Institutions, and Reform under Frederick II, 1760–1785* (Cambridge, 1987), p. 122.
9. Landgrafschaft Hessen-Kassels, *Sammlung Fürstliche Landesordnungen und Auschreiben* (Kassel, 1760–1816), 1–8. Cited as HLO, 16.12.1762.
10. Sabean, *The Power in the Blood,* pp. 20–27.
11. Peter K. Taylor, *Indentured to Liberty: Peasant Life and the Hessian Military State, 1688–1815* (Ithaca, 1994), pp. 75–77.
12. Ibid., pp. 78–82.
13. Ibid., p. 179.
14. Ibid., pp. 171–73.
15. Ibid., p. 176.
16. Ibid., p. 189.
17. Ibid., pp. 190–94.
18. Rodney Atwood, *The Hessians: Mercenaries from Hesse-Cassel in the American Revolution* (Cambridge, England, 1981), pp. 19–21.
19. Hessisches Staatsarchive Marburg, *Akten,* 4h Kriegsachen, 5 Geheimenrat, Bestände 17e Ortsrepositur, 49d Steuerrektificationskommission. Cited as StaM, Best. 4h 4072.
20. StaM, Best. 5 16,622, Lahnstrom.
21. Eugen von Frauenholtz, *Entwicklungsgeschichte des deutschen Heerwesens: Die Heerwesen in der Zeit des Absolutismus* (Munich, 1940), pp. 4, 13–14.
22. James C. Scott, *Domination and the Arts of Resistance* (New Haven, 1990), pp. 4–7.
23. Wolf, *Europe,* p. 289.
24. Sabean, *The Power in the Blood,* pp. 31, 94–95.
25. Wolf, *Europe,* p. 289.
26. Hermann Rebel, "Why Not 'Old Marie' . . . or Someone Very Much Like Her?" *Social History* 13, no. 1 (1988): 1–20.
27. Jakob and Wilhelm Grimm, *Die Kinder und Hausmaerchen der Brüder Grimm* (Berlin, 1812, 1815), nos. 25, 9, and 119. Cited as KHM.
28. Peter Taylor and Hermann Rebel, "Hessian Peasant Women, Their Families, and the Draft: A Social-Historical Interpretation of Four Tales from the Grimm Collection," *Journal of Family History* 6, no. 4 (Winter 1981): 257–367.
29. Ibid., p. 365.
30. Orlando Patterson, *Slavery and Social Death: A Comparative Study* (Cambridge, Mass., 1982), p. 39.
31. Taylor and Rebel, "Hessian Peasant Women," pp. 364–65.
32. Heinz Röllecke, "Die stockhessischen Märchen der 'alten Marie': Das Ende eines

Mythos um die früheste Aufzeichnung der Brüder Grimm," *Germanische-Romanische Monatschrift* 25 (1975): 341.

33. "The Seven Ravens," KHM, 1819, no. 25.
34. Taylor and Rebel, "Hessian Peasant Women," p. 365.
35. Ibid., p. 360.
36. Jack Goody, *The Development of the Family and Marriage in Europe* (Cambridge, England, 1983), p. 136. The head was the most ancient common ancestor, and digits of hands and feet were the most recent generation.
37. Sabean, *The Power in the Blood,* pp. 140–43.
38. David Warren Sabean, "Aspects of Kinship Behavior and Property in Rural Western Europe before 1800," in Jack Goody, Jean Thirsk, and E. P. Thompson, eds., *Family and Inheritance: Rural Society in Western Europe, 1200–1800* (Cambridge, England, 1976), p. 108.
39. Peter K. Taylor, "The Household's Most Expendable People: The Draft and Peasant Society in 18th Century Hessen-Kassel" (Ph.D. diss., Univ. of Iowa, 1987), p. 316.
40. StaM, Protokolle I and II, Gerichtsprotokolle Kaldern und Reizberg, Eheprotokolle 1783–1820.
41. Mack Walker, *German Hometowns: Community, State, and General Estate, 1648–1871* (Ithaca, 1971), p. 145.
42. Wolf, *Europe,* pp. 88–93.
43. Hermann Rebel, "Re-imagining the *Oikos,* Austrian Cameralism in Its Social Formation," in Jay O'Brien and William Roseberry, eds., *Golden Ages, Dark Ages: Reimagining the Past* (Berkeley, 1991), p. 17; Walker, *German Hometowns,* 149; Karl Pribram, *A History of Economic Reasoning* (Baltimore, 1983), p. 90; and Mark Raeff, *The Well Ordered Police State* (New Haven, 1983), pp. 60–65.
44. Rebel, "Re-imagining the *Oikos,*" pp. 34–35.
45. Ibid., p. 35.
46. Ibid., p. 37.
47. HLO, 28.8.1736, 11.6.1739; Johann Jacob Rambach, *Kirchen Gesangbuch* (1733).
48. Karl Strippel, *Die Währschafts- und Hypothekenbücher Kurhessens* (Marburg, 1914), pp. 49–53; Kersten Krüger and Klaus Greve, "Steuerstaat und Socialtruktur," *Geschichte und Gesellschaft* 3 (1982): 295–305.
49. E. P. Thompson, "The Grid of Inheritance: A Comment," in Jack Goody, Joan Thirsk, and E. P. Thompson, eds., *Family and Inheritance: Rural Society in Western Europe, 1200–1800* (Cambridge, England, 1976), pp. 328–60.
50. Krüger and Greve, "Steuerstaat und Socialtruktur," p. 303.
51. Rebel, "Re-imagining the *Oikos,*" p. 17; Otto Brunner, "Das 'Ganze Haus' und die Alteuropäische Ökonomik," *Neue Wege der Verfassungs- und Sozialgeschichte* (Göttingen, 1968), pp. 102–23.
52. Hessisches Staatsarchive Marburg, *Kadaster I,* Niederweimar, 1747. Cited as StaM, Kadaster I, etc.
53. StaM, Kadaster I, Allna 1747.
54. StaM, Best. 40d, Rubr. 29, nr. 26, Kaldern und Reizberg, 1732.
55. Hessisches Staatsarchive Marburg, *Protokolle I and II, Gerichtsprotokolle,* Reizberg und Kaldern, 1780–1820. Cited as StaM, P I and II, etc.
56. Wolf, *Europe,* pp. 91–92.

57. Gunther Thies, *Territorialstaat und Landesdefensionswerk* (Marburg, 1973), p. 35.
58. Hans Georg Boehme, *Die Wehrverfassung Hessen-Kassels* (Kassel, 1954), pp. 30, 34–35.
59. StaM, Best. 4h, 4053.
60. HLO, 16.12.1762.
61. HLO, 6.1.1698.
62. Raeff, *The Well Ordered Police State,* p. 74.
63. HLO, 24.4.1702.
64. Gunther Franz, *Der Dreißigjährige Krieg und das deutsche Volk* (Stuttgart, 1979), p. 41; George Thomas Fox, "Studies in the Rural History of Upper Hesse" (Ph.D. diss., Vanderbilt University, 1976), p. 268; and Ottfried Dascher, *Das Textilgewerbe in Hessen-Kassel vom 16. bis 19. Jahrhundert* (Marburg, 1968), pp. 166–86.
65. HLO, 19.5.1741.
66. HLO, 18.1.1734.
67. HLO, 19.5.1741.
68. Otto Könnecke, *Rechtsgeschichte des Gesindewesens in West- und Süddeutschland* (Marburg, 1912), p. 59.
69. Ibid.
70. HLO, 11.6.1739.
71. Könnecke, *Rechtsgeschichte,* pp. 372–78.
72. StaM, Best. 49d Marburg 556, Niederweimar 1738.
73. HLO, 29.8.1747.
74. HLO, 26.7.1755.
75. Philip Losch, *Soldatenhandel* (Marburg, 1976), p. 25; Boehme, *Die Wehrverfassung Hessen-Kassels,* pp. 27–8.
76. HLO, 16.12.1762.
77. HLO, 16.12.1762.
78. HLO, 10.4.1767.
79. HLO, 16.12.1762.
80. HLO, 19.11.1773.
81. Hermann Rebel, "Peasant Stem Families in Early Modern Austria," *Social Science History* 2, no. 3 (1978): 255–80.
82. Thompson, "The Grid of Inheritance," pp. 329–34.
83. Hermann Rebel, *Peasant Classes: The Bureaucratization of Property and Family Relations under Early Hapsburg Absolutism, 1519–36* (Princeton, 1983), pp. 194–98.
84. Wilhelm Holtzapfel, "Oberlandsbezirk Kassel," in Max Sering, ed., *Die Vererbung des ländlichen Grundbesitzes* (Berlin, 1899–1910); Teodor Mayer-Edenhauser, *Untersuchen über Anerbenrecht und Güterschluß in Kurhessen* (Prague, 1942); Fox, "Studies in the Rural History," pp. 368–71; Ingrao, *The Hessian Mercenary State,* pp. 116–20.
85. HLO, 19.11.1773.
86. HLO, 19.11.1773.
87. HLO, 19.11.1773.
88. Krüger and Greve, "Steuerstaat und Socialtruktur," pp. 302–3.
89. Ingrao, *The Hessian Mercenary State,* p. 120.
90. Ibid., 134.
91. HLO, 19.11.1773.

92. Taylor, *Indentured to Liberty,* pp. 172–76.
93. Michel Foucault, *The History of Sexuality* (New York, 1980), pp. 81–85.
94. T. Jackson Lears, "The Concept of Cultural Hegemony: Problems and Possibilities," *American Historical Review* 90, no. 3 (June 1985): 590.
95. Fredric Jameson, *The Political Unconscious: Narrative as a Socially Symbolic Act* (Ithaca, 1981), pp. 286–91.

References

Primary Sources

Grimm, Jakob and Wilhelm. *Die Kinder und Hausmaerchen der Brüder Grimm.* Berlin, 1812, 1815, 1819. Cited as KHM.

Hessisches Staatsarchive Marburg. *Akten,* 4h Kriegsachen, 5 Geheimenrat, Bestände 17e Ortsrepositur, 49d Steuerrektificationskommission. Cited as StaM, 4h, etc.

Hessisches Staatsarchive Marburg. *Kadaster I,* 1747. Cited as StaM, K I, etc.

Hessisches Staatsarchive Marburg. *Protokolle I and II, Gerichtsprotokolle.* Cited as StaM, P I and II, etc.

Landgrafschaft Hessen-Kassels. *Sammlung Fürstliche Landesordnungen und Auschreiben.* Kassel, 1760–1816, 1–8. Cited as HLO, [date].

Rambach, Johann Jacob. *Kirchen Gesangbuch.* Darmstadt, 1733.

Secondary Sources

Anderson, Fred. *A People's Army: Massachusetts, Soldiers, and Society in the Seven Years War.* Chapel Hill, 1984.

Atwood, Rodney. *The Hessians: Mercenaries from Hesse-Cassel in the American Revolution.* Cambridge, England, 1981.

Berge, Otto. "Die Innenpolitik Landgraf Friedrich II." Ph.D. diss., Mainz, 1950.

Boehme, Hans Georg. *Die Wehrverfassung Hessen-Kassels.* Kassel, 1954.

Brunner, Otto. "Das 'Ganze Haus' und die Alteuropäische Ökonomik." In *Neue Wege der Verfassungs- und Sozialgeschichte.* Göttingen, 1968.

Büsch, Otto. *Militärsystem und Sozialleben im alten Preußen, 1713–1807: Die Anfänge der sozialen Militärisierung der Preußisch-deutschen Gesellschaft.* Frankfurt,1981.

Dascher, Ottofried. *Das Textilgewerbe in Hessen-Kassel vom 16. bis 19. Jahrhundert.* Marburg, 1968.

Foucault, Michel. *Discipline and Punish: The Birth of the Prison.* New York, 1979.

———. *The History of Sexuality.* New York, 1980.

Fox, George Thomas. "Studies in the Rural History of Upper Hesse." Ph. D. diss., Vanderbilt, University, 1976.

Franz, Gunther. *Der Dreißigjährige Krieg und das deutsche Volk.* Stuttgart, 1979.

Frauenholtz, Eugen von. *Entwicklungsgeschichte des deutschen Heerwesens: Das Heerwesen in der Zeit des Absolutismus.* Munich, 1940.

Goody, Jack. *The Development of the Family and Marriage in Europe.* Cambridge, England, 1983.

Greenough, Paul. "Indian Famines and Peasant Victims: The Case of Bengal in 1943–44." *Asian Studies* 2 (1980).

Holtzapfel, Wilhelm. "Oberlandsbezirk Kassel." In Max Sering, ed. *Die Vererbung des ländlichen Grundbesitzes.* Berlin, 1899–1910.

Ingrao, Charles W. *The Hessian Mercenary State: Ideas, Institutions, and Reform under Frederick II, 1760–1785.* Cambridge, 1987.

Jameson, Fredric. *The Political Unconscious: Narrative as a Socially Symbolic Act.* Ithaca, 1981.

Könnecke, Otto. *Rechtsgeschichte des Gesindewesens in West- und Süddeutschland.* Marburg, 1912.

Krüger, Kersten, and Klaus Greve. "Steuerstaat und Socialtruktur." *Geschichte und Gesellschaft* 3 (1982).

Lears, T. Jackson. "The Concept of Cultural Hegemony: Problems and Possibilities." *American Historical Review* 90, no. 3 (June 1985).

Losch, Philip. *Soldatenhandel.* Marburg, 1976.

Mayer-Edenhauser, Teodor. *Untersuchen über Anerbenrecht und Güterschluß in Kurhessen.* Prague, 1942.

Melton, James Van Horn. "Absolutism and 'Modernity' in Early Modern Central Europe." *German Studies Review* 8 (1985).

Oestreich, Gerhard. *Geist und Gestalt des frühmodernen Staates.* Berlin, 1978.

Patterson, Orlando. *Slavery and Social Death: A Comparative Study.* Cambridge, Mass., 1982.

Pribram, Karl. *A History of Economic Reasoning.* Baltimore, 1983.

Raeff, Mark. *The Well Ordered Police State.* New Haven, 1983.

Rebel, Hermann. "Peasant Stem Families in Early Modern Austria." *Social Science History* 2, no. 3 (1978).

———. *Peasant Classes: The Bureaucratization of Property and Family Relations under Early Hapsburg Absolutism, 1519–36.* Princeton, 1983.

———. "Why Not 'Old Marie' . . . or Someone Very Much Like Her?" *Social History* 13, no. 1 (1988).

———. "Cultural Hegemony and Class Experience: A Critical Reading of Recent Ethnological-Historical Approaches (Part 1 and 2)." *American Ethnologist* 16, nos. 1–2 (1989).

———. "Re-imagining the *Oikos:* Austrian Cameralism in Its Social Formation." In Jay O'Brien and William Roseberry, eds., *Golden Ages, Dark Ages: Reimagining the Past.* Berkeley, 1991.

Röllecke, Heinz. "Die stockhessischen Märchen der 'alten Marie': Das Ende eines Mythos um die früheste Aufzeichnung der Brüder Grimm." *Germanische-Romanische Monatschrift* 25 (1975).

———. *Die Älteste Märchensammlung der Brüder Grimm.* Constance, Germany, 1976.

Rosenberg, Rainer Freiherr von. *Soldatenwerbung und militärisches Durchzugsrecht im Zeitalter des Absolutismus.* Berlin, 1973.

Sabean, David Warren. "Aspects of Kinship Behavior and Property in Rural Western Europe before 1800." In Jack Goody, Joan Thirsk, and E. P. Thompson, eds., *Family and Inheritance: Rural Society in Western Europe, 1200–1800.* Cambridge, 1976.

———. *The Power in the Blood: Popular Culture and Village Discourse in Early Modern Germany.* Cambridge, 1984.

Schoof, Wilhelm. *Zur Enstehungsgeschichte der Grimmischen Märchen.* Hamburg, 1959.

Scott, James C. *Domination and the Arts of Resistance.* New Haven, 1990.

Siddle, D. J. "Inheritance Strategies and Lineage Development in Peasant Society." *Continuity and Change* 1 (1986).

Strippel, Karl. *Die Währschafts- und Hypothekenbücher Kurhessens.* Marburg, 1914.

Sutherland, D. M. G. *France, 1789–1815: Revolution and Counterrevolution.* New York, 1986.

Taylor, Peter K. "The Household's Most Expendable People: The Draft and Peasant Society in 18th Century Hessen-Kassel." Ph.D. diss., Univ. of Iowa, 1987.

———. "'Patrimonial' 'Bureaucracy and 'Rational' Policy in Eighteenth-Century Germany: The Case of Hessian Recruitment Reforms, 1762–93." *Central European History* 22, no. 1 (1989).

———. "Military System and Rural Social Change in Eighteenth-Century Hesse-Cassel." *Journal of Social History* (Spring 1992).

———. *Indentured to Liberty: Peasant Life and the Hessian Military State, 1688–1815.* Ithaca, 1994.

Taylor, Peter, and Hermann Rebel. "Hessian Peasant Women, Their Families, and the Draft: A Social-Historical Interpretation of Four Tales from the Grimm Collection." *Journal of Family History* 6, no. 4 (Winter 1981).

Thies, Gunther. *Territorialstaat und Landesdefensionswerk.* Marburg, 1973.

Thompson, E. P. "The Grid of Inheritance: A Comment." In Jack Goody, Joan Thirsk, and E. P. Thompson, eds., *Family and Inheritance: Rural Society in Western Europe 1200–1800.* Cambridge, 1976.

Tilly, Charles, ed. *The Formation of National States in Western Europe.* Princeton, 1975.

Vogel, Hans, and Wolfgang von Both. *Landgraf Friedrich II von Hessen-Kassel: Ein Fürst der Zopfzeit.* Munich, 1973.

Walker, Mack. *German Hometowns: Community, State, and General Estate, 1648–1871.* Ithaca, 1971.

Wolf, Eric R. *Europe and the People without History.* Berkeley, 1982.

CHAPTER TWO

Dark Events and Lynching Scenes in the Collective Memory: A Dispossession Narrative about Austria's Descent into Holocaust

HERMANN REBEL

Before we get to the story promised in the title about a dispossession figure moving through Austria's historical experience with the Nazi Holocaust, I have to make such a story possible by, first, considering briefly some current objections to writing any histories about the Holocaust at all and, second, attempting to conceptualize a historical anthropology capable of narrating *specific* "long-duration" cultural histories that are punctuated by periodic collapses into concealed—but known—mass murders.

Eric Hobsbawm's concession to what he calls "serious" historians' acceptance of the Holocaust as inexpressible tragedy makes sense in terms of his concern that some Holocaust histories have indeed come to rely on legitimations that are themselves implicitly the ground for wholesale, annihilating formulas about classes of people.[1] Thus Yehuda Bauer characterizes the often-cited October 1943 speech at Poznan in which Himmler claimed that mass murder, performed as "duty towards our people," brings "no harm to our inner being, our soul," with a scapegoating formula about this "clear enunciation of a petit-bourgeois morality."[2] However, I find troubling Hobsbawm's failure to recognize his own exclusionary devaluation of a class of presumably "not serious" historians. These historians cannot apparently assume the pose of irreducible "incomprehension" in face of occurrences of genocide, as has been fashionable among some of the most visible participants (including Dan Diner, Arno Mayer, Charles Maier, even Raoul Hilberg) in the current debates.[3] A somewhat more interesting reading of Himmler's Poznan speech is offered by Saul Friedlander, who notices "dissonance" between Himmler's "innocent" genocide and his hedging injunction that it must never appear in the historical record. Friedlander's discussion promises a turn toward paying attention to "rhetorical" dimensions and to Freud's idea of the "uncanny" but disappoints when he, too, follows the intellectual fashion by declaring the Holocaust past as "fundamentally irrelevant for the history of humanity."[4]

Subsequently, Friedlander made an elegant recovery by attaching himself to a particular emergent "postmodern" reading of the Holocaust and by drawing on language by LaCapra, Lyotard, Blanchot, and others. With Yosef Yerushalmi he recognized that the Holocaust was an unprecedented trauma in Jewish experience. Unlike prior comparable horrors, it had not been closed in the collective memory by a redemptive myth and therefore offered an opportunity for cultural change, for an actual "working through" by means of memory work in which historians reveal themselves in the "commentary" they manage to attach to their Holocaust stories, commentary designed to go on resisting closure, to go on taking advantage of and contributing to the "excess" of memories attached to the Shoah (before time and the public desire for closure diminish it) in order to gain access to places that yet remain unopened and to integrate in unprecedented ways narratives of "background" and of Holocaust events.[5]

The outline offered below means to tell such a story, but one about perpetrators and bystanders whose redemptive myth never fully broke down before, during, or after the Holocaust.[6] The myth of *Heimat,* the hometown in its countryside, a space of gentle caring that remains outside of and far away from History, was as alive in 1943 as it was in 1985 when Edgar Reitz's epic film by that title appeared. It is one thing to sneer at the present turn toward ethnographic languages in attempts to historicize this myth; it is quite another to come to terms with the sources of this ascribed redemptive power of the home villages and with why, to return to Friedlander's justified puzzlement about Himmler's dissonances, the morally clean, necessary-for-the-village killings nevertheless had to be kept a secret.[7]

The story presented below seeks to build a history around information that reveals the fraudulent qualities of the structural and discursive formation called "village." The opening for such a story appears in Theodor Adorno's discovery, via Kogon, that many of the Auschwitz guards were dispossessed peasant sons whose exclusion stabilized their villages' social appearance sufficiently to continue to reassure urban white-collar workers that there was a secure rural life at the core of things.[8] To follow this lead and to find out what else connected village to Holocaust experiences is not as simple a task as historians have tended to make for themselves. Even the barefoot historian with tape recorder is in the end just another circulating signifier that the villagers have seen or heard stories about before and know how to "handle." To make progress, we have to confront some of the conceptual difficulties that stand in the way of such work, not the least of which is that historical anthropologists of villages are deeply divided into two opposing camps that cannot debate each other. The next section will explore aspects of that unwaged debate between what appear to be system-manipulating logician-grammarians, on one side, and system-deconstructing rhetoricians, on the other. The kind of story I want to tell remains impossible as long as these traditional divisions of the trivium divide cultural analysis. The first task is to make logic, grammar, and rhetoric circulate and serve us as the analytical rock-paper-scissors game they could be.

Anthropological History or Historical Anthropology?

It is useful to begin with Clifford Geertz's quip about anthropologists not studying villages but studying in villages, a charming bon mot that effectively covers its author's tracks.[9] By denying that anthropology is about any particular villages, it begs the question why anthropologists are precisely there and not here. They must be expecting to find something that they cannot find somewhere else, even if they think it is only a "simpler" version of some totality, something reproduced on a lesser scale, some perhaps more "authentic" enactments of "everyday life" perceived as "closer to the ground," as "local knowledge" living with "effects of the totality," and so on: all of which means they are indeed studying the villages they are in. The question is, why do some analysts have to misrecognize what they are doing by looking down from the height of a "comparative science of human culture"?[10] Does that mean that those who are there meaning to study the villages they are in are forever too immersed in particular conjunctures to have anything more than researched "facts" to offer up—no doubt with a respectful tug of the lock—to those earnestly suffering under a more or less private pathos of "doing science"?

It might be more interesting to have questions about divergent, possibly irreconcilable (and in mutual regard pathetic) purposes for doing "cultural science." Why are some content with achieving endlessly comparable taxonomies of the logics and grammars of possible cultural systems, aiming toward an all-controlling hermeneutics? Conversely, why do others find primary satisfaction in calculating the relative human advantages and damages incurred by the historical workings of particular systems? My own stake in the second approach seeks a connection between doing science and mobilizing a will for change by deconstructing, if only to destabilize it, the ongoing "edifying conversation" about purportedly systemic "necessities," by disclosing how such discourses permit and even requisition human destruction for the alleged "survival" of particularly favored systems. It seems obvious that both rationales for "doing science" would benefit from including something of the other, but debates between the two have had a hard time getting started.

Referring to one instance, we may point to Joan Vincent's critique of the ruling place given to static systems analyses over what she sees as analyses of particular historical processes. Her initiative was affirmed in spirit by Renato Rosaldo, who, however, immediately reabsorbed it into the terms of an internal debate among generational schools of formalist anthropologists upon whom he urged the study of what he calls "nonorder" within order as a superior research focus for understanding "consciousness, collective mobilization and improvisation in everyday life practices."[11] In his construction of "process," we see an attempt to modernize Durkheim by the conceptualization of a "space between order and chaos," a kind of social dream realm that exists in the unremarkable moments of so-called everyday life, lived in the expectation of a flow of natural time passing to allow for improvisation, for living change spontaneously, pre-systematically.[12] Is that to say, pre-linguistically? And do all participants in such moments(?) of everyday life share equally in an experience of unconstructed, unpressured, non-contingent time that some seem to feel exists? It is difficult to judge whether this was an opportunity missed for having a debate. Vincent's account of "process" analysis ended with a discussion of Gerald Sider's Newfoundland villages as particular locations for experiencing the manifold historicities of the world-system's various dis- and re-articulations and transformations of social-cultural formations. Sider's work without a doubt opens a window that one would suppose could have shown much to Rosaldo. The debate remained unwaged, however, with Rosaldo's essay finishing with a narrow perception of social microcosms (dinner party guests! and with Bunuel so far ahead in that game!) as locations for reenactments of general cultural "processes" that he reduces to variations in timing, in perceived breaches of etiquette, in gossip, and so on (that is, in the "informal" forms) and that then can appear recombined with more traditional "control" function concerns.

In the next section we shall return to consider the analytical usefulness of a more evolved version of processes in an everyday life space "between chaos and order," and move on to seeing several such spaces in a darker light as they simultaneously move about below, ahead of, and inside historical change. To help us do that, it is useful to think further, by means of the following example, about why a debate between the two approaches to the location of villages in larger systems' "processes" might not be wageable at all.

During the past three decades, some of the most significant historical analyses of village-level societies have focused on family and household forms and related demographic behaviors. One particularly prominent direction, calling itself "experimental history," emerged during the 1970s out of the Cambridge Group to follow Peter Laslett's lead toward creating mathematical models of statistically ascertainable boundaries and parameters surrounding and contained within possible historical family forms. Among the original intentions of the Cambridge Group's efforts had been the creation of standards for testing comparisons between different villages and between villages and their modeled counterparts.[13] Whereas one cannot but find such calculations useful and look for more, Laslett's more ambitious project for mining historical databases to create a universal mathematical family model with predictive powers relative to different choices of dependent and independent variables seems to have receded into the background and all but disappeared.[14] One can argue that it encountered and failed to solve a logical flaw in its model building: as a first step toward achieving a "universal" model, it had to posit an impossible primary village, one whose vital linkages with an "outside" world were disallowed in the model.

It was apparent early that the Laslett project might lose itself in obsessive taxonomy constructions accompanied by endless definitional tinkering, all to get rid of a so-called stem family thesis of already questionable analytical value. The portentous search "to find for us the world lost to time"—to be done only in terms of very few databases deemed sufficiently "trustworthy"—premised a kind of natural-historical "pre-industrial" European village-world of many fractal (that is, closed, replicable, and not scale bound) communities acting harmoniously in national systems.[15] In Laslett's words, "The whole pattern must therefore be thought of as a reticulation rather than as a particulation." He envisaged a village, however, that was, despite "perpetual negotiations with its neighbours," itself composed of self-contained webs of household and family relations and obligations that made modern economic rationality impossible.[16] His insistence on a closed pre-industrial village experience stands in sharp contrast to how other anthropologists and historians also currently involved in "village" analyses view, instead, the same systemically frozen villages as fluid, temporary, transitional locations where shifting and specific economic rationalities could not be avoided. In this view, Laslett's static late medieval European villages were actually undergoing "reinvention" to solve structural impasses in different European political economies, their historical appearance forming merely "a shell that was finally cast off with enclosure."[17]

In their account of the unfolding of the experimental history project, Wachter and Hammel reveal a crisis that occurred when they had to decide whether to model an open or closed village; that is, whether to require their marriage search algorithm to supply partners from an existing "closed" village population or from an open-ended village/world interaction where partners from outside were supplied "whenever the random numbers decreed marriage for an individual in the computer population."[18] While others developed this second strategy or went back to aggregate demographic statistics, Laslett's insistence throughout on a closed village model turned out to be self-defeating not only in terms of practical application but also in terms of the model's replicability in historical processes. Several years before Laslett began to formulate his "lost world," Eric Wolf was developing an antithetical social vision of rural villages that rejected "the then prevalent notion of community as a closed system" and instead "conceptualize[d] community as a nodal point of intersection among overlapping relational fields in open systems."[19] Wolf's was a widely debated model and still resurfaces periodically in different contexts; however, it obviously never reached the learned perambulations along the Cam. It seems all of a piece: the search for closure in a final, controlling algorithm, the closed village model, the autism of a transatlantic, computer-linked, interdisciplinary "science." When the metaphysical pleasures of the model, the pleasures of grammar and logic games, of putting more or less subtle forms through their paces, displace historical-anthropological analysis, then the latter, with its commitment to the pleasures of specific texts, is put on the defensive, even to the extent of returning the refusal to debate when any anti-historical withdrawal into the model threatens to breed monsters.[20]

The debate remains unwageable as long as the metaphysical commitments separating the two sides remain hidden in an unspoken opposition between transcendent integral selfhood (*logos*) and transcendent atomized willing (*nous*), forming an apparently irreconcilable divide whose intellectual history is traceable to the ancient Greeks' first entanglements with the Persian empire.[21] On the logocentric, logician-grammarian side, we hear such, from a rhetorician-Epicurean perspective somewhat naive, questions as "Is it still possible to organize this indistinguishable plurality of individual acts according to shared regularities?" or "How can we consider at one and the same time the irreducible freedom of readers[!] and the constraints meant to curb this freedom?"[22] Or, Peter Burke: "Was Montaillou . . . a typical medieval village? A typical Mediterranean village? . . . Not typical at all?" One might leave Burke to ask such questions in peace were he not adding confusion by claiming a ground he calls historical anthropology to construct something looking more like anthropologized history, seeking "to write the history of the 'grammar' of a culture and relate it to the messages emitted by individuals using this grammar, and the perception and interpretation of these messages by their recipients."[23] Against such closed historical semiotics claiming to be *the* interdisciplinary ground between history and anthropology it seems useful to insist on a heuristic distinction between "anthropological history" and "historical anthropology" to refer to types of analysis suitable, respectively, to closed or open system-models.

Reading the Logics and Grammars of "Dark Events" and "Lynching Scenes" in Historical Villages

My sympathies are with the historical anthropologists. At the same time, it will become obvious that the narrative outlined below owes a considerable debt to notions about the historical devolution of intertwined forms and structures. The next section's presentation of a narrative outline is bound to favor the structural side of the story and to underplay the power of the rhetorical analysts' revelations of the devil in the details. Nevertheless, the "formal" elements of the story are sufficiently strong to warrant a debate with anthropological history about the concepts that serve such analysis.

There is no denying the power of ritual performances and games as dimensions of historical processes, but such forms do not necessarily have to be seen as *the* source of order and social cohesion in particular systems. The naturalized identification of ritual with order misdirects analysts to look, with Rosaldo (or Victor Turner), for an imagined, "nonordered," "liminal" buffer zone of "everyday life" between chaos outside and order at the formal core. For historical analysis concerned with discovering and tracing processes leading to historical instances of mass murder, such an innocent representation of everyday life remains uselessly enchanted by its necessary sense of paralogical selves (however "decentered") performing entirely on this side of the pleasure principle. It misses the points (1) that it is particularly in games and performances governed by rules, not laws, and perceived as somehow outside formal historical processes that "lives can be at stake" and (2) that ritually disconnected human experience is most easily reducible to terrorized solitude whose occurrence enters, nevertheless, into historical "readings" by other performers.[24] In this light, it seems unacceptably ingenuous to see repetitions in ordinary life as relief from uncertainty and as the source of an underlying stability that makes it possible for a few to cross over into the non-ordinary to effect change.[25] Need one point out that the compulsions that often hide behind ordinary repetitions may have pathological qualities and be the greatest obstacles to change? It gives me goose flesh to read *German* historians discussing "communal" life as a "secret codex," a separate "mentality sphere" where it is possible to bury in the daily life of "common sense" and "real practice" publicly inadmissible and unbridgeable conflicts, where the "common man" may be relied on to have the necessary "discipline" to endure in his psychological spaces those unresolved public conflicts that cannot be admitted to public discourse without endangering the stability of the "higher complexity."[26] What continues to threaten the wageability of the debate for me is that, as I have reason to suggest via the narrative below, such formalist, neo-stoic formulas are deeply implicated in the intellectual-psychological construction and historical transmissibility of German-speaking holocaust forms.[27]

To undermine these undebatable differences and to discover how the disorder-necessary-for-order that has been displaced into spaces imagined variously as "non-order," "everyday life," and so on enters remembered experience, historical anthropologists need to stay within poaching distance of the middle-range theory precincts of the anthropological historians. One such hunting ground may be found, for example, in Benedict Anderson's somewhat mechanistic modernization narrative. He relegates the kind of time thinking caught by Auerbach's description of the temporal contiguities of promises and fulfillments contained in Biblical sacrificial memory to some pre-modern realm that contrasts with modern "homogeneous, empty time."[28] Anderson's argument about a revolution in temporal perception undergirding a newly imagined shift toward national "community" experience is too simple but nevertheless elicits two possibly creative observations. First, we have to understand and try to avoid his dependence on a tradition of thinking about collective memory that is grounded in Halbwachs and Durkheim and in Leibniz before that. In this unacknowledged, even overtly rejected functionalism, historical experience is lost in a distinction between "memory" and "history." "Memory" appears, in this view, only in a living conscious community acting out an imaginary, plotted, and more or less understood drama through succeeding generations; "history," meanwhile, is perceived as something different, as a discontinuously transformable substance of time references and data that do not depend on a lived continuity of communal experience but rather on its opposite, on the separation of a "reality" different from and outside of communal remembrance.[29] Such a distinction poses a couple of difficulties in that it burdens professional historians with the impossible obligation to become the transcendent self that is capable of existing outside the historical process. Not only are historians accordingly required to exclude themselves from communal processes of remembering, but they are denied the possibility that doing history could be an educated form of "memory work." Historians can reappear as members of the community ensuring that historical "realities" that are not admitted to exist in the community's conscious enactments of the memory play may be woven back into the action in ways that rewrite the play itself and that can, in this sense, constitute change.

One can make a second observation about Anderson's recollection of Auerbach's time of sacrificial fulfillment. Auerbach's perceived *contiguities* of historical experience, shorn of Anderson's simplistic vision of "modernized" time experience, offer a so-far-unexplored opportunity for holding evidently unconscious and pathologically repetitious elements present in Nazi violence against untenable current rationalizations of Nazi actions as "evil" irruptions of unpredictable willfulness into otherwise orderly, rational, linear time.[30] To explore further the historical-analytical potential of notions of sacrificial time, I draw on work by René Girard concerning the paradoxes of violence in community formation rituals. Girard offers a description of collective pathology as a dimension of the unease arising from known but unacknowledged, unconfronted collective crimes.[31] He envisions a perpetual historical malaise of experience and memory driving communities toward repeated and apparently spontaneous and unanimous selections of both perpetrators and victims for sacrificial acts that suppress an increasingly visible but unfathomable and therefore ineffable "original" crime that will not bear scrutiny for fear of endangering the existing order as it becomes too fragile under the growing mass of the unspeakable.

Girard's insight concerning prior and collectively enacted "dark events" that repeatedly authorize subsequent "lynching scenes," which he sees as displaced reenactments of those "forgotten" prior crimes by, I would say, historically (not "mythically") selected perpetrators and victims, cries out for transposition to a reconsideration of Nazi violence. It is out of no disrespect for the dead and the survivors of the Nazi Holocaust that my story argues, from Girard's perspective, that the terror they suffered was a kind of cover-up for domestically concealed forms of violence that had for centuries been tacitly accepted as necessary for maintaining the immanent integrity of a family ideal that was perceived as the central value of the Austrian-German community.[32]

Girard's reluctance to turn his vision toward historical analysis rests on his distinction between ordinary and extraordinary—that is, "ritual"—states. This seems a somewhat forced distinction when we are confronted by violent historical events and actions that do not respect its boundaries.[33] It would appear that the lived oscillations in everyday life between ordinary and ritual actions are too rapid for any but theoretical separation; in this sense we have to recognize that there are differences of occasion where the weight given to readings of expressly ritual events gains special importance but also that the spectrum and durations of, and the number of participants in, expressly ritual moments vary enormously. A considerable part of the following story is about such oscillations, about unceasing ritual displacements ingredient to everyday experience and necessary to the normal operations of Austria's provincial family life, simultaneously retaining and suppressing what Girard terms "the sheer proximity of knowledge" about the concealed crime.[34] It is to comprehend the paradoxical necessities of the continuous, repeated rhetorical subversions of the logic and grammar of communalist forms, extending into the most intimate spheres of everyday life, that we need to draw together the methods of anthropological history and historical anthropology.

A further theoretical element in our search for language about "necessary" violence may be found in Pierre Bourdieu's specific notion of "misrecognition," which is close to Girard's idea of a "successfully misunderstood victimage." Bourdieu: "I call *misrecognition* the fact of recognizing a violence which is wielded precisely inasmuch as one does not perceive it as such."[35] Bourdieu's formulation concretizes Girard's implicit challenge to historians to recover historical experiences of such dark events, to link them to their respective lynching scenes and to write conscious memory narratives outlining what is in effect a specific and collectively unconscious knowledge about both foundational and related "redemptive" acts of violence. One may see a *testable* quality in misrecognition and find historical moments when violence is simultaneously committed and denied by means of languages, institutions, and acts whose historical presence and fields of action remain to this day unaccounted for. One also has to take seriously Bourdieu's admonition that such narratives of misrecognition be embedded in "objective structures (price curves . . . etc.)" and in "social formations," and this, to be absolutely clear, is not to pay homage to any alleged material determinations in some metaphysics of a "last instance" but to allow us, rather, to deconstruct historical and communal discursive formations about *necessity* that required some to pay the price of a misrecognizable victimage.[36] The following narrative is an attempt at engaging with and putting to use this "formalist" research program in a history concerned with misrecognizing and annihilating tropes. The story that follows does not intend to reveal some absurd cause of or meaning for fascist murders but assembles rather a number of historical narratives, in themselves conventional, that in juxtaposition and in concert point toward unspeakable conjunctures, toward an interplay of historical contingencies, modal articulations, and unstable rhetorical erasures that finally required *this* fascism, *this* organized mass murder to be put into words and acted out.

Austrian Family Formations, Dispossession, and Descent into Holocaust

The present story is a sequel to and harvest of a previous story that focused on specific "modernizing" fiscal restructurings of the Hapsburg state during the sixteenth and the early seventeenth centuries and on the adjustments to these changes by Austrian family economies as they were revealed in peasant household inventories.[37] In the new revenue sharing and administrative partnership between Crown and Estates that emerged in the German provinces of the Hapsburg monarchy by 1650, peasants' "houses" had been "equalized" legally as the lowest tier of administrative entities in the state. Peasant farmers, on finding themselves in charge of these impartible tribute-producing "firms," subject to accounting and eminent domain supervision by various levels of public and private corporate authorities, reoriented their family operations toward the new conditions of tenure and impartibility. They developed a tax-sheltering system of *inter vivos* property transmission to advantage designated heirs, incumbents, and "retired" stem-elders. Their families were increasingly divided internally into a class of heir-incumbents and a class of effectually dispossessed family members who had to part in childhood from the home farm with only promises of future payments of residual inheritance portions that consisted of equal shares based on half of the farm's assessed net value.

The intent now is to follow the course of some of the historical experiences set in motion after 1650 by this new social contract. It concerns the historical devolution of the necessarily suppressed conflicts between the two social classes contained in peasant families and of corollary conflicts among the different economic and emotional calculations required by participants in the new family formations to absorb, replicate, and resist the strategies of the corporate and tribute authorities above them. The authorities' repeatedly restructured tribute demands reproduced themselves in peasant family life in the form of intensified competition over the allocation of family debt and the distribution and management of inheritance, all requiring difficult choices about inclusion and exclusion among parents, children, and siblings.

The first narrative focus is on the so-called reform period beginning around 1750 under Empress Maria Theresa. My argument is that the Theresian-Josefinian reforms sought to expand the Austrian dynastic-corporate tribute empire by entering into the inner workings of family and communal life at all social levels in order to make new and growing tribute claims on private fiscal arrangements. My sources for this part of the story are the account books of aristocratic estates and of ecclesiastical foundations as these latter were secularized. I also look at the minutes of business meetings and accounts of parish and artisan guilds as Estates and Crown increasingly pressured and absorbed their trust funds. Finally, I examine various authorities' trust fund accounts detailing the administration of the residual inheritance portions (so-called *Pupillengelder*) of the dispossessed peasant children. All these private funds were increasingly forced into public paper securities that functioned, in turn, as credit instruments circulated by the ruling dynasties and corporations to allow them to participate in the new global market competition that had begun in earnest with the War of the Spanish Succession. Already implicit in the mid-seventeenth-century arrangements, now, a century later, it was a more tightly intertwined articulation of family economies at several social levels, with tribute forms of public finance whose actively concealed operations were experienced privately and invisibly and, so far, ineffably in a destructive calculus of necessary alliances within and exclusions from families and communities.[38]

An analysis of peasant household inventories reveals basic adjustments that peasants made in their family economies. If we treat the peasant farms and households revealed in the inventories as business firms, we can assess the management of their assets by means of statistical analyses of the changing relationships among such selected components of the house economy as equity, debt, liquidity, inheritance allocations, and so on. Moreover, by sorting the inventories into overlapping legal status, wealth, and occupational groupings and by gathering them into four temporal blocs (1649–69, 1710–30, 1770–85, and 1790–1802), we can project a multifaceted survey of the peasant householders' strategic and tactical adaptations over time, and we can detect emerging through the detail a clear mid-eighteenth-century dividing line between different directions and consequences of family business strategies.

It is my sense from the inventories that the Theresian-Josefinian reforms were experienced, first, as a destabilization of the heir incumbents' sense of control over their family economies and, second, as a costly restructuring when the heirs had to develop a new fit with the increasingly limited trade and capital markets that accompanied the further elaboration of the tariff and trade monopoly walls that surrounded the Hapsburgs' "free trade" empire. There is not the space here to indicate what I find about the "reagrarianization" of the peasantry, the tendencies to hold equity rather than debt, the changes in the transport sector, and so on. These and other changes indicate that the peasantry was turning away from market transactions in a larger radius and turning toward further sheltering and concealing their more limited incomes within family and communal accounts. Most significantly, they protected themselves against the incursions of the tribute state by increasing the rate of dispossession, thereby reducing the pressures of state-supported residual inheritance claims on incumbent heirs. In this regard a central narrative pivot occurs in the inventory analysis when we learn that at a time when, as a result of Theresian-Josefinian inheritance "reforms," the number of those with claims should have been increasing, in practice this number actually decreased sharply and fell even below seventeenth-century levels.[39] How this was done and what resulted is the subject of the second half of the story, whose focus is on the historical experience of the dispossessed in late imperial Austria.

The focus for this second part is on the private and public institutions that managed both the practices and results of this general increase in the rate of dispossession. To grasp the hegemonic ideology operating in these institutions, we need to rethink the intellectual history of the textual sources, logical constructions, and implications of the neo-stoic and cameralist languages in which the Austrian tribute empire conducted its business and social transactions.[40] We can make an argument that the Austrian bureaucratic aristocracy's imposition of a dangerously modified neoclassical household utopia and of its accompanying ideal personality type required the heads of households to overcome any moral dilemmas they might have faced in the exercise of their duties, including the forced selection-for-disinheritance and worse among their own children, all for the sake of maintaining the ongoing viability of the tribute-producing quality of their house. Woven into the ethical and, above all, *disciplining* precepts that informed intertwined systems of structurally (and therefore "practically") necessary exclusions from households of, successively, origin, employment, and welfare, was a philosophically elaborated subordination of family attachments to the requirements of a police state seeking to realize a metaphysical human transformation that explicitly devalued the disinherited even to the point of their physical annihilation.

Looking at the "life course" of the dispossessed, at the endangering and annihilating experiences of that great majority of children who did not inherit or who were already children of previously disinherited children, we are, arguably, looking at the concealed "dark events" in Girard's equation. From the time of their birth, those who had to be necessarily selected for dispossession were in constant mortal danger in ways that were not simply the "natural" result of "poverty" but were intrinsic to the designs of families and institutions that, on one hand, required the selection of family members for terminal, often deadly expulsion while, on the other, they espoused redeeming family and community values. Part of the experience of dispossession in this system was not to have the means to defend oneself against highly theorized and pervasive institutions and practices that were conducting successfully misrecognizable forms of human destruction visited primarily on those who, for structural reasons, were separated from and could not found families.

In outline, I focus on the successive life-course phases of birth, coming of age, employment, illness, old age, and dying, and I find, in each of these areas of experience, particular inversions, negations, reversals, and cover-ups by which expressly positive public languages and institutions affirming social membership and caring became in practice attached to and designated exclusion, neglect, and death. Thus: state-founded gynecological institutions became agencies for infanticide; corporatively and publicly managed trust funds consisting of residual inheritance portions of the dispossessed were "necessarily" eroded by aristocratic fund managers and state treasury officials; employment became migrant unemployment and progressively further disconnection from family and home communities; unemployed migration became capture and forced conveyance by police authorities who practiced attrition rather than repatriation. A life of dispossession often ended in hospitals, asylums, and other welfare installations that did not so much dispense care as serve as collection and holding stations for persons on the way to the charity death wards. All these institutions were in the hands of a hierarchy of academics, administrators, medical and legal professionals, police officials, and service staffs whose fundamental values and practices, grounded in "modernized" neo-stoic languages, became the standards—which recursively entered the daily experience of private family life—by which everyone who entered the "process" was routinely and repeatedly classified as worthwhile or expendable.

The stage prop dominating the background for the last part of the story, covering the period from the mid-nineteenth century until after World War II, is this Austrian (imperial, republican, and fascist) social machine grinding out surplus dispossessed children for absorption by destructive communal-state processes and institutions. To construct a narrative bridge between these ongoing dark events and the lynching scenes that followed during the 1940s, we can trace the planning, founding, and operation after 1850 of one of the institutions that supported the social management of the necessary dispossession, the provincial mental hospital Niedernhart in Linz. Reading the planners' documents and debates and observing the institution in action, one is led to the conclusion that its primary function was not to care for and possibly heal mentally ill people but to remove and restrain unmanageables who, for one reason or another, were too "costly" to continue in their communities. We finish by considering Niedernhart's subsequent liaison with the euthanasia station at nearby Hartheim castle and with the operations of the concentration camp Mauthausen. The featured performers in this final story about the devolution of a psychiatric hospital into an assembly ward for wholesale murders of the "unfit" are the same judicial and medical officials of the province who carry on "normal" operations while they are making their institutions, talents, and authority available for the physical extermination of persons whose lives were judged to be, in the correlate sociobiological language, "unworthy of living."

At this point, the historical circle connecting dark events and lynching scenes closes in at least two respects: first, in terms of familial agreements about inheritance and labor economies that became dependent on correlate institutional-ideological and professionalized processes of (misrecognizable) murders that serviced the larger social contract between kin and tribute; second, in the Nazis' pathological, impossible appeal to a community of blood claims informing a vengeful, paranoid, and violent politics of racial dis- and repossession, all incongruously aiming at realizing a positive, benevolent European empire on foundations of necessary and politically acceptable dispossessions, deportations, and mass murders.

I want to finish by coming back briefly to the life-course experiences of the dispossessed in this system, to make clearer the key idea that the human destruction that accompanied the centuries of dark events was not visited simply on a small minority of particularly unfortunate people but happened on a scale so large that one can only marvel at the effectiveness of the rhetorical displacements and concealments of these practices. One has to be shocked by the brutal clarity of the historical traces and even by the, in effect, open admissions of murder and other forms of destruction that remain clearly visible in the sources, if one chooses to look at specific levels of detail. It took a couple of readings of the earliest (1795–99) protocols kept at the Linz gynecological hospital for me to pay sufficient attention to the relatively small crosses that indicated a death and to begin to count and combine figures only to discover that in the period the first-year death rate among children born at and left "in care" of the institution was 85 percent, reaching 95 percent in 1797. One has to read court interrogations of infanticidal mothers repeatedly and in the varied contexts of the interrogations of other family members to recognize that infanticide was often not just a matter of poor women killing to avoid shame or to be able to continue working but was also in the domain of fraudulent deals being struck within families, in which a pregnant daughter already selected for dispossession agreed to carry out the necessarily violent removal of a potential additional inheritance claim in return for a continuation of her own claim against the patrimony—a tragic bargain in which not only a child had to be murdered but the infanticide herself carried all the remaining risks and was in mortal danger throughout.

We find languages of care and of social membership commanding, simultaneously, terminal neglect and expulsion, reoccurring at every "stage" in the life course of the dispossessed. A look behind the scenes into the various corporate and state authorities' account books detailing the handling of the trust funds of minors (who were called orphans whether they had family or not) reveals high rates of attrition and outright nonpayment of these funds, as even the best-managed tribute corporations holding, collecting interest and fees from, and trading in the bonds that secured the system went through cycles of insolvency, "conversion," monetary devaluations, and so on. This second "procedural dispossession" was experienced individually, privately against a hegemonic public chorus of endless protestations of paternal trustworthiness even as it marked for many young adults the loss of a last connecting link to families of origin and was a further step along a road of social and often physical annihilation.

Many descended into a world of lifelong bachelorhood (itself a mark of congenital incompetence, mandating lifelong wardship and tutelage) and of temporary labor that contained its own misrecognizable forms of life-threatening violence.[41] Languages about steady and communally respectable employment covered actual practices of unrenewable one-year labor-service contracts—whose day-to-day status could be read in everyday rituals of silencing and exclusion—which forced job searches increasingly far from the home community and segued into a life of "vagabondage," which often ended, when aging and illness rendered laborers uncompetitive, in the hands of the "conveyance" authorities (in the so-called *Schubsystem*). These authorities' job was to return persons to more or less fictitious "home communities," and their practices, observed in local police authorities' extradition agreements, arrest and transport records, and other administrative communications, in effect "selected" those who would survive by keeping them out of the system. The rest would be shunted from prison to detention center and back until they either escaped and disappeared—their "wanted" circulars reveal such practices as the experimental tattooing of identification numbers on their arms—or until they could be delivered to the charity death wards. There, the sources reveal, service staff went through ritual-yet-ordinary motions of conducting post-mortem testamentary closures (*Sperren*), inventorying "property" and properly inquiring after possible heirs—all in all, a bizarre inversion considering who had just died, under what circumstances, and with what pitiful heirlooms. During a dispossessed person's entire life-course experience, he or she had ample opportunity to observe the inheriting siblings and their bureaucratic partners engaging in "correct" rituals, featuring in particular various gestures of caring-after-death as a kind of readmission of the deceased dispossessed to the world of social membership, redemptive tropes that primarily served to exonerate the mechanisms of dispossession that had required and hastened these deaths.

The imperial Austrian state's particular articulations of tribute and family relations had resulted in a social contract that had to be realized by institutions ready to receive, process, and quietly eliminate those who could not be allowed to press inheritance or welfare claims against their families or communities. In time, the flow of Nazi directives for specific exterminations simply initiated a shifting by these interlocking institutions, personnel, and practices into higher, as yet untried gears for an accelerated and infinitely expansile circulation of fatal signifiers. From this perspective, the well-known gray buses with blacked-out windows that combed the countryside around Linz looking for "life unworthy of living" or delivering cargoes of those identified as "living corpses" from the railheads to Mauthausen and from there to Niedernhart and Hartheim, had at least as much to do with the old "conveyance system" as with the, for some, exciting new fascist science of "racial hygiene." The latter is also implicated, of course, when it speaks in German about "racial" features as inheritance (*Erbgut*) and unconsciously, linguistically, connects blood-membership with a claim to inheritance. Finally, in the light of the theoretical effort outlined above at reuniting logic, grammar, and rhetoric in historical anthropology, we can now view the buses that helped implement this "scientific" program as rhetorical constructs in the memory performances by which Nazis simultaneously revealed and concealed the everyday-life betrayals and destructions of those who had been, for more than three centuries, selected both for dispossession and for implementing the dispossession.

Acknowledgments

An earlier version of this chapter appeared under the title "Dispossession in the Communal Memory: An Alternative Narrative about Austria's Descent into Holocaust," in *Focaal* 26–27 (1996). Even earlier versions were presented in Budapest and Salzburg during the spring of 1995 and in New Haven during the fall of 1995. Many thanks go to Julia Szalai, Janos Bak, Sepp Ehmer, and James Scott.

Archival research for this project was done at the Upper Austrian Landesarchiv in Linz, Austria, and was funded by the John Simon Guggenheim Foundation, the American Philosophical Society, and the University of Arizona. Release time for writing was made possible, in 1991, by the Wenner-Gren Foundation and by the Harry Frank Guggenheim Foundation.

Notes

1. Eric Hobsbawm, in *Die Zeit* 37 (Sept. 16, 1994): 13–14.
2. Bauer, *Holocaust and Genocide,* p. 38.
3. See Habermas, "Die zweite Lebenslüge der Bundesrepublik," and other contributions in the same volume by Dubiel, Frank, and Beck; see also Maier, *The Unmasterable Past,* pp. 34–39 and passim. Does the next Great Coalition lurk in the fundamental agreement between Ernst Nolte and Arno Mayer that the Nazis were primarily motivated by anticommunism as well as in their corollary explanation that if the Nazis had been allowed to take Eastern Europe and Russia there would have been no Judeocide, only some, presumably acceptable, deportations to Africa? These ideas resonate curiously with the acceptability of *containable* ethnic cleansings and genocides found in the formulations of leading policy planners in the present moment (see Nolan, *Global Engagement*); a ground-clearing pragmatist description of Nazism as an ultimately "inexplicable" ir-

ruption of ahistorical willfulness into otherwise orderly communal processes may be found in Putnam, *Reason, Truth, and History,* pp. 170–71, 203–16.

4. Friedlander, "The 'Final Solution,' " pp. 23–35.
5. Friedlander, "Trauma, Memory, and Transference."
6. It is interesting to find it, for example, told with a straight face from the author's personal experience in the best current survey of Austrian social history (Bruckmüller, *Sozialgeschichte Österreichs,* p. 527).
7. Maier, *The Unmasterable Past,* pp. 118–19. If one really were to begin a critique of the Heimat syndrome, one could say that other films show a completely broken mythology of "the country," one for which there has yet to be a redemptive reclosure. I am thinking of films like Helma Sanders-Brahms' *Germany, Pale Mother.* (See the discussion in Kosta, *Recasting Autobiography,* pp. 121–52).
8. For a discussion of Adorno's point, see Rebel, "Cultural Hegemony," 118–19.
9. Clifford Geertz, quoted in Kalb, Marks, and Tak, "Historical Anthropology: The Unwaged Debate," programmatic statement for a special issue of *Focaal: Journal of Anthropology* 26–27 (1996).
10. Clifford, *Person and Myth,* p. 24; as far as Geertz is concerned, one has to add in all fairness that he seems to have come around to a view that admits that "it has become harder and harder to separate what comes into science from the side of the investigator from what comes into it from the side of the investigated" (Geertz, "Disciplines," p. 102).
11. Rosaldo, "Putting Culture in Motion," 105; Vincent, "System and Process."
12. If perhaps not *the* intellectual ancestry of Rosaldo's notion of a "space between," then at least its modernist intellectual ground might be found in Michel Foucault's peculiar appropriation of Binswanger's dream theory and its conflation with conceptions by Broch, Blanchot, Bataille. Rosaldo finally loses himself in trivia, perhaps to avoid this dangerous intellectual neighborhood where the free space of an imaginary existing, allegedly, between form and chaos allows the recognition and actual enactment of rhythms of violent destruction and sacrifice. That it is a dangerous space would appear from an observation that these confusions (persuasively told in Miller, *The Passion of Michel Foucault*) could explain why Foucault ended his life as a murder-suicide.
13. Wachter, Hammel, and Laslett, *Statistical Studies,* p. 12 and passim.
14. See Rebel, "Peasant Stem Families," p. 258; this article surveys some of the early criticism, especially by Lutz Berkner, of the Laslett project. And there are, of course, other ways to read Wachter's and Hammel's absorbing account, "The Genesis of Experimental History."
15. Wachter and Hammel, "The Genesis of Experimental History," p. 406. This is to take nothing away from the gains claimed by the title of this Festschrift for Peter Laslett. They are substantial, without question, but whether they are inroads into a "lost world" (as the authors seem to conceive loss) is open to question.
16. Laslett, *The World We Have Lost,* pp. 59–60, 80–83, and passim.
17. The quote is from anthropologist W. Roseberry ("Potatoes, Sacks, and Enclosures," p. 44). It is interesting to find a somewhat comparable perspective appearing in the same year, 1991, in a completely different part of the intellectual landscape in historian Wunder ("Die ländliche Gemeinde").

18. Wachter and Hammel, "The Genesis of Experimental History," p. 392.
19. Ghani, "Writing a History of Power," p. 37; Ghani's reference is to several essays that Eric Wolf wrote during the late 1950s, the best known of which is "Closed Corporate Peasant Communities in Mesoamerica and Central Java."
20. Rebel, "The Austrian Model for World Development."
21. See Nietzsche on Parmenides and Anaxagoras in his *Die Philosophie im tragischen Zeitalter der Griechen.* The subsequent divisions of these metaphysical positions are best approached through the excellent compilation and commentary by A. A. Long and D. N. Sedley, *The Hellenistic Philosophers.* A key text is Kant's discussion of the paralogism of the self (*Critique of Pure Reason,* p. 328 ff). Windelband's tracing of the histories of "learning" and "will" in *Die Geschichte der neuren Philosophie* is part of the story. The explosion of "schools" in the present century is beyond any note, but see, for example, the quality of the current discussion on one side by Schulthess (*Am Ende Vernunft*); and, on the other, by De Man (*The Resistance to Theory*) and also Schmidt ("Hegel und Anaxagoras").
22. Chartier, "Texts, Printing, Readings," p. 156.
23. Burke, *Historical Anthropology,* pp. 4–5. My distinction between "historical anthropology" and "anthropological history" accords with Kalb, Marks, and Tak, "Historical Anthropology." A useful discussion of this terminology may be found in various contributions to Silverman and Gulliver, *Approaching the Past,* most notably the essay by Rogers, "The Anthropological Turn in Social History."
24. These two points are made, respectively, by Baudrillard (*Seduction,* pp. 132–37) and Lefebvre (*Everyday Life,* pp. 165 ff., 179–88, 192–93); see my critique from this perspective (Rebel, "Cultural Hegemony," pp. 124–25 and passim).
25. Lüdtke, "What Is the History of Everyday Life," pp. 5–6 and passim.
26. Mörke, "Die städtische Gemeinde"; similar sentiments are found in contributions by Blickle, Press, Hauptmeyer, Kaschuba, Schmidt in the same volume. For a review that expands on these thoughts see *Journal of Modern History* 67, no. 1 (1995): 203–6.
27. Also relevant to this line of thinking is Sluga, *Heidegger's Crisis,* and the unfairly maligned Farias, *Heidegger and Nazism.*
28. Anderson, *Imagined Communities,* pp. 28–40.
29. Halbwachs, *The Collective Memory,* pp. 78–83 and passim.
30. Putnam, *Reason, Truth, and History,* pp. 170–71, 203–16.
31. Girard, "Generative Scapegoating"; also useful in this regard is Nancy, *The Inoperative Community.*
32. Rebel, "Reimagining the *Oikos.*" Relevant also are Adorno's Auschwitz guards; see Rebel, "Cultural Hegemony," pp. 118–20. The distinctive features of *German* peasant family values in Austria appear in Rebel, "Peasantries under the Austrian Empire."
33. Nancy, *The Inoperative Community,* p. 46.
34. Girard, "Generative Scapegoating," p. 100.
35. Bourdieu and Wacquant, *An Invitation to Reflexive Sociology,* p. 168.
36. Bourdieu, *Outline of a Theory of Practice,* pp. 21–22.
37. Rebel, *Peasant Classes.*
38. This seems the appropriate spot to acknowledge the debt for these and other formulations that I owe not only to Eric Wolf's *Europe and the People without History* (1982)

but also to his entire body of writing, particularly his early brief summary of a currently resurgent, and to my mind now unnecessarily impoverished, conception of "cultural ecology" analysis (Wolf, "The Study of Evolution"; we now also have his further thoughts along these lines in *Envisioning Power*).

39. Rebel, "Peasants against the State," p. 19.
40. See Rebel, "Reimagining the *Oikos.*"
41. Important comparative data and analyses in this direction may be found in Ehmer, *Heiratsverhalten.*

References

Anderson, B. 1983. *Imagined Communities: Reflections on the Origin and Spread of Nationalism.* London: Verso.

Baudrillard, J. 1990. *Seduction.* New York: St. Martin's.

Bauer, Y. 1991. "Holocaust and Genocide: Some Comparisons." In Peter Hayes (ed.), *Lessons and Legacies: The Meaning of the Holocaust in a Changing World.* Evanston: Northwestern University Press.

Bourdieu, P. 1977. *Outline of a Theory of Practice.* Cambridge: Cambridge University Press.

Bourdieu, P., and L. Wacquant. 1992. *An Invitation to Reflexive Sociology.* Chicago: University of Chicago Press.

Bruckmüller, E. 1985. *Sozialgeschichte Österreichs.* Munich: Herold.

Burke, P. 1987. *The Historical Anthropology of Early Modern Italy.* Cambridge: Cambridge University Press.

Chartier, R. 1989. "Texts, Printing, Readings." In L. Hunt (ed.), *The New Cultural History.* Berkeley: University of California Press.

Clifford, J. 1992. *Person and Myth: Maurice Leenhardt in the Melanesian World.* Durham, N.C.: Duke University Press.

Ehmer, J. 1991. *Heiratsverhalten, Sozialstruktur, ökonomischer Wandel: England und Mitteleuropa in der Formationsperiode des Kapitalismus.* Göttingen: Vandenhoeck and Ruprecht.

Farias, V. 1989. *Heidegger and Nazism.* Philadelphia: Temple University Press.

Friedlander, S. 1991. "The 'Final Solution': On the Unease in Historical Interpretation." In P. Hayes (ed.), *Lessons and Legacies: The Meaning of the Holocaust in a Changing World.* Evanston, Ill.: Northwestern University Press.

———. 1994. "Trauma, Memory and Transference." In G. H. Hartman (ed.), *Holocaust Remembrance: The Shapes of Memory.* Oxford: Blackwell.

Geertz, C. 1995. "Disciplines." *Raritan* 14 (3).

Ghani, A. 1995. "Writing a History of Power: An Examination of Eric R. Wolf's Anthropological Quest." In J. Schneider and R. Rapp (eds.), *Articulating Hidden Histories.* Berkeley: University of California Press.

Girard, R. 1987. "Generative Scapegoating." In R. G. Hamerton-Kelly (ed.), *Violent Origins.* Stanford: Stanford University Press.

Habermas, J. 1993. "Die zweite Lebenslüge der Bundesrepublik: Wir sind wieder 'normal' geworden." In S. Unseld (ed.), *Politik ohne Projekt? Nachdenken über Deutschland.* Frankfurt: Suhrkamp.

Halbwachs, M. 1980. *The Collective Memory.* 1950. Reprint, New York: Harper.

Kant, I. 1965. *Critique of Pure Reason.* N. K. Smith edition. New York: St Martin's.

Kosta, B. 1994. *Recasting Autobiography: Women's Counterfictions in Contemporary German Literature and Film.* Ithaca: Cornell University Press.

Laslett, P. 1973. *The World We Have Lost.* New York: Scribner's.

Lefebvre, H. 1971. *Everyday Life in the Modern World.* New York: Harper.

Long, A. A., and D. N. Sedley. 1987. *The Hellenistic Philosophers.* Vol. 1. Cambridge: Cambridge University Press.

Lüdtke, A. 1995. "What Is the History of Everyday Life and Who are Its Practitioners?" In A. Lüdtke (ed.), *The History of Everyday Life: Reconstructing Historical Experiences and Ways of Life.* Princeton: Princeton University Press.

Maier, C. 1988. *The Unmasterable Past.* Cambridge: Harvard University Press.

Man, P. de. 1986. *The Resistance to Theory.* Minneapolis: University of Minnesota Press.

Miller, J. 1993. *The Passion of Michel Foucault.* New York: Doubleday/Anchor.

Mörke, O. 1991. "Die städtische Gemeinde im mittleren Deutschland, 1300–1800: Bemerkungen zur Kommunalismusthese Peter Blickles." In P. Blickle (ed.), *Landgemeinde und Stadtgemeinde in Mitteleuropa: Ein struktureller Vergleich.* Munich: Oldenbourg.

Nancy, J.-L. 1992. *The Inoperative Community.* Minneapolis: University of Minnesota Press.

Nietzsche, F. 1994. *Die Philosophie im tragischen Zeitalter der Griechen.* Stuttgart: Reclam.

Nolan, J. E., ed. 1994. *Global Engagement.* Washington: Brookings Institution.

Putnam, H. 1981. *Reason, Truth and History.* Cambridge: Cambridge University Press.

Rebel, H. 1978. "Peasant Stem Families in Early Modern Austria: Life Plans, Status Tactics and the Grid of Inheritance." *Social Science History* 2, no. 3.

———. 1983. *Peasant Classes: The Bureaucratization of Property and Family Relations under Habsburg Absolutism, 1511–1636.* Princeton: Princeton University Press.

———. 1989. "Cultural Hegemony and Class Experience." *American Ethnologist* 16, no. 1.

———. 1991. "Reimagining the *Oikos:* Austrian Cameralism in its Social Formation." In J. O'Brien and W. Roseberry (eds.), *Golden Ages, Dark Ages: Imagining the Past in Anthropology and History.* Berkeley: University of California Press.

———. 1992. "The Austrian Model for World Development: A Neoclassical Excitation." *East Central Europe/L'Europe du Centre-Est* 19, no. 1.

———. 1993. "Peasants against the State in the Body of Anna Maria Wagner: An Austrian Infanticide in 1832." *Journal of Historical Sociology* 6, no. 1.

———. 1999. "Peasantries under the Austrian Empire, 1300–1800." In T. Scott (ed.), *European Peasantries.* London: Longman.

Rogers, N. 1992. "The Anthropological Turn in Social History." In M. Silverman and P. H. Gulliver (eds.), *Approaching the Past: Historical Anthropology through Irish Case Studies.* New York: Columbia University Press.

Rosaldo, R. 1989. "Putting Culture in Motion." In *Culture and Truth: The Remaking of Social Analysis.* Boston: Beacon Press.

Roseberry, W. 1991. "Potatoes, Sacks and Enclosures in Early Modern England." In J. O'Brien and W. Roseberry (eds.), *Golden Ages, Dark Ages: Imagining the Past in Anthropology and History.* Berkeley: University of California Press.

Schmidt, G. 1990. "Hegel and Anaxagoras." In M. Riedel (ed.), *Hegel und die antike Dialektik.* Frankfurt: Suhrkamp.

Schulthess, P. 1993. *Am Ende Vernunft—Vernunft am Ende? Die Frage nach dem logos bei Platon und Wittgenstein.* Sankt Augustin, Germany: Academia.

Sluga, H. 1993. *Heidegger's Crisis: Philosophy and Politics in Nazi Germany.* Cambridge: Harvard University Press.

Silverman, M., and P. H. Gulliver (eds.), 1992. *Approaching the Past: Historical Anthropology through Irish Case Studies.* New York: Columbia University.

Vincent, J. 1986. "System and Process, 1974–1985." *Annual Review of Anthropology* 15.

Wachter, K.W., and E. A. Hammel. 1986. "The Genesis of Experimental History." In L. Bonfield, R. M. Smith, and K. Wrightson (eds.), *The World We Have Gained.* Oxford: Blackwell.

Wachter, K. W., E. A. Hammel, and P. Laslett. 1978. *Statistical Studies of Historical Structure.* New York: Academic Press.

Windelband, W. 1911. *Die Geschichte der neuren Philosophie.* 2 vols. Leipzig: Breitkopf and Hart.

Wolf, E. 1957. "Closed Corporate Peasant Communities in Mesoamerica and Central Java." *Southwestern Journal of Anthropology* 13.

———. 1964. "The Study of Evolution." In S. Tax (ed.), *Horizons of Anthropology.* Chicago: Aldine.

———. 1982. *Europe and the People without History.* Berkeley: University of California Press.

———. 1999. *Envisioning Power.* Berkeley: University of California Press.

Wunder, H. 1991. "Die ländliche Gemeinde als Strukturprinzip der spätmittelalterlich-frühneuzeitlichen Geschichte Mitteleuropas." In P. Blickle (ed.), *Landgemeinde und Stadtgemeinde in Mitteleuropa: Ein struktureller Vergleich.* Munich: Oldenbourg.

PART II *Agricultural Production and the Peasant Experience*

CHAPTER THREE

Agrarian Issues during the French Revolution, 1787–1799

PETER JONES

There is no comprehensive agricultural history of France, nor is it likely that one will ever be written. According to Lucien Febvre, "la France se nomme diversité" (France can be called diversity), and agronomists tend to agree. For statistical purposes modern France is divided into 473 discrete *pays,* or micro-agricultural regions, holding characteristics in common that distinguish them from their neighbors.[1] Nonetheless, it *is* feasible to study the agricultural, or rather the agrarian, history of France on a more confined scale. Thanks to a vigorous tradition of local and regional monograph writing, there exists today a rich and accessible collection of primary and secondary source material. This material varies in quality, and also in inspiration. But the best of it flows from the "genre de vie" school of geographer-historians founded in the earlier part of the twentieth century by Vidal de la Blache.[2] This, of course, is the main inspiration of *Annales* historiography.

So, it is possible to mount a serious investigation of a temporally or spatially defined slice of French history. Such is the purpose of this chapter. It brings together a great deal of fragmentary research into the agrarian experience of France at a momentous time in that country's history. As a rough-and-ready principle of organization, I have adopted a triple perspective: that of the state, that of the landlord class, and that of the peasantry. It scarcely needs emphasizing these days that the state (that is, rulers and their bureaucracies) was an important actor on the stage of agrarian reform. Ancien régime France was a large, territorially united country ruled by an absolute monarch whose bureaucracy was the envy of Europe. Moreover, this powerful state machine radically refashioned its priorities after 1789, which lends added interest to the whole question of the state as an "autonomous actor."

The landlord class consisted of the nobility, institutional owners of land belonging to the Catholic Church, and wealthy commoners. They, too, had a well-defined agenda of agrarian reform by the time the ancien régime drew to a close. More significantly, they also possessed appropriate vehicles for the articulation of these objectives in the political arena. In practice, though, the changes that landlords envisaged coincided with those of the state (in its pre-1789 configuration) to a large degree. By contrast, the peasantry had no consciously worked-out agenda for agrarian reform before 1789, and for two very good reasons. To start with, they lacked the means of making themselves heard, or at least such means as were available tended to be prohibitive (foot-dragging, "collective bargaining by riot," and so on) rather than affirmative. Second, the 19 or 20 million agricultural producers (67 percent of the total population in 1789) whom contemporaries would have recognized as peasants were incapable of formulating a joint program. The anxieties preoccupying landless laborers were necessarily different from those of tenant farmers, just as the worries of sharecroppers differed from those of self-sufficient polyculturalists or cash-crop vinegrowers.

That said, there was also an irreducible minimum that united nearly all echelons of the peasantry. Issues of subsistence, of surplus extraction, of culture, and of custom had a wide import. It is also worth noting that certain agrarian issues mobilized all the actors in the play. The debate over the commons is a good example. Rulers, administrators, landlords, and peasants—large or small—all had an interest in how they should be exploited. Because it is not possible to explore each and every agrarian issue within the space of a brief chapter, I shall concentrate on those topics where the interests of the protagonists overlapped. Such overlaps often generated tensions, and hence testimony as to what peasants really thought about the land, the rural community, the proper targets of collective action, and so forth.

One final remark about the approach adopted here. It is that of a historian raised on a substantial diet of Marxist social history in the late 1960s, and Annaliste structuralism and cultural anthropology in the 1970s and early 1980s. Subsequently, I have rediscovered political history, so to speak, as a result of a meditation on the role of what post-Braudel Annalistes would call "matrix-events." These are events of such potency (like the French Revolution or the collapse of Soviet communism?) that they triggered processes on their own.

None of these influences have prompted me to indulge in much cross-cultural speculation. For I remain a historian of France, and an archive-based one at that. My angle of comparison is internal rather than external. In recognition of the interdisciplinary thrust of this volume, however, I have extended that angle wherever possible. At least, I present the abundant data on the French peasantry in a way that is meaningful to social scientists and agricultural economists.

Reform Agendas

As far as eighteenth-century rulers were concerned, "agriculture" and the "peasantry" were not self-defining objects of interest. In a war-mongering century, the state viewed agrarian reform first and foremost as a tax-raising exercise. Ameliorative considerations (what Enlightenment writers dubbed "bienfaisance") came second. The pre-1789 French government had no Ministry of Agriculture or Bureau of Internal Affairs; instead, proposals for institutional change (partition of common land, drainage schemes, curtailment of collective rights, the standardization of weights and measures, and so on) were handled by the Controller-General of Finances. As for the minutiae of leases and modes of tenure, they were a matter of private treaty between individuals and therefore sacrosanct. No government in the period covered by this study was prepared to intervene in deals "freely" struck between landowners and tenants. It is true that when the revolutionary government started to dismantle the feudal regime after 1789, it discovered "unfree" forms of tenure that were more feudal than civil in character. These were abolished by and large.

Nevertheless, there can be no doubt that agrarian concerns were intruding increasingly upon government thinking during the second half of the century. This was true not merely of France but of a number of western European states. The motivation was still more fiscal than benevolent, although rulers were adept at disguising their calculations in the humanitarian language of the Enlightenment. Also, the language of political economy was starting to make an impact: specifically the equation of a wealthy peasantry with a wealthy state. In the case of France, we can summarize the growing apprehension of the importance of agriculture to the national economy under two headings: recognition of the need to maximize productive capacity, and the determination to enforce a redistribution of peasant surplus.

The land underproduced for all sorts of reasons. In certain regions, large areas lay waste for want of adequate population to farm them; large areas were designated as common and used mainly as rough pasture; large areas remained waterlogged for much of the year; and finally, between a third and a half of arable land lay fallow at any one time. The government reasoned that if more land was brought into regular cultivation, tax receipts would rise. Equally, fiscal capacity would improve if institutional restraints on the activities of existing cultivators were removed and market opportunities were enhanced. Such considerations prompted a spate of legislation in the 1760s and 1770s, encompassing land-clearance measures (*défrichement*), schemes to divide up the commons, and projects to curtail collective rights so as to facilitate enclosure. There were even a few beacon attempts to promote land consolidation (*remembrement*) in order to minimize the productivity losses involved in strip and plot farming. The point to note about this "top-down" approach to agrarian reform, however, is that it largely failed. Most of the legislation was permissive rather than prescriptive, and in any case the Bourbon monarchy, notwithstanding its claim to absolute power, lacked the coercive capacity to enforce the reforms.

Besides, there was an alternative. Why not renegotiate the contract that made the state, the church, and the nobility joint partners in the exploitation of the peasantry? If ecclesiastical surplus extraction (the tithe) could be diminished, the fiscal elasticity of the rural population would increase in proportion. Better still, the "dead hand" of the Catholic Church could be removed altogether, and its entire landed wealth (estimated at 10 percent of the soil surface) brought within the fiscal net. The fief-owning nobility were vulnerable to similar statist calculations, for they, too, extracted surplus from the peasantry in the form of harvest dues and monopoly rights.

Yet both maneuvers were difficult to perform at the political level, for any move to curtail the defining privileges of the church and the aristocracy threatened to undermine the very foundations of absolute monarchy. It is one of the paradoxes of the French Revolution that it began as a revolt of the privileged against the leveling intentions of absolute monarchy. Nevertheless, there is ample evidence that the government had committed itself to a redistribution of peasant surplus by the end of the ancien régime. The tithing prerogative of the clergy was under attack from many quarters, for in this sphere the priorities of state and the virulent anti-clericalism of the *philosophes* coincided neatly. The monarchy remitted its rights over vassals on royal demesne land in a bid to encourage lay and ecclesiastical seigneurs to do likewise. Meanwhile, ministers studied schemes for a general redemption of feudal dues similar to the operation carried through in Savoy by Charles Emmanuel III in 1772. In the provinces, agents of the Crown were even prepared to incite rural communities to contest the more onerous features of the seigneurial privilege in the courts.

Landlords also responded to the physiocratic vision of an unfettered agriculture, albeit from a rather different perspective. Those who were lay or ecclesiastical seigneurs might view the unrestrained fiscalism of central government with alarm. But the bulk of their income derived from rent, and anything that helped to maximize the resources of the soil was good news. They approved the reluctance of legislators to become involved in matters of contract; they applauded attempts to deregulate the grain trade (that is, to allow prices to fluctuate in keeping with supply and demand); and they naturally supported moves to curtail the losses that collective rights imposed on efficient farmers. Landlords were a relatively compact group whose objectives are captured in a memorandum that the Royal Society of Agriculture presented to the National Assembly in October 1789.[3]

Among the twenty or so items enshrined in that document, we find calls for:

- recognition of the right of landowners to grow whatever they pleased, howsoever they pleased, on their land,
- recognition of the right of enclosure and the concomitant right to withdraw land from community use,
- removal of impediments to the partition of common land and the draining of marshes,
- removal of restrictions on the planting of fodder crops,
- removal of taxes bearing on agriculture, such as the *gabelle* (salt tax),
- improvements to road communications, and
- uniform weights and measures in order to facilitate internal trade.

What of the peasant agenda for agrarian reform? To judge from the grievance lists (*cahiers de doléances*) compiled by country dwellers on the eve of the Revolution, there was overarching agreement on two issues: state fiscality and seigneurialism. Late eighteenth-century peasants, unlike their seventeenth-century forebears, did not challenge the taxing function of the state, but they did articulate a sense that the burden no longer corresponded to the benefits. Thus, the problem of taxation was less one of legitimacy than one of reciprocity. The same was true of the feudal regime. Seigneurs no longer provided the level of services that might justify the degree of surplus extraction. Even so, peasants did not challenge the legitimacy of seigneurialism directly. They tended to call for the commutation or reimbursement of harvest dues and monopoly rights. The frontal challenge came later, that is to say, once revolutionary politics began to alter definitions of the "legitimate."

Beyond these axioms, it is difficult to generalize, for the political emancipation of country dwellers in 1789 and subsequent years brought to light not one, but several agrarian reform agendas. Well-to-do tenant farmers (*coqs de village*) wanted security of tenure (that is, long leases), which would enable them to profit from the secular trend in grain prices. Indeed, they were not averse to using community pressures in order to restrict the market for tenancies. Yet, at the same time they did not hesitate to challenge the most basic instincts of the rural community by withdrawing land from *vaine pâture* (stubble, fallow, and meadow grazing) and by transporting "surplus" grain stocks from market to market. We also find complaints against sales taxes and the tithe, and demands that landlords pay compensation for improvements.

Smaller peasants who were neither commercial growers nor even self-sufficient farmers relied heavily on the retention of collective rights. They railed against the "abuse" of these rights by their larger neighbors and resisted most of the monarchy's agrarian reform initiatives of the 1760s and 1770s. Yet they were also desperately short of land and did not hesitate to dig and squat if the opportunity presented itself (as it did in 1789). Plot farmers also complained loudly about the activities of engrossers bent on constructing compact farms out of dozens of small tenures, and they objected to thc practice of subtenanting, too. When landlords brought in principal tenants (*fermiers-généraux*), it invariably resulted in a worsening of conditions for the renting peasantry.

Sharecroppers and landless laborers came at the bottom of the heap. The former were mostly wretched, dependent, and vulnerable. They formed the most benighted group in the French countryside, and attempts to organize them, whether in the eighteenth, nineteenth, or twentieth century, rarely took root. It is just possible to catch the sound of their voices at the height of the revolutionary Terror (1793–94), when rural hierarchies were momentarily turned upside down. And their call was overwhelmingly for lease reform: longer contracts (one year was the norm) and a renegotiation of the share system. Landless laborers, by contrast, were not always wretched. Their condition was determined by the types of employment available, the supply of common land, and the degree to which a community welfare ethic prevailed. However, they were always vulnerable to the harvest cycle and to price spikes. They demanded land, access to the land of others (through the mechanism of collective rights), and price controls.

Issues

Social and political scientists have long quarried the French Revolution for material from which to construct theories of modernization. And ever since the bicentennial it has become fashionable and "correct" to describe 1789 as the birth moment of "political modernity."[4] Yet it should not be forgotten that France provided the stage for one of the great peasant revolutions of modern history. What lessons can students of agrarian change distill from this experience? Georges Lefebvre, the historian who first brought to light the rural dimension of the French Revolution, expressed himself thus: "In spite of appearances, the [peasants'] influence was as much conservative as revolutionary: they destroyed the feudal regime but consolidated the agrarian structure of France."[5] This slightly deflating formulation of what was supposed to be a left-wing interpretation has always caused a certain amount of historiographical discomfort. Even experts on the peasantry have raised doubts about some of Lefebvre's judgments, monumentally researched though they were. In this section I want to demonstrate the relevance of the Revolution to fellow scholars working in the field of agrarian studies. But in order to do this, it is necessary to recast some of the agenda items mentioned above into a more manageable form.

Individual and community. The relationship between these two entities is a conundrum for all of us. Having started my career as a researcher with a longitudinal study of an allegedly "backward" peasantry, I encountered this problem early on.[6] It seemed to me that the rural society of the Southern Massif Central was predicated on the utter subjugation of the individual to family and, by extension, community norms. Yet it behooves the historian to explain how these norms came into being, which is often quite difficult. My solution was to postulate a set of "atavisms" (the imperatives of subsistence agriculture, household structure, inheritance strategies, religion, fiscality, and so on) that operated to influence, and very largely to determine, responses to such questions as what crops to grow, whether to divide up the village stock of common land, and whom to vote for. However, I remember well how one critic put his finger on the deficiency of the Annales approach when it comes to probing individual motivation. My peasants, he pointed out, found themselves locked in a structural cage with no input over their voting choices.[7]

Peasantries differ, of course. And in France, at least, we know of the existence in the eighteenth and nineteenth centuries of discrete regional cultures, even if we are hard put to explain how they came into being. I thus conducted a more thorough comparative investigation of the abundant literature on rural France. It left me perplexed, but by no means ready to abandon the field to the "rational choice" theorists. Not that there is anything wrong with rationality as a working hypothesis. But I remain skeptical of arguments that rethink behavior in the past according to modern notions of the rational, or which identify rational individuals but not rational groups. As for the sense of community, I have learned that it does *not* presume the existence of an autarkic, isolated, or "backward" peasantry. That elusive bonding existed even within highly differentiated and mobile peasantries, as several monographs have demonstrated.[8] This has prompted me to evolve the concept of the cultural community, which transcends physical space and manifest disparities of social condition.[9]

Land, commons, and common rights. The rhetoric of the French Revolution was individualistic, libertarian, and, progressively, egalitarian. As such, it was bound to clash with the corporate structuring of ancien régime society. This dictum holds for the clergy, the nobility, and bourgeois professionals, as well as the peasantry. Individual peasants tended to function within three more or less superimposed collectivities: the community of the faithful (parish), the fiscal community (known as the *communauté d'habitants*), and the agrarian community. The Revolution did violence to all three, but for present purposes we need concentrate only on the agrarian community.

The agrarian community is a well-understood concept, I think. It derived from a mode of agricultural production in which land was owned and farmed individually for the most part, but under the proviso of a large number of collectively determined restraints. Nearly every agricultural activity was subjected to regulation: planting and harvesting, cropping cycles, pasturing and stock holding, irrigation, scavenging, and so on. The reasons for organizing cultivation in this fashion seem to have been twofold: to optimize *gross* production and to provide minimal subsistence guarantees for the poor. Of course, this system—in all its regional variants—was predicated on the limitation of freehold property rights. Depending on what was being grown, a proprietor might enjoy sole possession of the fruits of his land for no more than four months in the year. It is true that villages consisting almost entirely of large tenant farmers working compact holdings existed, on the outskirts of Paris, for example. Such farmers could often afford to ignore planting and cropping norms.[10] But agricultural enterprises on this scale were unusual; they cannot be numbered among the "original characteristics" of French rural history.

Yet even in the juridical sense, freehold ownership scarcely applied before the Revolution. The right of "eminent property" belonged to the monarch, and all land was held from him on the basis of nominally feudal tenure. One of the first tasks of the revolutionaries was therefore to vindicate the concept of private property in law. This principle was easy to legislate but difficult to embody in day-to-day practice, for it implied the concomitant abolition of all those collective restraints that prevented a proprietor from doing as he saw fit with his land. In this respect, the early revolutionaries took over a large part of the reform agenda of the old monarchy. "The land must be as free as those who farm it," if the libertarian rhetoric of the *Declaration of the Rights of Man* was to mean anything at all.[11] But what if "men" preferred use-rights to freehold rights; or worse, what if they pursued the property-owning ambition while refusing to give up use-rights over the property of others? The revolutionaries never found a way of resolving these tensions between the theoretical applications of "liberté" and the grassroots reality of peasant revolution.

Georges Lefebvre's solution was to formulate a dichotomy.[12] The Revolution, he argued, brought two rival definitions of property into sharp alignment. For the bourgeoisie, 1789 brought a historic opportunity to vindicate absolute or freehold property rights, whereas the peasantry perceived the Revolution as an opportunity to retrieve those rights of use or access that had been sacrificed on the altar of agrarian individualism in the 1760s and 1770s, and generally to shore up the edifice of custom. This almost works, but not quite. Certainly, landlords (whether noble or bourgeois) moved smartly to set the agenda for reform. If peasants were going to be allowed to buy out feudal harvest dues (legislation of 15–28 March and 3–9 May 1790), why should proprietors have to put up with servitudes that made their fields, meadows, and forests public property for part of the year? Where Lefebvre's formulation misses the mark is in its estimation of peasant needs and ambitions. Having waded through many boxes of documents, I no longer think it sufficient to argue that the poorer peasantry always attached greater significance to rights of access than to rights of ownership. The mistake lies, perhaps, in trying to freeze or imprison a mobile and politically alert peasantry. What appeared to be the solution to a small peasant's subsistence problem in the summer of 1789 was not necessarily the solution he would favor four invigorating years later, when Jacobin politicians started to carve up the commons and the assets of the church and émigré nobles.

The issue of the commons brought the competing interests of government, landlords, and peasants into sharp focus. No other agrarian reform caused more upset and dismay at the grass roots, for it brought into play all the considerations mentioned so far and many more: the rights of the individual vis-à-vis the community, the role of the household in agricultural production, calculations of short- and long-term rationality, and differential concepts of property.

In my experience there are few safe generalizations to be made on this subject. Before 1789 fiscal exigencies were paramount, and they translated into a policy favoring partition, although the en bloc leasing of village commons was also promoted. After 1789, and during the Terror particularly, fiscal, humanitarian, and party political imperatives blurred together to produce a confusing, frustrating, and ultimately inoperable government policy. Before 1789 landlords and seigneurs had often rallied to the cause of partition because prevailing legislation gave them the lion's share of the land in question. After 1789, with different laws in force, they showed much less enthusiasm. Likewise, for the peasantry a long lease on a household plot appeared a more attractive solution than an exiguous strip allocated per capita on quasi-freehold terms (the egalitarian option enshrined in the decree of 10 June 1793). The former solution seemed to offer the best means of reconciling the immediate needs of land-hungry households with the long-term need for flexible community controls. By contrast, peasant proletarians beset with subsistence worries and fired up with Jacobin ideology might regard an exiguous strip of arable land as unquestionably preferable to continued collective exploitation. Owners of stock—that is to say, the larger peasants—nearly always derived disproportionate benefit from commons left as pasture.

So, the first clue to look for when assessing the socioeconomic dynamics of common land partition is the *mode de partage.* No section of the peasantry had an unalterably fixed position on this issue, nor did the state, nor did landlords. Even the radical decree of 10 June 1793 (the legislative inspiration for most of the partitions of the revolutionary era) was shot through with ambiguities. It provided for a per capita distribution of equal-sized plots to every man, woman, and child on the condition of a one-third vote in favor. However, the plots could not be sold for ten years, nor could they easily be grouped to form compact household holdings because they were allocated by ballot. And, in any case, all the strips remained liable to the usual array of collective rights. This was to put ideology before agrarian reform.

Nevertheless, many poor and semi-landless peasants did try and make use of the legislation. Equally, it is true that many tenant farmers and landowners did their utmost to defeat the egalitarian intentions of government. My guess would be that pressure to divide was most sustained in those villages where the reciprocities underscoring the rural community had broken down. A sense of resentment, perhaps, against larger farmers who were devouring the common pasture with many head of stock and who were even buying in stock that could then be fattened under the cover of common rights. But two points are worth noting. Even peasant proletarians found the 1793 legislation frustrating to implement, for it was designed to reward a social abstraction: the individual country dweller. Unless the plots could be combined to form family-sized holdings, the economic payoff was likely to be pretty minimal. Yet maybe this is to miss the object of the exercise, for the agrarian legislation of 1792–94 tended to blur economic and political rationalities. A full partition of the commons might bring swift retribution in terms of plummeting stock levels and declining soil fertility, as a number of villages would subsequently confess. But in the upside-down world of the Terror, it taught landlords and would-be engrossers an unforgettable political lesson.

Many, many more villages did not divide up their commons, in spite of all the inducements. Silence is difficult to interpret, for peasants rarely petitioned to say why they were not taking a particular course of action. Foot-dragging by village elites unquestionably played a part; often the land in question was of meager extent; often there were strong ecological arguments against the breaking open of the soil in upland regions. But a certain distaste for an operation that offended community norms seems also to have operated. Many villagers were prepared to grant to households and even individuals a temporary usufruct right over portions of common, but the Jacobin decree of 10 June 1793 proposed to alienate the commons to certain individuals or households for all time. This contradicted the deep-rooted belief that common land was held in trust for the benefit of each succeeding generation.[13]

Collective Action. For a historian who has tended to emphasize the localist, not to say atavistic, wellsprings of peasant political action, the crowd scenes of the French Revolution represent something of a challenge. Between 1788 and 1793 large numbers of country dwellers acted in unison to achieve certain objectives, which encompassed the demise of the ancien régime. Whether the consequences of that action were intended and willed is a matter for debate. Georges Lefebvre always insisted that the peasantry mounted their own "autonomous" revolution in these years, but this judgment needs to be considered in the historiographical context that produced it. Until Lefebvre's seminal study *Les Paysans du nord* appeared in 1924, it was axiomatic that 1789 inaugurated a "bourgeois revolution": nothing more and nothing less. Strictly speaking, the peasant revolution was not autonomous; it was independent (which is what Lefebvre meant). It cannot be described as autonomous because it was interactive; that is to say, mobilized peasants and bourgeois legislators nudged each other forward in a complex political dialectic.

But why did peasants mobilize in the first place? Here I find the "moral economy" approach to be most persuasive. As the gravity of the harvest crisis became apparent in the late summer of 1788, traditional subsistence anxieties were translated into violent crowd actions against the holders and producers of grain. Price-fixing (*taxation populaire*) in the public marketplace—the characteristic restorative action—was much in evidence. But so, too, were expeditions to the gates of monasteries, chateaux, and the barns of rich tenant farmers. This extension of the popular politics of subsistence from the point of distribution (the market) to the point of production (the farm) foreshadowed the more focused and dirigiste crowd actions of the revolutionary climacteric. Indeed, evidence indicates a shift in the pattern of subsistence-motivated protest as early as the 1770s.[14] Add to this the elite political crisis of 1787–89, and the royal government's public act of repudiation of its traditional subsistence responsibilities (deregulation of the grain trade in 1787), and we begin to understand the dynamics of collective action in 1789 and years beyond.

The conspicuous anti-seigneurial animus of rural revolt from the summer of 1789 onward had its roots in the ancien régime, then. But it owed more, I would argue, to the revolutionary process itself. The subsistence crisis, the drafting of cahiers de doléances, the election of deputies, and finally the convocation of the Estates-General itself worked a massive *prise de conscience* among country dwellers. A sense of the potentialities of crowd action (against seigneurial stewards, tithe proctors, engrossers, usurers, excise men, customs officials, and so on) developed—a sense that peasants might at last have the opportunity to make, or remake, the world around them.

The interactive politics of the peasant revolution flowed directly from this realization, for country dwellers soon learned how to coax concessions from legislators. Indeed, the whole story of the dismantling of the feudal regime in France between 1789 and 1793 is an object lesson in how the nominally weak and supposedly illiterate get their own way in the end.

Rural rebellion played a role in this process, but noncompliance (the refusal either to pay or to redeem dues) and obstructionism counted for more. Another factor was the need to buy the political support of the peasantry as Counter-Revolution beckoned. All this can be construed using standard archival techniques, but the patterning of collective actions requires a more sophisticated methodology, such as that pioneered by John Markoff at the University of Pittsburgh. By means of rigorous content analysis and quantification (of the cahiers, and of crowd "events"), he is able to demonstrate that "major clusters of legislation . . . follow, with some variation in lag, major bouts of rural activism."[15] On such issues as seigneurialism, tithe reform, the sales of national property (*biens nationaux*), and the partition of common land, then, peasants were as aware of debates taking place in the Legislative Assemblies as legislators were sensitive to news coming in from the countryside.

Subsistence Ethic. Anyone raised on the *Marxisant* economic history of Ernest Labrousse and the stagnationist social history of Emmanuel Le Roy Ladurie will find it difficult to conceptualize peasants in anything other than subsistence-oriented terms.[16] This includes nearly all the rural historians of France, not least the author of this chapter. As far as Labrousse was concerned, agriculture acted as the pacemaker for the entire eighteenth-century economy: proto-industrialization, urban consumer demand, credit networks, and overseas trade scarcely signified. We now know better. In particular, the link between proto-industrialization and the agrarian economy has received close attention.[17] Yet the articulations between peasant neo-subsistence (or would-be subsistence), agriculture, and the marketplace are still poorly understood. As grain price series indicate, France did not possess an integrated economy in the eighteenth century. Commercial agriculture was practiced in a few privileged regions, it is true, and signs of a specialist fatstock industry are evident from mid-century. But market pull was weak (urban growth of 2 percent at most across the century), and it deteriorated sharply with the onset of the Revolution.

If the economic character of the peasant household remains equivocal, there can be no doubt about the social, psychological, and political applications of the subsistence ethic. By 1789 the "famine plot" had become a staple of popular culture. Twice before, in 1763–64 and 1776, the monarchy had tried to relinquish its traditional responsibility for food supply, only to be forced by panic-stricken crowds into reimposing controls on the movement of grain. A third attempt was made in 1787, and some of the consequences have been mentioned above. As economic conditions worsened, the Revolution occasioned a general retreat from market exposure. Instead, the subsistence ethic became a symbol of political orthodoxy. Millers, merchants, and bakers were threatened with death for speculating in the people's foodstuffs, while the use of land for purposes other than the production of cereals risked accusations of *lèse-nation.* Menaced by famine and encircled by enemies, republican France adopted a primitive version of the siege economy. Commercial agriculture suffered sorely: in some regions farmers were forbidden to plant such oil-seed crops as colza; in others, recently planted vines were grubbed out. The resounding freedoms of 1789 (notably, the freedom to plant at will) were temporarily forgotten in a desperate bid to feed the armies defending the frontiers.

It was in these conditions that the subsistence ethic acquired an ideological underpinning. Famished consumers of town and country sought to alter the definition of property laid down at the start of the Revolution and proclaimed instead the "right to exist" (*droit à l'existence*). In the face of this primitive right of survival, all freehold conceptions of property were forced to give ground. The merchant's property in a warehouse full of sacks of corn must yield to the subsistence rights of the hungry crowd hammering on the gates. Likewise the farmer with a barnful of grain. This was both a natural extension of the raiding-party habits that greeted every harvest dearth, and an intellectual refinement of moral economy logic. In the hands of seasoned campaigners like the Paris sans-culottes, the *droit à l'existence* became a formidable weapon.

Capitalism and the "voie paysanne." The ultimate paradox of the French Revolution lies in its legacy. It lit a beacon of political radicalism that shone across Europe, while appearing to bequeath to France a legacy of unremitting social and economic conservatism. If the purpose of "the great bourgeois revolution of the end of the eighteenth century," as Soviet scholars used to label the events of 1789–94, was to prepare the way for capitalism, something must have gone seriously awry. According to Lefebvre the error lay in the premise, for the "autonomous" peasant revolution had never been capitalist in intent. On the contrary, it had set out to defend the communitarian system of agriculture from a statist obsession with political economy, and from the freelance attentions of improving landlords. Lefebvre reached this conclusion on the basis of extensive field research, and he adhered to it despite the misgivings that such a variant interpretation provoked among Marxist intellectuals. In fact, the tension between the "bourgeois revolution" of classical Marxism and Lefebvre's peasant revolution was ironed out only in the 1970s, when a younger generation of Lenin-inspired researchers reached maturity.

Chief among these was the Soviet historian Anatoli Ado. In a book published in 1971 and until recently available only in Russian, he formulated a vision of the "peasant route" to capitalism that rescued country dwellers from the ideological cul-de-sac into which Lefebvre had driven them.[18] If agrarian capitalism failed to take off on the morrow of the Revolution, he declared, it was not because small peasants obstinately clung to the rural community and collective rights, but because they failed to force a wholesale division of estates and thereby to terminate the rentier lifestyle of large landed proprietors. Thus the responsibility for France's tardy capitalist development is transferred from the peasant masses to an ill-defined proprietorial elite. In the process, the congruence of bourgeois and peasant revolutions is largely restored.

Endowed with freehold plots of land, liberated from feudal restraints, small and medium-sized peasants would rejuvenate the stalled agrarian capitalism of the eighteenth century through the mechanisms of petty commodity exchange. The trouble with this scenario, of course, is that it did not happen. A lot remains unclear about the functioning of the land market in the 1790s and the final destination of those properties (biens nationaux) put up for sale by the state. However, my own research soundings indicate that plots were sought, and used, mainly for subsistence purposes. It is true that nearly all echelons of the peasantry benefited to some degree from the easing of surplus extraction (reduction in land taxes, formal abolition of the tithe and of seigneurial harvest dues). But this seems to have facilitated enhanced auto-consumption rather than enhanced market participation. Consumer markets were in disarray by the end of the 1790s, in any case, and urban demand for agricultural products was severely depressed. Obviously, it is important to know where precisely to look for Ado's rural capitalists within the differentiated ranks of the French peasantry. An agricultural worker who acquires a plot of land in order to lessen his dependence on the marketplace is not the same quantity as a middling independent peasant. The numbers of these latter probably did increase as a result of the Revolution; even so, they would have constituted a slender base from which to mount a capitalist transformation of agriculture.

Many agrarian issues were in play during the French Revolution, and this chapter has aired a number of them. Few, if any, were resolved, whether in the 1790s or in subsequent decades. It has often been glibly remarked that 1789 liquidated the political ancien régime but left the economic ancien régime largely intact, and there is much truth in the statement. The fiscal needs of state were eventually solved by expropriation and bankruptcy, rather than by agricultural reform. Landlords emerged from the crisis with a more rational system of direct and indirect taxation, but little else. In many regions, enclosure became harder to achieve, not easier. It is true that rents rose, but so did labor costs. And large landowners bore the political cost of the Revolution (requisitions, harassment, imprisonment, and worse) to a considerable degree.

For peasants there were real benefits, although these were distributed unevenly. Proletarians profited from labor scarcities to push up wage rates; they successfully defended collective rights, and a few even managed to acquire parcels of land. Plot farmers shared in these benefits while profiting from the abolition of the tithe and seigneurial dues. They also won temporary relief from the threat of engrossment, for the Revolution had made this antisocial practice politically unacceptable. The middling peasant farmer, meanwhile, continued his stealthy conquest of the soil as the lands of the church and émigré nobles were divided up and sold on. They often ended up taking over the subsistence plots of their lesser neighbors, too. Sharecroppers were the one group to emerge from the Revolutionary cycle with negligible gains. The period of price inflation (1794–96) produced some windfall benefits, to be sure, but the root cause of their poverty and insecurity was tenurial. Not one of the political groups competing for power in the 1790s was prepared to grip the nettle of lease reform. This issue was tackled only in the Statut de Fermage of 1946—a long-overdue act of justice toward France's dwindling band of sharecroppers.

Notes

1. See F. Braudel, *L'Identité de la France: Espace et histoire* (Paris, 1986), p. 29.
2. See P. Vidal de la Blache, *Tableau de la géographie de la France* (Paris, 1903); P. Vidal de la Blache, *Principles of Human Geography* (London, 1926).
3. Printed in *Archives Parlementaires de 1787 à 1860: Recueil complet des débats législatifs et politiques des chambres françaises,* première série, 1787–99 (92 vols., Paris, 1862–1980), vol. 9, pp. 523–52.
4. "Interpreting the French Revolution: An Assessment of the Writings of Francois Furet" (debate at the annual meeting of the American Historical Association, San Francisco, 1989).
5. G. Lefebvre, "La Revolution française et les paysans,"reprinted in *Etudes sur la Révolution française* (Paris, 1954), p. 249.
6. See P. M. Jones, *Politics and Rural Society: The Southern Massif Central, c. 1750–1880* (Cambridge, England, 1985).

7. See W. Brustein, "The Politicisation of the Peasantry of France's Southern Massif Central," *Peasant Studies* 13 (1986): 245.
8. See, in particular, G. Dallas, *The Imperfect Peasant Economy: The Loire Country, 1800–1914* (Cambridge, England, 1982); L. Vardi, *The Land and the Loom: Peasants and Profit in Northern France, 1680–1800* (Durham, N.C., 1993).
9. P. M. Jones, "Parish, Seigneurie and the Community of Inhabitants in Southern Central France during the Eighteenth and Nineteenth Centuries," *Past and Present* 91 (1981): 74–108.
10. See J.-M. Moriceau and G. Postel-Vinay, *Ferme, entreprise, famille: Grande exploitation et changement agricoles: Les Chartier, XVIIe–XIXe siècles* (Paris, 1992), pp. 177–218.
11. For this gloss, see P. M. Jones, *The Peasantry in the French Revolution* (Cambridge, England, 1988), p. 131.
12. See G. Lefebvre, *Les Paysans du nord pendant la Révolution française* (2 vols., Paris, 1924); Lefebvre, "La Révolution française et les paysans."
13. In this connection, see P. M. Jones, "Common Rights and Agrarian Individualism in the Southern Massif Central, 1750–1880," in G. Lewis and C. Lucas (eds.), *Beyond the Terror: Essays in French Regional and Social History, 1794–1815* (Cambridge, England, 1983), pp. 135–36.
14. See C. A. Bouton, "Solidarities and Tensions in Ancien-Régime France: Rural Society Confronts the Subsistence Crisis of 1775" (Ph.D. diss., State University of New York, 1985). See also Bouton, *The Flour War: Gender, Class, and Community in Late Ancien Régime French Society* (University Park, Pa., 1993).
15. J. Markoff, *The Abolition of Feudalism: Peasants, Lords and Legislators in the French Revolution* (University Park, Pa., 1996), p. 487.
16. See C. E. Labrousse, *Esquisse du mouvement des prix et des revenus en France au XVIIIe siècle* (2 vols., Paris, 1933, reprinted 1984); Labrousse, *La Crise de l'économie française à la fin de l'Ancien Régime et au début de la Révolution* (Paris, 1944, 2nd edition 1990); E. Le Roy Ladurie, "L'Histoire immobile," *Annales E.S.C.* 29 (1974), pp. 673–92.
17. See Vardi, *The Land and the Loom.*
18. A. Ado, *The Peasant Movement in France during the Great Bourgeois Revolution of the End of the Eighteenth Century* (in Russian) (Moscow, 1971). Translated into French as A. Ado, *Paysans en révolution: Terre, pouvoir, et jacquerie, 1789–1794* (Paris, 1996).

References

Ado, A. *Paysans en révolution: Terre, pouvoir, et jacquerie, 1789–1794* (Paris, 1996).

Blache, P. Vidal de la. *Tableau de la géographie de la France.* Paris, 1903.

———. *Principles of Human Geography.* London, 1926.

Bouton, C. A. "Solidarities and Tensions in Ancien-Régime France: Rural Society Confronts the Subsistence Crisis of 1775." Ph.D. diss., State University of New York, 1985.

———. *The Flour War: Gender, Class, and Community in Late Ancien Régime French Society.* University Park, Pa., 1993.

Braudel, F. *L'Identité de la France: Espace et Histoire.* Paris, 1986.

Brustein, W. "The Politicisation of the Peasantry of France's Southern Massif Central." *Peasant Studies* 13 (1986).

Dallas, G. *The Imperfect Peasant Economy: The Loire Country, 1800–1914.* Cambridge, England, 1982.

"Interpreting the French Revolution: An Assessment of the Writings of Francois Furet." Debate at the annual meeting of the American Historical Association, San Francisco, 1989.

Jones, P. M. "Parish, Seigneurie and the Community of Inhabitants in Southern Central France during the Eighteenth and Nineteenth Centuries." *Past and Present* 91 (1981).

———. "Common Rights and Agrarian Individualism in the Southern Massif Central, 1750–1880." In G. Lewis and C. Lucas (eds.), *Beyond the Terror: Essays in French Regional and Social History, 1794–1815.* Cambridge, England, 1983.

———. *Politics and Rural Society: The Southern Massif Central, c. 1750–1880.* Cambridge, England, 1985.

———. *The Peasantry in the French Revolution,* Cambridge, England, 1988.

Labrousse, C. E., *Esquisse du mouvement des prix et des revenus en France au XVIIIe siècle.* 2 vols. Paris, 1933 (reprinted 1984).

———. *La Crise de l'économie française à la fin de l'Ancien Régime et au début de la Révolution.* Paris, 1944 (2nd edition 1990).

Lefebvre, G. *Les Paysans du Nord pendant la Revolution française.* 2 vols. Paris, 1924.

———. "La Révolution française et les paysans." Reprinted in *Etudes sur la Révolution française.* Paris, 1954.

Le Roy Ladurie, E., "L'Histoire immobile." *Annales E.S.C.* 29 (1974).

Markoff, J., *The Abolition of Feudalism: Peasants, Lords and Legislators in the French Revolution.* University Park, Pa., 1996.

Moriceau, J. M., and G. Postel-Vinay. *Ferme, entreprise, famille: Grande exploitation et changement agricoles: Les Chartier, XVIIe–XIXe siècles.* Paris, 1992.

Vardi, L. *The Land and the Loom: Peasants and Profit in Northern France, 1680–1800.* Durham, N.C., 1993.

CHAPTER FOUR Imagining the Harvest in Early Modern Europe

LIANA VARDI

In the minds of many, peasant identity is indelibly linked to the past. Not only is food production the oldest profession, but peasants, especially in Latinate definitions of the term, are associated with a region, a *pays*. In the nationalist revivals of the nineteenth century, peasants thus came to embody the nation, its traditions and mores, unsullied by modern concepts. What the Right seized on with glee, the Left rejected as reactionary. Karl Marx never wavered from his conclusion that peasants, attached to their plots of land, were a counterrevolutionary force. They could not accept the march of history. Both Right and Left agreed, then, on the timeless conservatism of the peasantry. Whereas other groups in society had a history, peasants were historical emblems.[1]

Treating peasants as repositories of traditional culture was a recent invention. Until the eighteenth century, Europeans viewed peasants as the very antithesis of culture. For medieval social theorists, peasants were bestial and servile by nature. Although Christian theology argued that the poor would inherit the earth, the approach toward them was the same: worthy or unworthy, the peasants' lot was to labor humbly and ceaselessly and find their reward in heaven. Steeped in these views, medieval culture contained peasants within the imaginary of serfdom, tying them to their betters.[2] Yet peasants were too important a segment in society to be dismissed consistently. As this essay will argue, views of peasants, inherited from the Middle Ages, were largely superseded in the early modern period. It was difficult to depict peasants as servile in an age that legally recognized their independence and no longer saw them as mere appendages to the manor. By the sixteenth century, moreover, cultural paradigms that sought to redefine the nature of man, stressing life in the world, also came to celebrate the peasants' daily activities. Peasants were shown in command of the fields and as members of independent and lively communities. Sixteenth-century art and poetry turned the countryside over to the peasants, while elites retreated to imaginary pastoral landscapes.

Whereas the peasants' cultural takeover of the land was particularly marked in the Renaissance, there was a strong reaction against their presence in the seventeenth century. Poets and painters restored the countryside to the aristocracy by depicting estates without laborers, by focusing, especially in the visual arts, on cultivated nature without people. The working peasant had become a threatening figure, the countryside a place of yearning and unease. An overall obsession with social order reinforced this process.[3] Decades of war intensified a personal search for safety and repose, which the countryside provided in theory but not in practice. Anxieties about peasant society pervade seventeenth-century depictions of the countryside. Eighteenth-century culture, on the other hand, reintegrated the peasant within the rural landscape. This enterprise necessitated the inclusion of the peasant within the ideals of virtue and civility.[4] The peasant community, composed of virtuous householders, was then believed to live in harmony with the elites, dedicated to a common pursuit of happiness.

This essay questions the standard interpretation of the place of peasants in European culture. Aristocratic cultures, we are told, abhorred the sight of manual labor and displays of crassness or vulgarity.[5] Peasants, deemed unsuitable subjects for the noble arts, were left out of high culture. At best, peasants appeared in art and literature only sporadically. Even then, their appearance was ritualized and circumscribed.[6] Peasants supposedly could be the subjects of broad farce but not of tragedy. They could figure as beasts of burden, tilling the fields, as emblems of Christian resignation, or embody sins of gluttony and sloth in depictions of the deadly sins. Any substantial departures from such a limited repertory, such as the evident importance of peasants in Spanish Golden Age theater or of peasant genre scenes in seventeenth-century Holland, are either set aside—as in the case of Spain—or else neatly fitted within the aristocratic framework.[7] Thus Protestant bourgeois Holland gave rise to genres that were disdained elsewhere, until they were taken up in post–Civil War England. The more common approach, in any case, is to treat each national phenomenon separately, viewing it essentially as an internal development.[8]

There is a related reading of the peasants' place in early modern European culture. Since contemporaries viewed them as ugly and unsuitable subjects for high art, once peasants were beautified, they necessarily ceased to be real. They became something other than themselves and could therefore be treated simply as metaphors.[9] This interpretation neatly applies to any dignified representation of the peasant. The most compelling example is the eighteenth-century Rococo, with its pretty rustic couples, which no one ever accepts as an actual depiction of residents of the French countryside. Yet, as I will argue, this genteel portrayal was connected to a highly significant drive to validate the peasant estate.

My approach will be broadly comparative, trying to capture what I perceive to be European-wide trends, and I will discuss the transfer of cultural images from one country to another. Despite my use of broad frameworks, I want to stress that I believe in the force of the images themselves, be they visual or written. Poets and painters, whether they mimicked ancient models or aimed at universal truths about the human condition, still chose at times to embody them in peasants. Viewers could interpret these images as they wished, aided by an arsenal of traditional and contemporary guides. Yet the image itself remained potent. No matter what allegorical messages the images contained, they must have reminded viewers forcibly of the real peasants around them. Moreover, at times, the images became more powerful than the discourse within which they were meant to be contained. This brought about crises in the representation of peasants. As I will demonstrate, the principal loss in this system of representation was the harvest, which had been one of the standard themes of European culture.

I have chosen to focus on the harvest theme for several reasons. As the culmination of the agricultural year, bringing in a good crop was the central concern of peasants, farmers, and landowners. Throughout the early modern period, even in the eighteenth century, wealth was based essentially on landownership, and taxes were raised principally on agricultural produce. Governments and individuals alike had a heavy stake in the outcome of the harvest, which consisted mainly of cereals, for bread was the basic food. Given its importance, the harvest had a permanent place in the representation of country life. Another reason for looking closely at depictions of the harvest is that representing this activity necessarily involved choices about what should be portrayed. Plowing was done with either horses or oxen, but it was a task that a man usually performed alone, and its depiction rarely varies, even if the setting of the field changes. In contrast, before mechanization in the late nineteenth century, the grain harvest required a huge amount of labor. The crop ripened all at once and had to be brought in very quickly before the plants dried and the grain fell to the ground. Peasants would therefore be on the fields from morning 'til night, extra hands would be hired for the season, and bands of harvesters would move from field to field. They cut the crop with sickles or scythes (although scythes were used primarily for haymaking), others gathered the fallen stalks into sheaves, and eventually the crop was carted back to the farm. These tasks were sometimes but not always gendered, with men doing more of the cutting and women more of the bundling and raking.

This process could never be fully rendered pictorially. There were simply too many people out in the fields. In the Middle Ages, when the natural world and its tasks were pared down to symbols, the harvest was depicted in vignettes showing a single harvester holding a single tool. During the Renaissance, painters amalgamated all the phases of the harvest into a single frame. Later, artists focused on one part of the harvest and left out the rest. In every period, therefore, the artist faced a choice about what to put in and what to leave out: about how to portray not so much the reality—because no artist intended that—as the spirit of the harvest. Although poets were more at liberty to suggest the massive use of labor that the harvest required, they, too, made peculiar choices about the ways they depicted this event.

Scenes of harvest, though part of the traditional painterly and poetic repertoires, are not a major art form. Some periods are richer than others; as I will argue, the sixteenth and eighteenth centuries are relatively abundant, while the second half of the seventeenth century has left us many fewer such scenes. Altogether, however, except perhaps for the nineteenth century, this is a minor corpus, one, moreover, where geniuses like Pieter Bruegel overpower their far feebler imitators. It would not be surprising, therefore, to learn that pictures of the seasons lie forgotten in attics or museum storage rooms or are fading on the walls of country houses and inns. In a similar vein, execrable poems on the seasons may well have been left out of modern-day editions and prove impossible to resurrect. Nonetheless, hundreds of prints and poems have been catalogued, and these allow for clear themes to emerge.[10]

Late medieval iconography included peasants in several settings. The first was as background figures in religious paintings depicting the Nativity, the Flight into Egypt, and various parables.[11] Peasants also illustrated Old Testament scenes, most notably the Book of Ruth, which typically included a scene of reaping. Although one can imagine thousands of possible biblical illustrations, the repertoire was in fact limited, and artists tended to work within an accepted canon of suitable representations. This was also true of the cycles of the seasons. The months had been illustrated since antiquity, and this motif was taken over by medieval Christianity. The ancients linked the cycle of the seasons to religious feasts; Christianity detached it from its pagan roots and focused instead on human activities, especially agriculture.[12] Man's lot, since Adam's curse, was to live by the sweat of his brow.

Medieval artists carved emblems of the months on church capitals and illustrated calendars in Books of Hours. Whereas early medieval versions had included a large variety of rural activities, from weeding to shoveling dung, Carolingian calendars fixed the themes more rigidly. The images were cleaned up and simplified. They focused more closely on the grain fields and vines that produced bread and wine, the twin symbols of the Mass.

Historians of technology have pored over these pictures looking for evidence of changing techniques.[13] The most significant development, until the invention of harvesting machinery in the nineteenth century, was the increase in the use of scythes for harvesting grains. Yet even this did not become a normal procedure for wheat until the late eighteenth century, because mowing with scythes was rougher on the plant and led to the loss of grain. But the degree of accuracy in the depiction is not my concern here. I am interested rather in the image of reaping itself, be it with sickle, scythe, or reaping hook, as the symbol for harvest *work.* Artists who wished to depict nature's abundance had other options, such as an image of the goddess Ceres.

Medieval illustrations were usually miniatures, focused on a single, stylized figure, and a rudimentary version of a field (Figure 4.1). Sometimes, the image was more complex, embodying a clear social hierarchy.[14] The peasants were squat and ungainly, bent over their tools, in submissive poses. More often, social hierarchy was simply taken for granted, and the lord or his steward walked amid busy laborers (Figure 4.2). In the duc de Berry's famous Book of Hours, scenes of peasant labor alternate with aristocratic pastimes: feasting, hunting, and Maying. The peasants work in the shadow of various oversized castles (Figure 4.3). The calendar genre was conceived as depictions of seasonal work and seasonal amusements, each performed by the appropriate class.[15]

Peasant work was significant in medieval pictures for yet another reason. The paintings symbolized man's troubled relationship to nature.[16] One way to conceptualize this relationship was in the form of a garden, representing paradise. This garden was enclosed, separated from the wilderness without, the wilderness to which man had been condemned after the Fall and which he could tame only by his labor. While the garden represented the ideal state, scenes of agricultural labor expressed man's fallen condition as well as his potential redemption through work. When artists wanted to convey this message, scenes of nature and of the countryside concentrated on working peasants.

The Renaissance, of course, extended the range of profane subjects. Artists, like philosophers, explored the relationship between man and nature and attempted to capture the essence of nature itself. The idyllic view represented by the medieval garden developed into a full-blown pastoral style, celebrating man in harmony with nature: mankind did not have to work but could simply enjoy

Figure 4.1. *An example of medieval miniatures, showing a single harvester.* August, *Gradual and Sacramentary (German, thirteenth century), courtesy of the Pierpont Morgan Library, New York, MS M. 711, f. 5v.*

nature's fruits. In this vision, leisure and play dominated, and idealized mythic figures pranced or lounged about. Although the message was timeless, contemporary figures were sometimes merged within the setting, reassuring the viewer that he or she belonged in this universal scheme.[17] The overall effect was cerebral and elevated, embodying what many in this period conceived as the proper function of art.

Yet this ideal version did not totally replace a grittier approach to the countryside. In some ways, interest in real agricultural work actually increased in the sixteenth century. This new interest derived partly from the medieval tradition of viewing work as an aspect of man's redemption. It also grew from the renewed influence of classical learning and in particular Virgil's *Georgics.*[18] Unlike Virgil's *Eclogues,* which along with writings of Theocritus, Horace, and Ovid evoked pas-

Figure 4.2. *Social hierarchy depicted at harvest time. Pietro di Crescenzi,* Livre des profits ruraux, *Bruges, ca. 1470, courtesy of the Pierpont Morgan Library, New York, MS M. 232, f. 201v.*

toral landscapes and pastoral poetry, the *Georgics* focused on agricultural work.[19] In fact, they come close to being an agricultural manual in verse. For Virgil, work is man's lot and, if performed with true forbearance, will produce a simple, fruitful, and peaceful life.[20]

Classical literature in the sixteenth century joined the biblical imagery that had dominated medieval representations of the countryside. Agricultural work thus penetrated the learned sixteenth-century panorama, armed with the twin

Figure 4.3. *Scenes of peasant labor in the shadow of aristocratic castles. Paul de Limbourg,* Juin *(1414–15), Très Riches Heures du Duc de Berry. Courtesy of Giraudon/Art Resource, New York.*

Figure 4.4. *An illuminated page from a Book of Hours shows peasants in a naturalistic setting, one of the artistic responses to peasants in the early modern period that led viewers to think of them as controlling the countryside.* August: Mowing Wheat, Binding Sheaves, *Da Costa Hours (ca. 1515), by Simon Bening and others, Bruges, courtesy of the Pierpont Morgan Library, New York, MS M. 399, f. 9v.*

stamps of classical virtue and Christian salvation. Inspired by Virgil, poets depicted seasonal tasks, including harvest work, as part of their repertoire. The setting, moreover, was always recognizably modern, invoking contemporary practices. Artists, on the other hand, placed the harvest in both modern and classical surroundings. One current of art breathed new life into the old calendar tradition by moving the laborer to a naturalistic village environment. Simon Bening's cycle of the months illustrates this trend (see Figure 4.4).[21] The second approach, informed by the Renaissance revival of classical art forms, portrayed the harvester

Figure 4.5. *The influential Pieter Bruegel the Elder depicts the muscular strength of the peasants and implies the energetic labor needed for the harvest.* Summer *(1568), courtesy of the Metropolitan Museum of Art, Harris Brisbane Dick Fund, 1926 (26.72.23).*

as an Olympian athlete, a Hercules laboring under a Greek sky. This idealized version, mostly confined to allegories, nonetheless reinforced the association of peasants with energetic exertions.

By the sixteenth century, pictures of rural occupations moved from illuminated manuscripts to full-sized panels, canvases, and prints.[22] Perhaps no painter was more influential than Pieter Bruegel the Elder (1525–1569) in effecting this transition. Bruegel transposed religious scenes to Flemish villages and turned the peasants' daily routines into moral exempla.[23] He also energized harvest scenes. Whatever his intentions,[24] we are left with images of vigorous, muscular peasants, working hard, drinking hard, or collapsing from fatigue.[25] The artist emphasized the impressive effort involved with a technique of arrested movement. No image expresses the resulting sense of energy more powerfully than Bruegel's print *Summer* (Figure 4.5). The first object that catches the eye is the huge scythe that the harvester has put down in order to refresh himself; it juts out of

Figure 4.6. *Several stages of work, including rest, are displayed in Pieter Bruegel the Elder,* The Harvesters *(1565). Courtesy of the Metropolitan Museum of Art, Rogers Fund, 1919 (19.164).*

the frame. Around him, men and women are mowing, gathering, carrying—convincing us that abstract concepts (work, heat) have been brought to life, even if not fully realistically.

His harvest scene housed at the Metropolitan Museum of Art presents a larger and more encompassing vision (Figure 4.6). It depicts the stages of the harvest from the reaping to the carting and even includes, in tiny background detail, summertime village recreations. The foreground scene contrasts work and rest by closely intertwining the two. Those eating are not enjoying a summer picnic but, like the sleeper, are taking a break from their exertions. Meanwhile, others are reaping and gathering. Significantly, the men mowing are positioned in ways that exhibit their scythes and the strength needed to wield this heavy tool.

Bruegel's versions of harvest work remained the most vibrant among similar undertakings by contemporaries and imitators over the next fifty years. Sixteenth-century representations of summer continued to focus on harvest work, with peasants engrossed in their tasks, pointedly swinging sharp tools. Maarten de Vos's print *Summer,* from the late sixteenth century, depicted a community at work, energetically cutting, bundling, and even drinking in a pose borrowed from

Figure 4.7. *Flemish imitators of Bruegel continue the vigorous depiction of harvest work. Maarten de Vos (1532–1603),* Summer *(n.d.), Courtesy of the Metropolitan Museum of Art, the Elisha Whittelsey Collection, the Elisha Whittelsey Fund, 1949 (49.95.2003).*

Bruegel (Figure 4.7).[26] I. A. Wierix's print of a Maerten Van Cleve harvest (circa 1580) focuses directly on the bundling and cutting, as a harvester brandishes a short-handled scythe (Figure 4.8). This style still obtained in France around the middle of the seventeenth century, when the court painter Jacques Stella produced harvesting scenes of a vigorous nature, with the lively motions reminiscent of Bruegel and of his own Flemish background (Figure 4.9).

Realism about peasant labor, of course, did not require that peasants be placed within recognizable modern landscapes. Humanist fashion encouraged some artists to present peasants in archaic costumes and settings. The mode originated in Italy, but it was applied to harvesters in Mannerist paintings in Flanders, Holland, and France. Perhaps the most striking version is an allegory engraved by Philips Galle, after Maerten van Heemskerck, *Summer* (1563). In this highly gendered picture, two women tend sheep on the right, while naked men and youths mow, rake, reap, bundle, and prominently sharpen their scythes (Figure 4.10). The alle-

Figure 4.8. *In the foreground, a display of the cutting and bundling during the harvest. I. A. Wierix after Maerten van Cleve,* La moisson *(ca. 1580), courtesy of Cabinet des Estampes, Bibliothèque Royale Albert Ier, Brussels.*

gorical figure is not the usual representation of Ceres, the goddess of the harvest, but a muscular male clutching wheat stalks.[27] Harvest work is here an excuse to display human anatomy and rippling muscles.

The classical treatment was not always so athletic. In France, around 1550, Etienne Delaune engraved a series on the labors of the months that combined fanciful characters with hard-working peasants. Once again we are in ancient Greece rather than the Ile de France, but the work represented is real. Men and women mow, reap, and carry heavy burdens (Figure 4.11). In the July scene that depicts haymaking, some people work and some frolic, apparently to show warm-weather pastimes, but the artist still concentrates on the physical aspects of harvesting. The antique setting, for all its silliness, merged lowly manual labor with lofty, classical aspirations, without completely losing touch with the present.

Italy would seem to be the logical place to produce classicizing images; instead, rural scenes there focused on contemporary peasants. Leandro Bassano's *Summer* (from the late sixteenth century) depicts seasonal tasks set in the countryside. The tending of sheep occupies most of the picture, and the carting and threshing

Figure 4.9. *French court painter Jacques Stella continues the Bruegel tradition of vigorous figures in harvest scenes with* Reapers, *from the collection* Les pastorales *(Paris, 1661). Courtesy of the Metropolitan Museum of Art, Harris Brisbane Dick Fund, 1937 (37.37.34).*

of the corn are more prominently displayed than the reaping itself, which is taking place in the background. In fact, Italian artists treated threshing as the quintessential summer labor.[28] Threshers beating their flails dominate Antonio Tempesta's *July,* dated 1599.[29] His rendering of reaping in *June* is less dynamic, and the foreground lunch break overshadows the harvesting scene behind it (Figure 4.12). Yet the picture leaves no doubt that it is the representation of a working community.

National variations were thus of some significance in the sixteenth century; Italian artists included a wider range of seasonal tasks, and their peasants were on the whole handsomer than those portrayed by northern artists. But, whatever their origin, depictions always showed that harvests involved a substantial number of laborers. People crowded around the fields, either working or resting, but never disconnected from the business at hand. These, of course, were not the only possible images for summer. The season was sometimes rendered in purely allegorical fashion or as aristocratic recreation. When the representation consisted

Figure 4.10. *An allegory of Summer in which muscular, almost naked peasant men and women appear in an antique setting. Philips Galle after Maerten van Heemskerck,* Aestas *(Summer) (1563), courtesy of the Metropolitan Museum of Art, gift of Mr. and Mrs. S. Bieber, 1953 (53.667.2).*

of seasonal work, however, the harvest remained central to the depiction, and harvest work consisted primarily of cutting.

Poetry and plays linked task and tool even more dramatically. Thomas Nashe's play *Summer's Last Will and Testament,* performed in 1592, openly identified harvest with harvest tools and brought them out on stage.

Harvest, by west, and by north, by south and southeast,
Show thyself like a beast.
Goodman Harvest, yeoman, come in and say what you can.
Room for the scythe and the sickles there!
[Enter Harvest with a scythe on his neck, and all his reapers with sickles . . .][30]

Figure 4.11. *Hard-working peasants in antique costume. Etienne Delaune,* Augustus *(ca. 1550), courtesy of the Prints Collection; Miriam and Ira D. Wallach Division of Art, Prints and Photography; the New York Public Library; Astor, Lenox and Tilden Foundation.*

In Shakespeare's *The Tempest,* "reapers, properly habited," perform a dance and are addressed as "You sunburned sickle-men, of August weary."[31] Although we can only imagine their costumes now, contemporary court ballets were sometimes more explicit in stage direction. In *The Vision of the Twelve Goddesses* by Samuel Daniel, a masque presented at Hampton Court near London in January 1604, Ceres is described "in straw colour and silver embroidery with ears of corn and a dressing of the same, present[ing] a sickle."[32] In a late sixteenth-century French "mascarade rustique," the nature of harvesting is made clear.

> They say that in this land the soil is very fertile,
> And that your harvest lasts a long time;
> If the wheat in your fields needs cutting,
> We are ready to use our sickles.[33]

Lyric poets expressed the same sentiments, especially in France, where the georgic blossomed in the sixteenth century.[34] The poems made remarkable use

Figure 4.12. *Antonio Tempesta,* June *(1599) shows the Italian interest in portraying the task of threshing, here overwhelmed by the focus on the noonday meal of the harvesters. Courtesy of the Prints Collection; Miriam and Ira D. Wallach Division of Art, Prints and Photography; the New York Public Library; Astor, Lenox and Tilden Foundation.*

of three images: physical exertion, sharp tools, and numerous harvesters. Jacques Peletier's 1547 poem *L'été* (Summer) describes a troop of harvesters ("la gaye troupe") and focuses on a man mowing, his sharpened scythe ("sa faux acérée"), and the powerful strokes with which he cuts the stalks.[35] Pierre de Ronsard's *Ode de la venue de l'été* also uses the term *troop* ("la diligente troupe") to refer to the harvesters and describes the sickles that bring down the crop. Scythes are sharpened in Rémi Belleau's *Le ver luisant,* in Philibert Guide's *Juillet,* and in Claude Gauchet's *Les moissons.*[36] This last poem follows a peasant from home to field, where

> Swinging around him his scythe with great force,
> He orders his work in such a way
> That a snail's head can be seen twirling in the air.[37]

In Pierre de Brach's *Son voyage en Gascogne* as well, a mower swings his scythe. The entire band of harvesters, the reapers and the rakers, personifies whirlwind activity. One last example hints that admiration of this force is tinged with nervousness. Germain Forget's *L'été* of 1584 describes a landscape filled with workers ("la campagne ja pleine de dispos Aousterons") and ends:

> Without delaying long, let each and everyone
> Be sure to return deep inside their humble lairs
> Their sharp instruments, their mercenary tools.[38]

Time and again, poets emphasized that harvesting brought together individual laborers of Herculean strength. And, as the last poem points out, this collectivity could be menacing. They were urged to limit their vigor to harvesting. The sixteenth century ended, therefore, with a powerful set of images about country people. Poetry, plays, and art offered a portrayal that accentuated the physical strength and prowess of the peasants and placed them within a large working community.[39] Peasants had traditionally been represented as cowardly, for bravery was the attribute of the noble warrior. At best, the peasant could be crafty and win the day through trickery. The revival of the georgic and the Renaissance fascination with the human body lifted the peasant to a heroic and martial status. This new status was not the sole representation of the countryside. Peasants could still be shown as pious or farcical, but the theme of the seasons and of the harvest, all appeared to agree, was best served by these vigorous, tool-wielding images.

Energetic harvest scenes clearly lost their appeal in the seventeenth century, although the theme of the seasons survived. A search for seventeenth-century pictures of harvest work quickly reveals the striking fact that, as the century progressed, these became rarer and rarer. The sixteenth century readily yields its series of representations, and so does the eighteenth century. In other words, the seventeenth century did not bring the demise of a visual genre so much as the replacement of its content. The same is true of poetry, where seasons were emptied of their georgic associations.

Scenes from the harvest continued to be part of the standard Flemish painterly repertoire in the first half of the seventeenth century. Yet something had changed. Pictures of summer, which had previously centered on the harvesting, now conveyed the impression that the work had been completed. The crop was being carted home, the work day was over. In paintings by Jan Brueghel the Elder (1568–1625), for example, a few stragglers continue to cut the grain, as the harvest cart, now at the center of the picture, rolls by them, surrounded by peasants resting after their labors.[40] Peter Paul Rubens renders his *Return from the Fields* (1630s) as a festive event. In his *Rainbow Landscape* (1636), the wagon is loaded and the field hands are going home (Figure 4.13).

In Dutch pictures of the same period, only a few token harvesters are still

Figure 4.13. *Seventeenth-century artists shift their depictions of the harvest scene to a time after the main tasks are over. In Peter Paul Rubens,* The Rainbow Landscape *(1636), the loaded wagon is shown ready to bring in the grain. Reproduced by permission of the Trustees of the Wallace Collection, London.*

shown in the fields. In Jan van de Velde's *August* (1616), the larger figures on the left have their noonday meal, while a man and a woman on the right continue to work. Behind them, a man follows a loaded wagon, and, further off to his left, a couple of idlers stroll through the fields (Figure 4.14). Jan van Almeloveen shows token male harvesters on the left, set among male strollers and rowers (Figure 4.15). Little remains of the realities of the harvest, with its massive use of labor. The picture no longer carries the message that the effort involved is magnificent. The focus has shifted to the presumed rewards of country life: harmless amusements like bathing and fishing, strolls and courtships, within a soothing, undulating rustic setting.[41]

Although paintings of the countryside were extremely popular in the Dutch Republic, by mid-century few if any echoed the sixteenth-century fascination with harvest work. Painters focused instead on the landscape, on scenes that showed peasants congregating around their houses (whether the hovels of Jan van Goyen or the prosperous cottages of Meindert Hobbema), walking down a country road, taking goods to market, or minding their cow. Pictures of the seasons, meanwhile, turned into scenes of leisure and repose.[42]

Art historians have noted this absence of agricultural work in paintings from the Dutch Golden Age. "Surprisingly few Dutch landscapes show farmers at work. For the most part, the view of the landscape is recreational rather than agri-

Figure 4.14. *Cutting the crop occupies only a few people, as others eat, cart, or stroll. The harvest has been marginalized. Jan van de Velde,* Augustus *(1616), courtesy of the Metropolitan Museum of Art, Harris Brisbane Dick Fund, 1933 (33.52.57).*

cultural," writes Christopher Brown.[43] The most common explanation is that such innocuous paintings answered city folk's nostalgic yearning for the simple life of the country during a period of rapid urbanization. The rise of landscape paintings that focused on broad vistas or on nature itself, with their tiny, often lone figures, similarly embodied the urban observer's desire for quiet seclusion. The countryside that had once teemed with workers, villages that Bruegel and his followers had shown bursting with boisterous, drunken revelers, were recast as depopulated, or underpopulated, places where a few villagers might come together for the occasional drink (see Figure 4.16).[44]

The new depiction seemed to validate country life but deprive it of its basic activities and, just as openly, deprive it even of its inhabitants. Perhaps the most extreme—and most beautiful—version of this position comes from Jacob van Ruisdael (1628–1682). He painted more than twenty pictures of empty wheatfields, all nearly or already ripened. Not one of them included harvesters, although the odd traveler might stroll by (Figure 4.17). Tiny figures populate his paintings of the Haarlem bleaching fields or trudge through the snow in his grim winterscapes. His wheatfields, however, were meant to seem untouched by human hands.

Figure 4.15. *Harvesters are shown among rowers and strollers. The countryside belongs to all, urban and rural alike. Jan van Almeloveen (active 1658–ca. 1678–83),* Landscape with Harvesters, *courtesy of the Metropolitan Museum of Art, the Elisha Whittelsey Collection, the Elisha Whittelsey Fund, 1951 (51.501.5254).*

Commentators have treated this approach as a reflection on nature's spontaneous beneficence and hope for man's liberation from the drudgery that had been taken as his lot.[45] Ruisdael expressed, in the strongest terms, a European-wide recoiling from the labor that food production demanded. This attitude is also noticeable in Spain, to which some Flemish artists migrated in the early seventeenth century.[46] Elsewhere in Europe, pure harvest scenes were evidently not being painted.[47] One Italian artist, Ludovico Mattioli (1662–1747), designed a set of seasons at the end of the seventeenth century showing handfuls of tiny peasants working in vast deserted landscapes (Figure 4.18). After the Peasant War of 1525, the Germans showed a singular reluctance to depict laborers.[48] In France, Jacques Stella's harvest scene mentioned above (Figure 4.9) included vigorous workers, but not a single painting by the Le Nain brothers—known for their realistic scenes—shows peasants working.[49] A fascinating exception is Nicolas Poussin's *Summer* (1660–1664), which used the biblical theme of Ruth gleaning in the fields of Boaz (Figure 4.19). The background, as was traditional in such pic-

Figure 4.16. *Dutch and Flemish artists now prefer to show peasants as social drinkers rather than drunken revelers. Adriaen van Ostade (1610–1685),* Peasants Drinking with a Violin Player *(1660s?), courtesy of the Metropolitan Museum of Art, the Elisha Whittelsey Collection, the Elisha Whittelsey Fund, 1951 (51.501.1087).*

tures, showed a group of busy harvesters cutting and bundling the crop. Despite the lofty classical rendition, Poussin had no imitators. The other pictures of Ruth and Boaz produced in the seventeenth century gradually followed Rembrandt's lead in focusing on the pair, removed from the fields and garbed in exotic costumes.[50] The seasons, when they were represented, tended to be allegorical, as in Charles Le Brun's tapestries at Versailles, *Les quatre saisons* (late 1660s). All over Europe, painters turned away from depicting harvesting and the mundane realities of rural life. They preferred imaginary landscapes, rocky seclusions, and broad vistas that made the countryside into a patchwork of colors without people.

Seventeenth-century poetry suggests the reasons for this turning away. Literary historians have stressed the role of aristocratic complacency. Country-house

Figure 4.17. *One of many empty grainfields painted by Jacob van Ruisdael,* Wheatfields *(ca. 1670). Courtesy of the Metropolitan Museum of Art, bequest of Benjamin Altman, 1913 (14.40.623).*

poems in the seventeenth century celebrated the landed estate while leaving out the workers who produced its riches.[51] At best, as in Robert Herrick's poem *The Hock-Cart,* the tenants joyfully carried their crop to their master's barn.[52] Poets, when they did address agricultural work, imbued it much more directly than in the sixteenth century with a moral, regenerative quality. It was worthy, peaceful, and gay. It was also performed for the benefit of the lord. Scenes of reaping, when they appear, lack the absorption with swing and sweat that had typified the sixteenth-century georgic:

> Here is no other case in law
> But what the sunburnt hat of straw
> With crooked sickle reaps and binds
> Up into sheaves to help the hinds.[53]

The harvest was a common symbol of plenty, and it had appeared in that guise in sixteenth-century French poems, such as Joachim du Bellay's *Ode à Cérès* (1558). Its symbolic association also extended to religious revelation, as in Germain Forget's *Les plaisirs et felicitez de la vie rustique* (1584).[54] These symbolic

Figure 4.18. *Italian painter Ludovico Mattioli also shows tiny harvesters swallowed up in the landscape.* July *(ca. 1700), Courtesy of the Graphische Sammlung Albertina, Vienna.*

usages became more pronounced in the seventeenth century. In Racan's *Stances sur la retraite,* written around 1618, the harvest stands for plenty, as work yields to leisure and the georgic gives way to the pastoral.[55]

The sixteenth century had also associated the harvest with feelings of loss, even devastation.[56] For Du Bellay, Rome fell the same way that "the peasant harvests . . . the golden fields."[57] This theme appeared more frequently in the following century. The harvest figured as a symbol for lost love in Gilles Ménage's *Le moissonneur:*

> The fields have fewer ears, and these ears fewer grains
> Than my sorrowful heart has mortal chagrins.[58]

Saint-Amant's *Sonnet sur la moisson d'un lieu proche de Paris,* written around 1620, expanded the themes of destruction and loss.

> The gold yields to the blades; already the harvesters
> Stripping the fields of their yellow standards
> Make the desolation both gay and beautiful.
> This useful cruelty works for the benefit of all.
> . . . .
> The greater the devastation, the sweeter I find it.[59]

Figure 4.19. *A favorite Old Testament theme, Ruth gleaning in the fields of Boaz, shows peasants harvesting, safely transferred to a different place and time. The predominant approach in the seventeenth century is to avoid depicting laborers. Nicolas Poussin (1594–1665),* L'été ou Ruth et Booz. *Courtesy of Giraudon/Art Resource, New York.*

The harvest had become a symbol of death and destruction.[60] Numerous other poems conflated field and battlefield. Thus Jean-François Sarasin, in the *Ode de Calliope:*

He breaks a thousand battalions
And the bristling pikes
Fall before him
Like wheat on the fields.[61]

And in an *Ode* by François de Malherbe:

Under the burning sun
There is less corn on the fields
Than the battalions
That teemed in this army.[62]

English poets made even more powerful use of the same ideas. In *Troilus and Cressida,* Shakespeare had Nestor say about Hector:

And there the strawy Greeks, ripe for his edge,
Fall down before him, like a mower's swath.[63]

Andrew Marvell used similar imagery:

The Mower now commands the Field;
In whose Traverse seemeth wrought
A Camp of Battail newly fought:
Where as the Mead with Hay, the Plain
Lyes quilted o'er with Bodies slain:
The Women that with forks it fling,
Do represent the Pillaging.[64]

In Richard Lovelace's poem *The Grass-hopper* (1649), the harvest stands for the king's execution:

But ah, the Sickle! Golden Eares are Cropt;
Ceres and Bacchus bid good night;
Sharpe frosty fingers all your Flow'rs have topt,
And what Sithes Spar'd, Winds shave off quite.[65]

And in James Shirley's *Dirge:*

Death lays his icy hands on kings;
Scepter and crown
Must tumble down
And in the dust be equals made
With the poor crooked scythe and spade.
Some men with swords may reap the field,
And plant fresh laurels where they kill.[66]

Like Virgil and Horace, seventeenth-century poets used familiar images contrasting war and agriculture, destruction and plenty.[67] Yet the experience of war overwhelms seventeenth-century poems, to the point that the apparent comparison slides into an equivalence: rather than a contrast between agriculture and violence, the poems suggest the presence of violence *within* agriculture. The vigorous mower had become the Grim Reaper.

A closer look at attitudes toward harvest tools reinforces this interpretation. In the Middle Ages, the sickle and scythe were viewed as symbols of both life and death, of fertility and the ravages of time. The sickle, unlike the scythe, was at times associated with Christ as an emblem of redemption. In the fifteenth century, however, while the sickle remained a symbol of rebirth and fertility, the scythe acquired a more lasting and sinister association with death.[68] We take this so much

Figure 4.20. *Death comes upon his victim with a nobleman's weapon, a spear.* Atropos *(French, ca. 1360), courtesy of the Pierpont Morgan Library, New York, MS M. 132, f. 140v.*

for granted that it is surprising to note that Death was first depicted pointing an arrow or a spear, that is, a nobiliar weapon, before it became overwhelmingly associated (in the late sixteenth century) with the scythe, the instrument wielded by the peasant (Figures 4.20 and 4.21).[69] This connection between death and the scythe, between the peasant and death, became extremely potent. Death became routinely depicted with a scythe, commonly seen in Baroque tombs and paintings.[70] In the middle of the nineteenth century, Jean-François Millet still used the image to great effect: a shrouded skeleton, holding a scythe, grabs a peasant on the open road (*Death and the Woodcutter,* 1859). Scythe-wielding peasants became equated with death and destruction; the harvest, with its bustling activity, its massing of the peasantry, its violence—elements that the sixteenth century had openly celebrated—now betokened danger and alienation.

The harvest could be dissociated from death only by detaching it from the instruments of destruction. One solution was to rid the fields of workers and treat the countryside as a place of refuge, where nature spontaneously yields its pro-

Figure 4.21. *Now Death sits with his scythe, a peasant tool, enthroned above his harvest.* The Triumph of Death, *Farnese Hours, Rome (1546), courtesy of the Pierpont Morgan Library, New York, MS M. 69, f. 79v.*

duce. This idea was especially attractive given the tensions of the late sixteenth century. By the 1550s, rising grain prices and declining real wages had stimulated the reconstitution of large landed estates in many parts of Europe, at the expense of an increasingly indebted and impoverished rural population.[71] More than in earlier centuries, large landowners between 1550 and 1650 devoted careful attention to the composition of their properties. This consolidation did not remove peasants from the countryside, since grain-growing estates still needed large pools of labor. But questions about who owned the countryside were in everyone's mind; landowners did not want images of crowded, united, armed villages. The emptied countryside of seventeenth-century art thus described not a reality so much as a wish.

Partly because of changing property relations and partly because of political and religious strife, the social order appeared especially fragile to upper-class Europeans between 1550 and 1650. During this period, European countries experienced civil war or severe social disruptions that culminated in the great rebellions of the mid-seventeenth century. Those countries that were spared observed the ravages among their neighbors. The need to ensure tranquillity and the respect for hierarchy seemed particularly acute after such long periods of civil disorder.[72]

These social and political conditions provide a context to the artistic changes that I have traced. They do not offer sufficient explanation. Similar events, after all, can be cited for other periods. Further, images could shape as well as express rural social relations. Nonetheless, seventeenth-century Europeans revealed a growing distaste for the images they had inherited. The international climate called for a new image of the countryside, with far fewer pictures of peasants and above all far fewer working peasants. The dominant solution was to empty the countryside. But alternative solutions were available. Spanish, Italian, and Dutch artists, for instance, replaced images of rustic boisterousness with pictures of bandits and highwaymen, transferring social anxieties onto clearly defined marginals. Only Spain found a way of including the peasants within its search for order, as symbols of national regeneration.[73] Sometimes harking back to a feudal past and sometimes pointing to a better future, playwrights Lope de Vega, Calderón de la Barca, and Tirso de Molina constructed an organic vision of society that linked peasants, farmers, and landowners in mutual respect and mutual dependency. The premise that social harmony could not be effected without including peasants would shape the eighteenth-century cultural construction of the countryside.

The eighteenth century opened under the influence of a literary event. In 1697, the English poet John Dryden had published his complete translation of Virgil's works. Joseph Addison's introduction to Dryden's translation, praising the *Geor-*

gics as the greatest work of literature, incited poets to revive the genre.[74] The interest in the *Georgics* is usually treated primarily as a cultural validation of work.[75] I wish to point to a peculiar aspect of the eighteenth-century georgic revival, the way it envisioned the harvest.

In Book 1 of the *Georgics,* Virgil describes a storm that interrupts the harvest. The modern Loeb Classical Library translation renders lines 316–17 as follows: "Often, as the farmer was bringing the reaper into big yellow fields and was now stripping the brittle-stalked barley . . ."[76] French and English translations of the same lines, in the sixteenth and seventeenth centuries, assumed the cutting of the crop must be with a sickle or scythe. For example, Thomas May's 1618 translation:

> I oft have seen when corn was ripe to mow
> And now in dry and brittle straw did grow.[77]

A 1691 French prose translation belabors the point: "I have often seen, at the time when the farmer brought his harvesters into the golden, ripe corn, and when they had put their sickles to the barley with its fragile stalk."[78] Michel de Marolles preferred an alternate reading in 1649, which took Virgil to mean hand-reaping: "When the harvester was in the golden plain and held in his hands the fragile stalks of corn."[79]

Dryden, however, took a radically different approach:

> Ev'n while the reaper fills his greedy hands
> And binds the golden sheaves in brittle bands.[80]

This reading of the action as "binding" would be taken up by other English and French translators. Binding one's sheaves ("lier ses gerbes") is used in two eighteenth-century French versions, including the abbé Delille's celebrated translation, and in a number of English ones.[81] A new image of the harvest had emerged. It moved the focus away from cutting and shifted it instead to the gathering of the crop. Dryden's and Delille's were by far the most popular of all Virgil translations and thus the versions everyone read. Dryden's choice of images solved the problem of how to represent the harvest without invoking force and violence and without falling into the other excess of treating it simply as a joyous occasion. Gathering, binding, stacking, and carting, were, in fact, important parts of the harvest, albeit tasks that required less skill and force than reaping and mowing. Thus the new focus on gathering reinstated work within country scenes, in a way that seemed entirely "realistic," that is georgic and descriptive, without dwelling on its darker sides. Poets, artists, and audience were going to have their cake and eat it, too.

In Flanders, David Teniers the Younger had developed a similar visual image.

Teniers (1610–1690) was son-in-law to Jan Brueghel the Elder, and he followed the family interest in peasant themes, further influenced by his Flemish contemporaries Adriaen Brouwer and Adriaen van Ostade, who were renowned for their peasant genre scenes.[82] Teniers had a long and extremely productive life and, if probate records can be trusted, became the most popular and most reproduced artist in eighteenth-century France.[83] Besides toned-down pictures of village fairs and other peasant festivities, Teniers produced several cycles of the seasons. They harked back to early seventeenth-century versions. The cutting of the crop is marginalized; instead, peasants stack the corn, rake the remains, and have their lunch. The fields are huge, but the implication is that this tiny crew of workers will manage to reap them without urgent exertion. Teniers's most influential picture (which spawned a whole genre in the eighteenth century) is that of a young man holding a huge sheaf of wheat, while two harvesters work in the distance (Figure 4.22). There he stands, shyly looking away, a harmless, poetic soul. Some may read this as an image of brute force rather than meekness. However, the contrast with the sixteenth-century peasant Hercules is impressive. Instead of the muscular peasant brandishing a scythe surrounded by equally hard-working villagers, we have a lone youth, gathering the fruits of the harvest.[84]

This was the image that would dominate eighteenth-century pictures of the harvest. It was widely imitated, most obviously in France. Nicolas Lancret's *L'été* from 1738 (part of a cycle of the seasons painted for Louis XV) shows a rustic dance and courtship next to two young men holding and binding large sheaves (Figure 4.23).[85] *The Seasons,* panels by Jean-Baptiste Pater (1676–1738) for the duc de Choiseul, have a summer scene that also portrays peasants burdened with oversized sheaves. In England, Thomas Gainsborough's 1749 portrait of Mr. and Mrs. Andrews represents the harvest simply with bundled sheaves stacked near the couple, without laborers, but this is an exception (Figure 4.24).[86] The whole point, in fact, was to find a way of reintroducing harvest work in genteel fashion. Pretending that nature was simply munificent could only go so far. It presumed too obvious a denial of social realities.[87] Better to suggest a convergence between nature's abundance and human activity. Nature provided sustenance; it only needed gathering. This was, moreover, fun and easy work—eighteenth-century elites delighted in this image, exemplified by George III's haymaking—something that the sweat-drenched images of the sixteenth century or the blood-stained ones of the seventeenth could not successfully project.

Jean-Jacques Rousseau popularized the idea of pleasant summer work by focusing on tedding, a minor aspect of haymaking that consisted of turning over the cut hay on the meadows. "The simplicity of a rustic or pastoral life is always touching. One need only look at the fields covered with people tedding and sing-

ıgure 4.22. *From Flanders, David Teniers the Younger depicts* Summer *(1664) as a youth holding sheaf. This picture will be widely imitated, starting in France, where Teniers became extremely ›pular in the eighteenth century. Courtesy of the Trustees of the National Gallery, London.*

Figure 4.23. *Nicolas Lancret painted a cycle of the seasons for one of Louis XV's castles. This part,* L'été *(1738), reveals a genteel peasant dance next to the wheatfields. At left, two young men are binding sheaves. Musée du Louvre. Courtesy of Alinari/Art Resource, New York.*

ing, and at the herds spread around as far as the eye can see."[88] The raking and gathering of the crops came to stand for the entire harvesting process.

Eighteenth-century georgic poetry made the point explicitly. One of the most influential poems of the eighteenth century, James Thomson's *The Seasons,* which first appeared between 1726 and 1730 to universal acclaim, describes as charming a harvest as one could wish:

> Soon as the morning trembles o'er the sky,
> And unperceived unfolds the spreading day,
> Before the ripened fields the reapers stand
> In fair array, each by the lass he loves,
> To bear the rougher part and mitigate
> By gentle offices her toil.
> At once they stoop, and swell the lusty sheaves;
> While through their cheerful band the rural talk,
> The rural scandal, and the rural jest

Figure 4.24. *Thomas Gainsborough's portrait* Mr. and Mrs. Andrews *(1749) sets the couple beside large sheaves without laborers. Most other eighteenth-century harvest scenes included nonthreatening binders, rakers, and gatherers. Courtesy of the Trustees of the National Gallery, London.*

> Fly harmless, to deceive the tedious time
> And steal unfelt the sultry hours away.
> Behind the master walks, builds up the shocks,
> And, conscious, glancing oft on every side
> His sated eye, feels his heart heave with joy.
> The gleaners spread around, and here and there,
> Spike after spike, their sparing harvest pick.
> Be not too narrow husbandman! but fling
> From the full sheaf with charitable stealth
> The liberal handful.[89]

There is no cutting of the crop. Just as paintings were defining the harvest as the bundling of sheaves (the version we saw Dryden adopting), so here, too, Thomson turns it into "swelling the lusty sheaves." The act of gathering is further underscored by the entry of gleaners, and Thomson ends his section on the harvest with a modernized version of the story of Ruth.

Thomson's poem was translated and imitated by French poets in the second half of the eighteenth century.[90] They sentimentalized labor and infused it with gaiety, lingering sententiously on gleaning.[91] This new treatment of the harvest also permeated Charles Simon Favart's opera *Les moissonneurs* (1768). The

Figure 4.25. *A castle in the background reminds the viewer that the harvest produce will go to the landowner. J. B. Lallemand,* Vue de Château de Daix *(ca. 1780), courtesy of the Musée de Brou, Bourg-en-Bresse.*

action centers on Rosine, a gleaner. The stage directions mention reaping, but the word itself is never spoken on stage, despite the play's title. Instead, the farmer gives the following orders:

> For fear that the wheat will sprout on the ground,
> Gather the sheaves into shocks:
> Let them be locked in the barns;
> And let the ricks on the farm
> Show passers-by your work.[92]

The harvest's renewed popularity as a cultural theme cannot be doubted, but it was clearly preferable to present it in truncated form.[93]

Eighteenth-century images of harvest thus revolved primarily around the gathering and carting of the crop. The produce, moreover, was restored to the landowners. Paintings, as in the Middle Ages, were likely to show harvesting around a castle (Figure 4.25). French poets in particular eulogized the harmony between grateful, paternalistic landowners and their diligent and devoted laborers. The surge of gleaning tales at the end of harvest poems, following Thomson, highlighted the importance of gathering on the one hand but also manifested on the other the fairy-tale union of the squire with his needy dependents.

The fascination with gleaning was essentially confined to literature. Meanwhile, eighteenth-century artists turned the loaded wagon into a primary sym-

Figure 4.26. *The depth of this eighteenth-century harvest scene shows just how many peasants were still needed to gather and cart the crop. J. P. Le Bas, after Michau,* Les moissonneurs *(1770s), courtesy of the Metropolitan Museum of Art, Harris Brisbane Dick Fund, 1917 (17.3.3275).*

bol of the harvest, along with raking or sheaf-binding. We find such pictures in England, France, Holland, and Germany.[94] Taken as a whole, these images suggested, far better than in the sixteenth century, how many people it took to perform a single task, such as the gathering or carting of the crop (Figure 4.26). The countryside was filled with working people.

Moreover, after decades of showing peaceful bundling and gathering, the virtuous nature of the harvesters had been so firmly established that peasants handling cutting tools could be reintroduced in the late eighteenth century. In L. J. Watteau de Lille's storm scene in the northern France of the 1780s, a family of harvesters, the husband with a scythe on his shoulder accompanied by wife and babes, make their way home as the last harvesters hurriedly gather and load the remaining—prominently displayed—sheaves before the storm hits (Figure 4.27).

Figure 4.27. *Before the storm, a virtuous family of harvesters hurries to shelter. The harvest tools may now be shown in the hands of peasants. Louis Joseph Watteau de Lille,* Les vêpres ou l'orage *(1780s), courtesy of Réunion des Musées Nationaux/Art Resource, New York.*

George Stubbs's *Reapers* of the same period (1785) shows cutters rather awkwardly bent over their sickles, while others stand, holdings their sheaves à la Teniers (Figure 4.28).[95] Docility, gentility, and domesticity dominate these representations. Harvest tools are of secondary importance.

For the peasantry had been civilized and domesticated to fit an elite's desire to integrate them within an Enlightenment vision of progress and happiness. Bestial peasants, however pathetic, simply could not evoke warm feelings. The eighteenth century wanted to send a clear message about the beauty of the peasants, their effeminacy—their incapacity, almost, to perform strenuous tasks—which required both sexes to be portrayed in this way.[96] Bruegel's rough peasant had been replaced by slim and dainty countrymen and women. The peasant stopped being a clear physical type, coarse and squat (until revived by Millet in the nineteenth century). The eighteenth-century peasant was defined by setting and dress, not by demeanor.[97]

Some of the Enlightenment recasting of the peasant is evident in the example of

Figure 4.28. *A docile group of reapers bend to their harvest work; the sheaves are prominently displayed. George Stubbs,* The Reapers *(1785), © Tate, London 2000.*

the poet and *salonier* Jean-François de Saint-Lambert. A provincial who moved to Paris in the mid-eighteenth century, Saint-Lambert wrote thirteen articles for Denis Diderot's *Encyclopédie* (1751–1776), and he read his poems to his fellow Encyclopedists for years before they were published.[98] His lengthy poem on the seasons, which appeared in 1769, had been eagerly awaited. We may take his vision of the peasantry as representing that of a much larger milieu.[99]

Like so many other eighteenth-century poets, Saint-Lambert stressed the joys of rural life. His peasants are happy, loving, familial, contented with their lot. "Happy people of the fields," he addresses them, "your labors are like feasts." The peasant, "happy to see the days of indolence draw to a close, wants to merit abundance through his labor."[100] Yet the poem also reveals darker sides in the peasant's situation. The people suffer grave injustices. The corvée reduces them to slavery; unspecified taxes and rents condemn them to poverty. Yet they are too weak to rebel and too brutalized to farm their lands effectively:

> In the most favourable climates, the country people
> Made stupid by the excess of its sufferings,
> Knows not at all how to second nature's work.
> Habit and instinct alone directing its labors,
> They invent nothing and tremble to imitate;

> They dare not attempt to rise from their poverty;
> And dragging along their miserable lives,
> They believe that the gods have condemned them to suffer.
> Go forth, people of the countryside, make known your voices
> Even in that refuge where dwell your kings.[101]

For Saint-Lambert, scientific know-how and upper-class benevolence will restore the countryside to abundance and ease. His emphasis on peasant docility and child-like goodness in fact forms part of a larger program. Decent, unthreatening, yet oppressed, the peasants need the help of a reforming state and benevolent, well-informed landowners. For Saint-Lambert, depicting the peasants' basic goodness was a necessary prelude to calls for political change. Physiocracy worked hand in hand with poetry.

The peasant family man, celebrated in art and literature, became the foundation of the state. To eighteenth-century would-be technocrats, peasants were uneducated and tradition-bound. The sentimental overlay, however, turned them into virtuous and innocent creatures, ready to be guided. A.-R.-J. Turgot's plan for municipal reform thus gave propertied peasants a far greater voice than propertied town dwellers.[102] Likewise, there were many more rural than urban electors for the Estates General. As Georges Lefebvre explained, "The peasants were favored in this respect because nobody was afraid of them."[103] Peasants represented the future welfare of the state, not the past.

Representations of peasants, though subject to developments in artistic genres, reveal elite responses to country people. These responses need not be viewed as transparent reflections of material or political conditions. In fact, the representations are especially meaningful because, for the most part, they were initially intended either to fulfill allegorical functions or to display ingenuity in adapting the classics. They were not meant as mirrors of the social world, but they did shape perceptions.

The production of images of independent peasants in the sixteenth century broke dramatically with traditional modes of depicting peasants as subservient and fundamentally helpless. Medieval peasant rebellions, which erupted sporadically, had been treated as contrary to nature, whether the authorities heeded the peasants' grievances or not. Assimilating peasants to classical heroes in the sixteenth century, ensconcing them within large communities—on canvases and in print—rendered their existence more potent, their autonomy more probable. Since social mobility was visibly increasing in this period, the social fabric seemed dangerously weakened, both in town and country. The changes in depictions of peasants proved disquieting because they appeared to exceed their mandate. They could no longer be read simply as Christian messages or clever stylistic exercises.

Owners did not have to burn their peasant canvases to express their disaffection from the genre. Seventeenth-century patrons expressed their preference for different images of the country. Peasants hardly figured in them. When they did, it was rarely as part of large groups or as agricultural laborers. Agricultural work, moreover, prompted thoughts of destruction and loss. Poets in particular delved into the dark recesses of country life. The peasant was once again the "Other," best avoided.[104] The seventeenth century thus left a cultural legacy of violence, one in which rural work could not be portrayed without arousing ire or dismay. Yet showing cultivated places without people, be it in paintings or rustic poetry, did not adequately convince anyone that the countryside was empty. If anything, the move of nobles to country houses, separated from the village, implied the existence, out there, of a band of rustics, left without adequate supervision. The countryside needed therefore to be rethought as a place of harmony and peaceful interaction. The gathering of the harvest seemed to provide the means for doing this. It offered a solution to the problem of how to show labor without showing its dangerous aspects.

The eighteenth century thus adopted a new vision of the peasant. As a laborer, he was harmless and piteous and therefore a natural object of charity and paternalist concern. As an independent farmer, he was virtuous, hard-working, and devoted to his family. Anxious to learn and to be guided, the peasant emerged as a fitting citizen of the state. By the end of the eighteenth century, this figure had become an emblem for mankind.

Acknowledgments

This chapter was previously published in *American Historical Review* 101, no. 5 (December 1996). Much of the research was conducted thanks to a fellowship at the Yale Program in Agrarian Studies. My special thanks for discussions and detailed comments to Jean-Christophe Agnew, Robert Baldwin, Jonathan Dewald, Paul Freedman, George Grantham, Irving Lavin, Timothy Le Goff, David Quint, Theodore Rabb, James C. Scott, and Christopher Wood.

Notes

1. Latin, *pagus,* or district; Old French, *païs,* region. The consensus about the nature of peasants has been challenged lately by a more sympathetic historiography, one that has sought to restore them agency, especially in defense of their rights.

2. For a recent synthesis, see Del Sweeny, ed., *Agriculture in the Middle Ages: Technology, Practice, and Representation* (Philadelphia, 1995).
3. Theodore K. Rabb, *The Struggle for Stability in Early Modern Europe* (New York, 1975).
4. Peasant movements during the French Revolution destroyed the growing consensus about the gentle, docile nature of country folk.
5. See, for example, Norbert Elias, *The History of Manners,* Edmund Jephcott, trans. (New York, 1978), 207.
6. Paul Freedman provides an excellent overview of medieval approaches to peasants in *Images of the Medieval Peasant* (Stanford, Calif., 1999).
7. A recent example is Dian Fox, *Refiguring the Hero: From Peasant to Noble in Lope de Vega and Calderón* (University Park, Pa., 1991).
8. Nowhere is this truer than for the Dutch Republic. See for example, Simon Schama, *An Embarrassment of Riches: An Interpretation of Dutch Culture in the Golden Age* (New York, 1987).
9. This is one of the positions taken by Noël Salomon, *Recherches sur le thème paysan dans la "comedia" au temps de Lope de Vega* (Bordeaux, 1965).
10. Art books and exhibition catalogues are an excellent source, but, since this project required that I locate broad trends, I also looked through the published compilations of prints and woodcuts (*The Illustrated Bartsch,* F. W. H. Hollstein), as well as non-illustrated guides such as A. P. F. Robert-Dumesni's *Le peintre-graveur français,* 10 vols. (Paris, 1835–50). This allowed me to collect a body of harvest scenes, while at the same time gauging their relative importance within the overall work of the artist. As for the poetry and drama, I began with twentieth-century compilations, used references in scholarly works, which the renewal of interest, especially in rustic poetry, made both fruitful and suggestive. Given the vastness of the potential sources, I am immensely grateful to those literary scholars who pointed me in the right direction, and, with their help, my collection of plays and poems continues to grow.
11. Derek Pearsall and Elizabeth Salter, *Landscapes and Seasons of the Medieval World* (London, 1973); Walter S. Gibson, *"Mirror of the Earth": The World Landscape in Sixteenth-Century Flemish Painting* (Princeton, N.J., 1989); Otto Pacht, "Early Italian Nature Studies and the Early Calendar Landscape," *Journal of the Warburg and Courtauld Institutes* 13 (1950): 13–47.
12. Georges Comet, *Le paysan et son outil: Essai d'histoire technique des céréales (France, VIIIe–XVe siècle)* (Rome, 1992); Perinne Mane, *Calendriers et techniques agricoles (France-Italie, XIIe–XIIIe siècles)* (Paris, 1983); James Webster, *The Labors of the Months in Antique and Medieval Art to the End of the Twelfth Century* (Princeton, N.J., 1938).
13. Comet, *Le paysan et son outil;* Michael Roberts, "Sickles and Scythes: Women's Work and Men's Work at Harvest Time," *History Workshop* (1979): 3–28.
14. Michael Camille, "Labouring for the Lord: The Ploughman and the Social Order in the Luttrell Psalter," *Art History* 10 (1987): 423–54.
15. For the argument that the calendar genre left out social relations, see Bridget Ann Henisch, "In Due Season: Farm Work in the Medieval Calendar Tradition," in Sweeny, *Agriculture in the Middle Ages,* 322. Harvesting peasants also appeared in illustrations to medical encyclopedias and in late medieval secularized renderings of moral dicta, for

example, Ambrogio Lorenzetti's good and bad government frescoes, painted in Sienna (1338–1339). Gisèle Lambert, "De l'espace sacré à l'esprit profane: L'apparition du paysage," in *Paysages, paysans: L'art et la terre en Europe du Moyen Age au XXe siècle* (Paris, 1994): 37–42.

16. This is the treatment in Emile Mâle, *The Gothic Image: Religious Art in France of the Thirteenth Century,* Dora Nussey, trans. (New York, 1958), 64–65; and in Pearsall and Salter, *Landscapes and Seasons of the Medieval World.*
17. This process is summarized in Robert C. Cafritz, Lawrence Gowing, and David Rosand, *Places of Delight: The Pastoral Landscape* (Washington, D.C., 1988). See also Margaretha Rossholm Lagerlöf, *Ideal Landscape: Annibale Carracci, Nicolas Poussin and Claude Lorrain* (New Haven, Conn., 1990).
18. The Middle Ages had viewed Virgil essentially as the author of the *Aeneid.* Ernst Robert Curtius, *European Literature and the Latin Middle Ages,* Willard R. Trask, trans. (1953; rpt. edn., Princeton, N.J., 1983), 36. See also R. D. Williams and T. S. Pattie, *Virgil: His Poetry through the Ages* (London, 1982); L. P. Wilkinson, *The Georgics of Virgil: A Critical Survey* (Cambridge, 1969).
19. Even in the *Eclogues* and Theocritus's *Idyls,* however, despite their emphasis on leisure, real work always lurks close to the surface, in images of sowing, plowing, and the real tasks of herding. Thus Virgilian poetry in general stimulated interest in images of work. See Annabel Patterson, *Pastoral and Ideology: Virgil to Valery* (Berkeley, Calif., 1987).
20. "Yes," the narrator tells us, "unremitting labour and harsh necessity's hand will master anything." And, "Oh, too lucky for words, if only he knew his luck, / Is the countryman who far from the clash of armaments / Lives, and rewarding earth is lavish of all he needs!" *The Eclogues and Georgics of Virgil,* C. Day Lewis, trans. (New York, 1964), Georgics, Book 1, ll. 145–46, Book 2, ll. 458–60.
21. Simon Bening, *August: Mowing Wheat, Binding Sheaves,* Da Costa Hours, Bruges, ca. 1515, M399, f.9v, Pierpont Morgan Library, New York City. Bening and his workshop produced a number of such illustrations. See his "August" in a Book of Hours at the Bayerische Staatsbibliothek München (Clm 23638), and at the Pierpont Morgan Library, M307 and M451.
22. In fact, most of the pictures that I will be discussing survived in the form of prints, meaning that they had a wide circulation and a broad appeal.
23. He did not invent the genre, for others (such as Hieronymous Bosch, fellow Netherlander) had already tied vices and virtues to ordinary folk. Bruegel, however, toned down Bosch's fantastic elements and created recognizable human scenes.
24. A continuing engagement with Bruegel's paintings rests on their ambiguity. They can be treated as humorous or censorious, but they are so vibrant that the message slides into a celebration of peasant life.
25. One recent analyst views him essentially in rhetorical terms. Margaret A. Sullivan, *Bruegel's Peasants: Art and Audience in the Northern Renaissance* (Cambridge, England, 1994).
26. Sixteenth-century prints overflowed with detail. William M. Ivins, Jr., *Prints and Visual Communication* (1953; Cambridge, Mass., 1969). Other examples of vigorous harvests include Adriaen Collaert's (1560–1618) medallions of the months, Jacob Grimmer's

Summer at the Dayton Art Institute, Johann Thomas de Bry's *August* in Vienna, and Frans Boels's *Summer* in Stockholm.

27. At approximately the same time, Cesare Ripa's *Iconologia* counseled a robust young woman for the representation of Summer (p. 502) and either a half-naked man holding a sheaf and sickle (p. 348) or a half-naked peasant holding his reaping and threshing instruments (p. 346) as symbols for the month of July (agricultural version): "Because the most important aspect of this month is the grain harvest . . ." The first version of *Iconologia* appeared in 1593; illustrated versions followed (Padua, 1611) (1644 edn., rpt., New York, 1976).
28. This attitude appears to have been established in the Middle Ages, when Italian illustrated Books of Hours already focused on threshing. There are several such examples at the Pierpont Morgan Library.
29. The obsession was transferred to France through followers of Nicolo dell' Abate who painted threshers on the walls of Fontainebleau, in the mid-sixteenth century. The painting also includes pitching hay, a rare example of what was to become a common motif in the eighteenth century.
30. Thomas Nashe, *The Unfortunate Traveller and Other Works* (Harmondsworth, 1985), 171.
31. William Shakespeare, *The Tempest,* act 4, scene 1, ll. 134–40.
32. *A Book of Masques in Honour of Allrdyce Nicoll* (Cambridge, England, 1967), 28.
33. *Mascarade rustique* (before 1600), in Paul Lacroix, *Ballets et mascarades de cour de Henri III à Louis XIV (1581–1652),* Vol. 1 (Geneva, 1868), 138.
34. See poets such as Jacques Peletier, Philibert Guide, Germain Forget, Rémi Belleau, Claude Gauchet, Pierre de Brach, and Pierre de Ronsard, inspired in part by Luigi Alamanni's modern adaptation of the *Georgics* in *La coltivazione* (1543). The poem was written while Alamanni resided in France and was dedicated to Francis I. *La coltivazione di Luigi Alamanni et Le api di Giovanni Ricellai* (Milan, 1804). Henri Hauvette, *Luigi Alamanni (1495–1556) sa vie et son oeuvre* (Paris, 1903), chap. 4.
35. Jacques Peletier, *L'été,* in Françoise Joukovsky, *La Renaissance bucolique: Poèmes choisis 1550–1600* (Paris, 1994), 29–33. "Et de son bras robuste / A grans traiz fait sa tasche."
36. Ronsard, Guide, and Gauchet, in Joukovsky, *La Renaissance bucolique,* 43–45, 59–60, 51–52; Belleau in M. Allem, ed., *Anthologie poétique française, XVI*e siècle, vol. 2 (Paris, 1965), 11.
37. Claude Gauchet, *Les moissons* (1583), in *Le plaisir des champs avec La venerie, volerie et pescherie,* Prosper Blanchemain, ed. (Paris, 1869), 127. Unless otherwise noted, all translations are my own. Other images in the same poem invoke the grain harvest, where the "scieur courbé, halletant, s'esvertuë / A mener, non oiseux, la faucille tortuë / Par les espicz dorez," 130.
38. Forget, in Joukovsky, *La Renaissance bucolique,* 63–64.
39. Paintings and poems also placed peasants within a festive community. See, for example, Svetlana Alpers, "Realism as a Comic Mode: Low-Life Painting Seen through Bredero's Eyes," *Simiolus* 8, no. 3 (1975–76): 115–42.
40. Brueghel's *The Corn Harvest,* in Stockholm, and also see Joos de Momper, *Die Kornernte,* private collection.

41. See, for example, Jan van Goyen's 1625 panels *Summer* and *Winter* at the Rijkmuseum, Amsterdam. Also see *Les plaisirs de l'été* by the Flemish painter Sebastian Vrancx in the mid-seventeenth century. Galerie Robert Finck, *Exposition de tableaux de maîtres flamands du XVe au XVIIIe siècles* (Brussels, 1963), Catalogue No. 28.
42. Jan van Goyen (1596–1656) and Meindert Hobbema (1638–1709) were Dutch genre and landscape painters. The proportional increase in landscape paintings has been well documented. See, for example, two essays: Jan de Vries, "Art History," 249–82, and John Michael Montias, "Works of Art in Seventeenth-Century Amsterdam: An Analysis of Subjects and Attributions," 331–72, both in David Freedberg and Jan de Vries, eds., *Art in History, History in Art: Studies in Seventeenth-Century Dutch Culture* (Santa Monica, Calif., 1991). Also see Alan Chong, "The Market for Landscape Painting in Seventeenth-Century Holland," in Peter Sutton, ed., *Masters of 17th-Century Dutch Landscape Painting* (Boston, 1987), 104–20. It is worth noting that many Dutch artists of the early seventeenth century had migrated north from Flanders.
43. Christopher Brown, *Dutch Landscape: The Early Years, Haarlem and Amsterdam 1590–1650* (London, 1896), 29–30. See also Sutton, *Masters of 17th-Century Dutch Landscape Painting,* 8; Svetlana Alpers, *The Art of Describing: Dutch Art in the Seventeenth Century* (Chicago, 1983), 147; Schama, *Embarrassment of Riches;* Ann Jensen Adams, "Competing Communities in the 'Great Bog of Europe': Identity and Seventeenth-Century Dutch Landscape Painting," in W. J. T. Mitchell, ed., *Landscape and Power* (Chicago, 1994), 35–76.
44. Significantly, peasant violence against soldiers disappeared as a genre of painting. Jane Susannah Fishman, *Boerenverdriet: Violence between Peasants and Soldiers in Early Modern Netherlands Art* (Ann Arbor, Mich., 1979); Alison McNeil Kettering, *The Dutch Arcadia: Pastoral Art and Its Audience in the Golden Age* (Totowa, N.J., 1983).
45. E. John Walford, *Jacob van Ruisdael and the Perception of Landscape* (New Haven, Conn., 1991). Wolfgang Stetchow, *Dutch Landscape Painting of the Seventeenth Century* (London, 1966), 129, emphasizes the serenity of Ruisdael's paintings. The work by Seymour Slive and H. R. Hoetink, *Jacob van Ruisdael* (New York, 1981), 94, is less concerned with iconographic interpretation and concentrates on the formal aspects of the painting.
46. As in the case of Jan Brueghel the Elder and Joos de Momper, who produced, among other scenes, the famous painting of the Infanta and her maids haymaking. See the *Excursion campestre de Isabel Clara Eugenis,* Prado Museum, Madrid.
47. Some English agricultural manuals contain pictures of tool-wielding harvesters, for example, M. Stevenson, *Twelve Months* (London, 1661). Allegorical or classical treatments were more common. Wenceslaus Hollar's illustrations to Virgil's *Georgics,* John Ogilby, trans. (London, 1652), has peasants in antique costume performing all of the farming tasks described in the book, except for reaping. The harvest scene focuses on dances to Ceres, with two reapers hidden behind the dancers. Hollar's illustrations were used for Dryden's translation of Virgil and in every eighteenth-century edition of Dryden.
48. Peasants display huge scythes (and less evident sickles) in Hans Sebald Beham's illustrations of the months in Martin Luther's prayer book of 1527. Beham, however, was a recognized rebel. Keith Moxey, *Peasants, Warriors and Wives: Popular Imagery in the Reformation* (Chicago, 1989), p. 30. Woodcuts from around 1525 show peasants men-

acingly armed with tools, including reaping instruments. Maurice Pianzola, *Bauern und Kunstler: Die Kunstler der Renaissance und der Bauernkrieg von 1525* (Berlin, 1961), 95–96. Then, for about a century, German representations shy away from showing harvest tools. The harvester is overshadowed by mounted lord and lady and his sickle is buried deep in the corn, in *Summer* by Jost Amman, and an adaptation of the Flemish painter Sebastian Vracx by Matthaeus Merian the Elder shows harvesters hard at work in the distance, while the gentry stroll or ride by and others bathe in the river. Tools return in a series of vignettes representing the labors of the months by Elias Holl (in the 1630s). There a single, rough peasant, carrying the requisite tool, symbolizes each of the months, in medieval fashion. See *Hollstein's German Engravings, Etchings, and Woodcuts 1400–1700,* vols. 8, 15, 26, 27, and 28 (Amsterdam, 1954–58).

49. Stella's harvests appeared among Arcadian scenes in a volume called *Pastorales,* published by his niece after his death in 1658 (Paris, 1661).
50. Seventeenth-century pictures of Ruth and Boaz survive by Jan Swart, Philips Galle (engravings), Pieter Verbeecq, Nicolas Berchem, Gerhard van den Eeckhout (who produced four versions in the 1650s), Jan Victors, Jacob Pynas, Lambert Jacobsz, Aert de Gelder, and Bernard Fabritius. *D.I.A.L.: Decimal Index of the Art of the Netherlands,* Marquand Library, Princeton University, Princeton, New Jersey.
51. Raymond Williams, *The Country and the City* (New York, 1973); Alastair Fowler, "Georgic and Pastoral: Laws of Genre in the Seventeenth Century," in Michael Leslie and Timothy Raylor, eds., *Culture and Cultivation in Early Modern England: Writing and the Land* (Leicester, 1992), 81–88; Anthony Low, *The Georgic Revolution* (Princeton, N.J., 1985); John Barrell, *The Idea of Landscape and the Sense of Place, 1730–1840: An Approach to the Poetry of John Clare* (Cambridge, 1972).
52. *The Complete Poetry of Robert Herrick,* J. Max Patrick, ed. (New York, 1963), 140–42.
53. Mildmay Fane, Earl of Westmoreland, *To Retiredness,* in *Ben Jonson and the Cavalier Poets,* Hugh Maclean, ed. (New York, 1974), 206–7.
54. Both in Joukovsky, *La renaissance bucolique.*
55. Honorat de Bueil, Seigneur de Racan, *Poésies,* Vol. 1 (Paris, 1930). The countryside appears most frequently in the seventeenth century as a place of retreat from war and worldly cares, in conscious imitation of Horace.
56. This association already existed in the Middle Ages, where Henry I of England was depicted frightened of peasants armed with spade, pitchfork, and scythe. Michael Camille, " 'When Adam Delved': Laboring on the Land in English Medieval Art," in Sweeny, *Agriculture in the Middle Ages,* 266–67. James Turner describes the horror that cutting tools inspired in seventeenth-century England: *The Politics of Landscape: Rural Scenery and Society in English Poetry, 1630–1660* (Cambridge, Mass., 1979), 167.
57. Joachim Du Bellay, *Les antiquités de Rome,* in *Les Regrets, Les antiquités de Rome,* S. de Sacy, ed. (Paris, 1967), 46.
58. Gilles Ménage (1613–1692) in Maurice Allem, ed., *Anthologie poétique française, XVIIe siècle,* Vol. 2 (Paris, 1966), 114.
59. Marc Antoine Girard, Sieur de Saint-Amant, *Les oeuvres,* 3 vols. in 1 (Rouen, 1668), 3: 36. Saint-Amant's more famous verses on the seasons set spring in Paris, summer in Rome, fall in the Canaries, and winter in the Alps; see 3: 11–14.
60. See also Turner, *Politics of Landscape,* 167–68.

61. An ode to the Grand Condé after the battle of Lens (1648), in Allem, *Anthologie poétique française, XVIIe siècle,* 2: 8.
62. François de Rosset, *Nouveau recueil des plus beaux vers de ce temps* (Paris, 1609).
63. William Shakespeare, *Troilus and Cressida,* act 5, scene 5, ll. 25–26.
64. Andrew Marvell, "Upon Appleton House" (1681), ll. 418–24, in *The Major Metaphysical Poets of the Seventeenth Century,* Edwin Honig and Oscar Williams, eds. (New York, 1969), 766.
65. *The Poems of Richard Lovelace,* C. H. Wilkinson, ed. (1930; rpt. edn., Oxford, 1953), 39. See Don Cameron Allen, *Image and Meaning: Metaphoric Traditions in Renaissance Poetry,* new enl. edn. (Baltimore, Md., 1968), 152–64.
66. *Ben Jonson and the Cavalier Poets,* 196.
67. Ralph Knevet, "The Vote": "Sharp pikes may make / Teeth for a rake; / And the keen blade, the arch-enemy of life, / Shall be degraded to a pruning knife; / The rustic spade / Which first was made / For agriculture, shall retake / Its primitive employment." In Alastair Fowler, ed., *The New Oxford Book of Seventeenth-Century Verse* (Oxford, 1991), 359.
68. Pamela Berger, *The Goddess Obscured: The Transformation of the Grain Protectress from Goddess to Saint* (Boston, 1985); Erwin Panofsky, "Father Time," in his *Studies in Iconology: Humanistic Themes in the Art of the Renaissance* (1939; rpt. edn., New York, 1965); Millard Meiss, *French Painting in the Time of Jean de Berry: The Limbourgs and Their Contemporaries,* 2 vols. (New York, 1974), 1: 30–32.
69. At times, Death was pictured with a spade, the agricultural tool also used to dig the grave.
70. Both Protestants and Catholics used the symbol.
71. There is a vast literature on this subject. See, for example, Wilhelm Abel, *Agricultural Fluctuations in Europe from the Thirteenth to the Twentieth Centuries,* Olive Ordish, trans. (London, 1980).
72. Massed peasants, harvesters with their sickles and scythes, suggested armies of insurgents. Yves-Marie Bercé describes peasants armed with muskets and scythes marching off to fight for their rights. *Histoire des Croquants* (Paris, 1986), 10. See also Thomas Robisheaux, *Rural Society and the Search for Order in Early Modern Germany* (Cambridge, England, 1989).
73. See Salomon, *Recherches sur le thème paysan.*
74. "But shall conclude this poem to be the most compleat, elaborate, and finisht piece of all antiquity." John Dryden, *The Georgics of Virgil,* reprinted from the First Folio (London, 1931), xiii. On the influence of Dryden's translation, see the books already cited by Anthony Low, John Barrell, and Raymond Williams, and see John Dixon Hunt, *The Figure in the Landscape: Poetry, Painting, and Gardening during the Eighteenth Century* (Baltimore, Md., 1976).
75. This is the approach of English literary studies of the eighteenth century. See, for example, Dwight L. Durling, *The Georgic Tradition in English Poetry* (New York, 1935).
76. Virgil, *Eclogues, Georgics, Aeneid I–VI,* H. Rushton Fairclough, trans. (1916–18; rpt. edn., Cambridge, Mass., 1935), 103. From the Latin: "saepe ego, cum flavis messorem induceret arvis agricola et fragili iam stringeret hodea culmo."
77. Thomas May, *Virgil's Georgicks* (London, 1628), 16.
78. *Nouvelle traduction des Bucoliques de Virgile avec des notes* (Paris, 1691), 46.

79. Michel de Marolles, *Les oeuvres de Virgile traduites en prose* (Paris, 1649), 56.
80. John Dryden, *The Works of Virgil* (1697; rpt. edn., London, 1931), 26. Dryden could describe reaping when he chose: "Your hay it is mowed and your Corn is reap'd, / Your barns will be full, and your hovels heap'd, / Come, my Boys, come / And Merrily Roar out Harvest Home." *King Arthur, or, The British Worthy: A Dramatick Opera* (London, 1691), 47. Note that the action has already taken place, and all that remains is to cart the harvest away.
81. Jacques Delille, *Les géorgiques,* in *Oeuvres* (Paris, 1950), 57; and Abbé Desfontaines, *Oeuvres de Virgile traduites en françois avec des remarques* (Paris, c. 1700s); *Virgil's Husbandry, or, An Essay on the Georgicks, Being the First Book Translated into English* (London, 1725), 32; James Hamilton, *Virgil's Pastorals Translated into English Prose as Also His Georgicks* (Edinburgh, 1742), 48.
82. Adriaen Brouwer (1606–1638) had a brief career in Flanders depicting rough tavern scenes of brutish, drunken, squabbling peasants. Critics agree that, although Brouwer had imitators, real success came to those artists who softened and civilized the genre, such as Adriaen van Ostade. See Walter A. Liedtke, *Flemish Paintings in the Metropolitan Museum of Art,* 2 vols. (New York, 1984), 1: 4; Eric Larsen, *Seventeenth-Century Flemish Paintings* (Dusseldorf, 1985); and Jane P. Davidson, *David Teniers the Younger* (Boulder, Colo., 1979).
83. Mireille Rambaud, *Documents du Minutier central concernant l'histoire de l'art (1700–1750)* (Paris, 1971), 2: 1008–10.
84. The seasons were often used as allegories of time and of the ages of man. In this series, however, Teniers specifically represented Summer as a frail youth, while Spring (and the other seasons) were all grown men.
85. The summer scene is described in a 1746 inventory as "L'été sous la figure de personnes qui font la moisson." Georges Wildenstein, *Lancret: Biographie et catalogue critiques* (Paris, 1924), 59–60. Lancret produced two series of the seasons, one now at the Louvre, the other at the Hermitage, and both were engraved.
86. See Ann Birmingham, *Landscape and Ideology: The English Rustic Tradition, 1740–1860* (Berkeley, Calif., 1986).
87. See John Barrell, *The Dark Side of the Landscape: The Rural Poor in English Painting, 1730–1840* (Cambridge, England, 1980). One possibility was to portray actual reaping but under the watchful eye of a superior, as in George Lambert's *Landscape with a Cornfield,* painted for the duke of Bedford in 1733. Elizabeth Einberg, *George Lambert, 1700–1765* (London, 1970).
88. Jean-Jacques Rousseau, *Julie, ou La nouvelle Héloïse,* Michel Launay, ed. (Paris, 1967), Part 5, letter 7, 456. The French term is *faner.*
89. James Thomson, *The Seasons and the Castle of Indolence,* James Sambrook, ed. (Oxford, 1972), "Autumn," ll. 151–74. On Thomson's life, see Sambrook, *James Thomson, 1700–1748: A Life* (Oxford, 1991).
90. [Mme Bontems], *Les saisons, poème de Thomson traduit de l'anglois* (Paris, 1759); "Les quatre saisons ou les géorgiques françaises," *Poésies diverses du Cardinal de Bernis,* Fernand Drujon, ed. (Paris, 1882); Jean-François de Saint-Lambert, *Les saisons, poème,* 3d edn. (Amsterdam, 1771); and Margaret M. Cameron, *L'influence des saisons de Thomson sur la poésie descriptive en France (1759–1810)* (Paris, 1927).

91. See Jean-Antoine Roucher's descriptions of August in *Les mois, poème en douze chants* (Paris, 1779). Roucher's actual treatment of the harvest, however, was more graphic than other contemporaries, which may explain why it was judged both tedious and vulgar.
92. *Théâtre choisi de Favart,* Vol. 2 (Paris, 1809), *Les moissonneurs,* act 2, scene 3. Any suggestion of cutting evaporates even in the English version, *The Reapers, or, An Englishman out of Paris* (London, 1770).
93. Paradoxically, while the actual cutting of the crop was usually omitted, the summer harvest could be symbolized by a sickle. Antoine Watteau's allegory *Summer* at the National Gallery, Washington, shows Ceres holding a sickle. In the 1769 production of "Les festes de l'hymen et de l'amour," the dancer representing a male peasant carried a sheaf, while the female peasant held a sickle. Nothing as obvious, a century earlier, when the ballet costumes for the 1651 "Fêtes de Bacchus" showed Autumn as a man, surrounded by men with garlands on their heads, sheaves, and corns of plenty, but no tools. Cyril W. Beaumont, *Five Centuries of Ballet Design* (London, 1939), 50, 61.
94. See Thomas Gainsborough's several versions of "The Harvest Wagon" (Paul Spencer-Longhurst and Janet M. Brooke, *Thomas Gainsborough: The Harvest Wagon* [Toronto, 1995]) or William Ashford's "Landscape with Haymakers and a Distant View of a Georgian House" (1780), at the Yale Center for British Art, with its hay wagons and workers raking and pitching hay. Late eighteenth-century prints show laborers carrying large sheaves and placing them on loaded wagons. Christiana Payne, *Toil and Plenty* (New Haven, Conn., 1993), 181. From Holland, see Hendrick Meyer, "Harvest near a Village," in J. W. Niemijer, *Eighteenth-Century Watercolors from the Rijkmuseum Printroom* (Alexandria, Va., 1993), 101; from Germany, Jacob Philip Hackert, *Sommer* [1780s?], in *Heroismus und Idylle: Formen der Landschaft um 1800* (Cologne, 1984), 102.
95. Given that Stubbs barely shows the reapers, and has no mowers at all in his picture of haymakers (they surround the haycart with pitchforks), Barrell's description of the painting is surprising: "Deceptively commonplace ingredients—men and women at work, swaths of hay and sheaves of corn, billhooks and scythes—are lifted high above the level of matters of fact by Stubbs's miraculously assured sense of design." Review of the Tate exhibition cited by Barrell, *Dark Side of the Landscape,* 26.
96. A striking example, besides François Boucher and Jean-Honoré Fragonard, is the engraving by Francis Bertolozzi after William Hamilton (London, 1793) (Yale Center for British Art), illustrating the passage of Thomson's *Seasons* cited earlier. Beautiful, stately women gathering sheaves and carrying lunch baskets are surrounded by children, and a single youth reaping. Thomson's image of male chivalry has been feminized.
97. Hence the debate about the peasants' clothing from those who objected to the over-ornate, "artificial" offenses of the Rococo.
98. Luigi de Nardis, *Saint-Lambert scienza e paesaggio nella posia del settecento* (Rome, 1961); and Cameron, *L'influence des saisons de Thomson,* chap. 2.
99. The finished product was judged to be less than the sum of its parts, and Saint-Lambert's poem was criticized for its lifelessness.
100. Saint-Lambert, "L'été," l. 497. and "Le printemps," ll. 148–49.
101. Saint-Lambert, "L'automne," ll. 505–14. Bernis and Roucher made similar pleas in their poems of the seasons.

102. Pierre Samuel Du Pont de Nemours for Anne-Robert-Jacques Turgot, *Mémoire sur les municipalités,* in Gustave Schelle, ed. *Oeuvres de Turgot,* 5 vols. (Paris, 1913–23), 4: 568–628.
103. Georges Lefebvre, *The Coming of the French Revolution,* R. R. Palmer, trans. (1947; rpt. edn., Princeton, N.J., 1967), 64.
104. Jean de La Bruyère played on this traditional disparagement when he likened the peasants to beasts. *Les caractères ou les moeurs de ce siècle,* 11: 128: iv, in Jean Lafond, ed., *Moralistes du XVII*[e] *siècle* (Paris, 1992), 866–67.

References

Abel, Wilhelm. *Agricultural Fluctuations in Europe from the Thirteenth to the Twentieth Centuries,* trans. Olive Ordish. London, 1980.

Adams, Ann Jensen. "Competing Communities in the 'Great Bog of Europe': Identity and Seventeenth-Century Dutch Landscape Painting." In W. J. T. Mitchell, ed., *Landscape and Power.* Chicago, 1994.

Allen, Don Cameron. *Image and Meaning: Metaphoric Traditions in Renaissance Poetry,* new enl. ed. Baltimore, 1968.

Allem, M., ed., *Anthologie poétique française, XVI*e siècle, vol. 2. Paris, 1965.

———. *Anthologie poétique française, XVII*[e] *siècle.* Paris, 1966.

Alpers, Svetlana. "Realism as a Comic Mode: Low-Life Painting Seen through Bredero's Eyes." *Simiolus* 8, no. 3 (1975–76).

———. *The Art of Describing: Dutch Art in the Seventeenth Century.* Chicago, 1983.

Barrell, John. *The Idea of Landscape and the Sense of Place, 1730–1840: An Approach to the Poetry of John Clare.* Cambridge, England, 1972.

———. *The Dark Side of the Landscape: The Rural Poor in English Painting, 1730–1840.* Cambridge, England, 1980.

Beaumont, Cyril W. *Five Centuries of Ballet Design.* London, 1939.

Bellay, Joachim Du. *Les Regrets, Les antiquités de Rome,* ed. S. de Sacy. Paris, 1967.

Bercé, Yves-Marie. *Histoire des croquants.* Paris, 1986.

Berger, Pamela. *The Goddess Obscured: The Transformation of the Grain Protectress from Goddess to Saint.* Boston, 1985.

Birmingham, Ann. *Landscape and Ideology: The English Rustic Tradition, 1740–1860.* Berkeley, Calif., 1986.

A Book of Masques in Honour of Allrdyce Nicoll. Cambridge, England, 1967.

Brown, Christopher. *Dutch Landscape: The Early Years, Haarlem and Amsterdam, 1590–1650.* London, 1896.

Cafritz, Robert. Lawrence Gowing, and David Rosand. *Places of Delight: The Pastoral Landscape.* Washington, D.C., 1988.

Camille, Michael. "Labouring for the Lord: The Ploughman and the Social Order in the Luttrell Psalter." *Art History* 10 (1987).

———. " 'When Adam Delved': Laboring on the Land in English Medieval Art." In Del Sweeny, ed., *Agriculture in the Middle Ages: Technology, Practice, and Representation* (Philadelphia, 1995).

Chong, Alan. "The Market for Landscape Painting in Seventeenth-Century Holland." In Peter Sutton, ed., *Masters of 17th-Century Dutch Landscape Painting.* Boston, 1987.

Comet, Georges. *Le paysan et son outil: Essai d'histoire technique des céréales (France, VIII*[e]*–XV*[e] *siècle).* Rome, 1992.

Curtius, Ernst Robert. *European Literature and the Latin Middle Ages,* trans. Willard R. Trask. 1953; rpt. edn., Princeton, N.J., 1983.

Davidson, Jane P. *David Teniers the Younger.* Boulder, Colo., 1979.

Delille, Jacques. *Les Géorgiques.* In *Oeuvres.* Paris, 1950.

Desfontaines, Abbé. *Oeuvres de Virgile traduites en françois avec des remarques.* Paris, c. 1700s.

Dryden, John. *The Georgics of Virgil.* Reprinted from the First Folio. London, 1931.

———. *The Works of Virgil.* 1697; rpt. edn., London, 1931.

Durling, Dwight L. *The Georgic Tradition in English Poetry.* New York, 1935.

Einberg, Elizabeth. *George Lambert, 1700–1765.* London, 1970.

Elias, Norbert. *The History of Manners,* trans. Edmund Jephcott. New York, 1978.

Fishman, Jane Susannah. *Boerenverdriet: Violence between Peasants and Soldiers in Early Modern Netherlands Art.* Ann Arbor, Mich., 1979.

Fowler, Alastair. "Georgic and Pastoral: Laws of Genre in the Seventeenth Century." In Michael Leslie and Timothy Raylor, eds., *Culture and Cultivation in Early Modern England.* Leicester, 1992.

Fowler, Alastair, ed. *The New Oxford Book of Seventeenth-Century Verse.* Oxford, 1991.

Fox, Dian. *Refiguring the Hero: From Peasant to Noble in Lope de Vega and Calderón.* University Park, Pa., 1991.

Freedberg, David, and Jan de Vries, eds. *Art in History, History in Art: Studies in Seventeenth-Century Dutch Culture.* Santa Monica, Calif., 1991.

Freedman, Paul. *Images of the Medieval Peasant.* Stanford, Calif., 1999.

Gauchet, Claude. *Les Moissons* (1583). In *Le plaisir des champs avec La venerie, volerie et pescherie,* ed. Prosper Blanchemain. Paris, 1869.

Gibson, Walter S. *"Mirror of the Earth": The World Landscape in Sixteenth-Century Flemish Painting.* Princeton, N.J., 1989.

Hackert, Jacob Philip. *Sommer.* Germany, 1780s(?).

Hamilton, James. *Virgil's Pastorals Translated into English Prose as Also His Georgicks.* Edinburgh, 1742.

Herrick, Robert. *The Complete Poetry of Robert Herrick,* ed. J. Max Patrick. New York, 1963.

Hunt, John Dixon. *The Figure in the Landscape: Poetry, Painting, and Gardening during the Eighteenth Century.* Baltimore, Md., 1976.

Ivins, William M., Jr. *Prints and Visual Communication.* 1953; Cambridge, Mass., 1969.

Kettering, Alison McNeil. *The Dutch Arcadia: Pastoral Art and Its Audience in the Golden Age.* Totowa, N.J., 1983.

Lacroix, Paul. *Ballets et mascarades de cour de Henri III à Louis XIV (1581–1652).* Volume 1. Geneva, 1868.

Lagerlöf, Margaretha Rossholm. *Ideal Landscape: Annibale Carracci, Nicolas Poussin and Claude Lorrain.* New Haven, Conn., 1990.

Lambert, Gisèle. "De l'espace sacré à l'esprit profane: L'apparition du paysage." In *Paysages, paysans: L'art et la terre en Europe du Moyen Age au XXe siècle.* Paris, 1994.

Larsen, Eric. *Seventeenth-Century Flemish Paintings.* Dusseldorf, 1985.

Lefebvre, Georges. *The Coming of the French Revolution,* trans. R. R. Palmer. 1947; rpt. edn., Princeton, N.J., 1967.

Liedtke, Walter A. *Flemish Paintings in the Metropolitan Museum of Art.* 2 vols. New York, 1984.

Lovelace, Richard. *The Poems of Richard Lovelace,* ed. C. H. Wilkinson. 1930; rpt. edn., Oxford, 1953.

Low, Anthony. *The Georgic Revolution.* Princeton, N.J., 1985.

Maclean, Hugh, ed. *Ben Jonson and the Cavalier Poets,* New York, 1974.

Mâle, Emile. *The Gothic Image: Religious Art in France of the Thirteenth Century,* trans. Dora Nussey. New York, 1958.

Mane, Perinne. *Calendriers et techniques agricoles (France-Italie, XIIe-XIIIe siècles).* Paris, 1983.

Marolles, Michel de. *Les oeuvres de Virgile traduites en prose.* Paris, 1649.

Marvell, Andrew. "Upon Appleton House" (1681). In *The Major Metaphysical Poets of the Seventeenth Century.* New York, 1959.

May, Thomas. *Virgil's Georgicks.* London, 1628.

Meiss, Millard. *French Painting in the Time of Jean de Berry: The Limbourgs and Their Contemporaries.* 2 vols. New York, 1974.

Moxey, Keith. *Peasants, Warriors and Wives: Popular Imagery in the Reformation.* Chicago, 1989.

Nardis, Luigi de. *Saint-Lambert scienza e paesaggio nella posia del settecento.* Rome, 1961.

Nashe, Thomas. *The Unfortunate Traveller and Other Works.* Harmondsworth, 1985.

Niemijer, J. W. *Eighteenth-Century Watercolors from the Rijkmuseum Printroom.* Alexandria, Va., 1993.

Pacht, Otto. "Early Italian Nature Studies and the Early Calendar Landscape." *Journal of the Warburg and Courtauld Institutes* 13 (1950).

Panofsky, Erwin. "Father Time." In Panofsky, *Studies in Iconology: Humanistic Themes in the Art of the Renaissance.* 1939; rpt. edn., New York, 1965.

Patterson, Annabel. *Pastoral and Ideology: Virgil to Valery.* Berkeley, Calif., 1987.

Payne, Christiana. *Toil and Plenty.* New Haven, Conn., 1993.

Pearsall, Derek, and Elizabeth Salter. *Landscapes and Seasons of the Medieval World.* London, 1973.

Pianzola, Maurice. *Bauern und Kunstler: Die Kunstler der Renaissance und der Bauernkrieg von 1525.* Berlin, 1961.

Rabb, Theodore K. *The Struggle for Stability in Early Modern Europe.* New York, 1975.

Rambaud, Mireille. *Documents du Minutier central concernant l'histoire de l'art (1700–1750).* Paris, 1971.

Roberts, Michael. "Sickles and Scythes: Women's Work and Men's Work at Harvest Time." *History Workshop* (1979).

Robisheaux, Thomas. *Rural Society and the Search for Order in Early Modern Germany.* Cambridge, England, 1989.

Rosset, François de. *Nouveau recueil des plus beaux vers de ce temps.* Paris, 1609.

Roucher, Jean-Antoine. *Les mois, poème en douze chants.* Paris, 1779.

Rousseau, Jean-Jacques. *Julie, ou La nouvelle Héloïse.* ed. Michel Launay. Paris, 1967.

Saint-Amant, Sieur de [Marc Antoine Girard]. *Les Oeuvres,* 3 vols. in 1. Rouen, 1668.

Sambrook, James. *James Thomson, 1700–1748: A Life.* Oxford, 1991.

Schama, Simon. *An Embarrassment of Riches: An Interpretation of Dutch Culture in the Golden Age.* New York, 1987.

Scott, Tom. *The Peasantries of Europe from the Fourteenth to the Eighteenth Centuries.* London, 1998.

Slive, Seymour, and H. R. Hoetink. *Jacob van Ruisdael.* New York, 1981.

Spencer-Longhurst, Paul, and Janet M. Brooke. *Thomas Gainsborough: The Harvest Wagon.* Toronto, 1995.

Stetchow, Wolfgang. *Dutch Landscape Painting of the Seventeenth Century.* London, 1966.

Stevenson, M. *Twelve Months.* London, 1661.

Sullivan, Margaret A. *Bruegel's Peasants: Art and Audience in the Northern Renaissance.* Cambridge, England, 1994.

Sutton, Peter, ed. *Masters of 17th-Century Dutch Landscape Painting.* Boston, 1987.

Sweeny, Del, ed. *Agriculture in the Middle Ages: Technology, Practice, and Representation.* Philadelphia, 1995.

Thomson, James. *The Seasons and the Castle of Indolence,* ed. James Sambrook. Oxford, 1972.

Turgot, Anne-Robert-Jacques, and Pierre Samuel Du Pont de Nemours. *Mémoire sur les municipalités.* In Gustave Schelle, ed., *Oeuvres de Turgot.* 5 vols. Paris, 1913–23.

Turner, James. *The Politics of Landscape: Rural Scenery and Society in English Poetry, 1630–1660.* Cambridge, Mass., 1979.

Virgil. *Georgics,* trans. John Ogilby. London, 1652.

———. *Nouvelle traduction des Bucoliques de Virgile avec des notes,* Paris, 1691.

———. *Virgil's Husbandry, or, an Essay on the Georgicks, Being the First Book Translated into English.* London, 1725.

———. *Eclogues, Georgics, Aeneid I–VI,* trans. H. Rushton Fairclough. 1916–18; rpt. edn., Cambridge, Mass., 1935.

———. *The Eclogues and Georgics of Virgil,* trans. C. Day Lewis. New York, 1964.

Walford, E. John. *Jacob van Ruisdael and the Perception of Landscape.* New Haven, Conn., 1991.

Webster, James. *The Labors of the Months in Antique and Medieval Art to the End of the Twelfth Century.* Princeton, N.J., 1938.

Wilkinson, L. P. *The Georgics of Virgil: A Critical Survey.* Cambridge, England, 1969.

Williams, Raymond. *The Country and the City.* New York, 1973.

Williams, R. D., and T. S. Pattie. *Virgil: His Poetry through the Ages.* London, 1982.

Wildenstein, Georges. *Lancret: Biographie et catalogue critiques.* Paris, 1924.

PART III *Agrarian and Environmental Histories: Case Studies from South Asia*

CHAPTER FIVE

Naturae Ferae: Wild Animals in South Asia and the Standard Environmental Narrative

PAUL GREENOUGH

Nature [is] red in tooth and claw.
—Alfred, Lord Tennyson, *In Memoriam*

In this chapter I shall examine two themes missing or only sketchily represented in the standard South Asian environmental narrative. These are, first, the hazards posed by wild animals to the security of rural communities in the past and present and, second, the defensive measures taken by these communities and by the state to ward off this threat. By "standard environmental narrative" I mean the emerging scholarly consensus that postulates a transition in rural India over the past 120 years from a condition of environmental harmony, distributive justice, and material abundance to one of ecological disruption, massive social inequity, and widespread misery.[1] This narrative is recounted by various authors, in long or short compass, to pinpoint the origin of contemporary environmental crises and to explain a correlated collapse of rural welfare in our time. The scholarly value of the standard environmental narrative (hereafter SEN) is that it gives to the emerging field of South Asian environmental history an overall trajectory of cause, effect, and sequence, while allowing new findings to be fitted into an agreed-on framework of events. Yet the standard South Asian environmental narrative is flawed, I believe, by its failure to take into account the historic reality of violence between humans and wild animals, and by its ignorance of the devastating victory that has been won by humans over wild animals in the past fifty years.

The SEN, which has come into focus only in the past twenty years, constructs its plot around disastrous policies that date from the period 1870 to 1880. The 1878

Indian Forest Act, which closed the forests to ordinary Indians and appropriated most of India's timber for the state, epitomizes these policies and marks the point of no return, after which the previously favorable linkages among cultivators, the arable, and the forest began to rupture, dooming the populace to resource scarcities of ever increasing severity. This arresting story, which clearly parallels a similar nationalist narrative of economic disruption and depletion under colonial rule, goes as follows:[2]

> In pre-colonial South Asia, peasant and tribal communities understood their intimate dependence on natural gifts of water, wood, fodder, and soil. Well aware that landlords and rulers could not protect them during droughts and floods, they strove for autarky and avoided market intrusions and technical innovations that promised higher income but threatened periodic famine. They embraced risk-averse cultivation and appropriate technologies that harmonized with natural cycles of regeneration and decay. They were recyclers and conservationists by habit, and their respect for natural processes was mediated by richly detailed knowledge of soils, seeds, crops, weather, medicinal herbs, and edible wild plants. This knowledge was codified in vernacular taxonomies, proverbs, and songs; it found its wider expression in rituals and myths that celebrated natural forces and enjoined worship of a goddess of nature, Prakriti. Prakriti assured the fertility of crops and domestic animals but also of women, and women were the principal repository of the nature cult. Women's contributions to community and family well-being were equally practical; they possessed the vital skills of foraging, threshing, cooking, dairying, and water- and seed-selecting, which complemented the masculine tasks of timbering, plowing, sowing, and reaping. This division of labor differed, however, from the distinction drawn in Brahmanic texts between a masculine realm of culture and a feminine realm of nature: instead, the labor of both men and women was invested directly in fields and forests, and both sexes honored the godlings of earth, forest, water, and seeds, all emanations of goddess Prakriti. Non-Hindu cultivators and swiddeners also held these assumptions.
>
> In this nearly vanished world of pre-colonial tradition, every village community commanded its adjacent forest, pasturage and arable, watercourse and well, and every household (despite undeniable practices of social exclusion and hierarchy) had access by custom to all that it needed from nature. Stretches of thick jungle isolated the lowland villages from each other, and where settlements were closely lodged—as in narrow sub-Himalayan valleys—the boundaries of the commons were maintained across forests and pastures without the intervention of external authority. Traditional rulers never questioned the right of village communities to use and maintain these natural reserves, which by custom were untaxed and conserved from generation to generation.

Colonial rule disrupted this system at nearly every point. By engaging in sustained military action against traditional rulers in all parts of the subcontinent; by collecting high taxes strictly, locally, and in cash; by meticulously creating and enforcing property rights in land and rents; by commercializing food crops and demanding the cultivation of indigo, opium, cotton, sugar, and jute for export; by creating an unprecedented demand for labor in cities, in the army, and on plantations in India and abroad; by introducing the powerful technologies of steam locomotion, mechanized processing of crops (such as jute), and hydraulic control over rivers and canals; by initiating a famine-relief system that worked well only during modest scarcities and that disrupted ancient practices of self-insurance and private charity; by appropriating commons-rights in forests, grasslands, and fisheries; and, above all, by introducing a radical ideology that regarded feminine Nature as an undisciplined opponent over which masculine Science must exercise control—by all these means British rule systematically eroded the political, legal, economic, moral, and ideological bases of popular subsistence, broke open communal solidarity, and ravaged soil, water, woods, and shoreline. Perversely, this entire process was named "progress" by colonial rulers and then "development" by postcolonial rulers, the arrogance of both being equaled only by their indifference to popular havoc. The long-term consequences, only now coming into focus, were environmental catastrophes, the extinction of numerous species, and the dissipation of once-common knowledge of sound farming, fishing, hunting, and forestry practices upon which the well-being of millions had depended for centuries. Only remnants of this knowledge base now remain; nonetheless, these remnants, held tenaciously in remote parts of Central India, Jharkhand, and Ladakh, are India's "ancient future" and offer the starting point for a restoration of rural welfare that will deflate the claims of Science and Progress. Further, only subaltern resistance can defeat an entrenched bureaucracy and its constituency of industrialists, contractors, and loggers; only direct action will force this unholy alliance to abandon the path of "development" that corrodes whatever it touches in nature and society. Dispersed examples of inventive opposition—preeminently the Chipko movement and the Save the Narmada Movement—offer the only hope and deserve widespread emulation.[3]

My rendition of the SEN may not capture all the subtleties, and what is meant to be a précis may read more like a parody, thereby giving offense to everyone. I take the risk to make the point that, although certain portions of the narrative are plausible, the whole has a Puranic character and requires amendment.[4] First and most obviously, the abruptly falling trajectory (¬) after 1880 is implausibly lapsarian: it suggests that South Asians, compelled to eat the fruit of colonial rule, were expelled from paradise. This version of the Fall is suspect: it is too sweeping (can all current ecological crises be traced to so narrow a set of distant causes?), its chronology is mistaken (colonial officials were India's first self-conscious environmentalists as early as 1800), its vision of the past is too nostalgic (were precolonial material conditions in all ways superior to contemporary ones?), and, conversely, its expectations for the future are too pessimistic (life expectancy in India continues to rise, as it has for nearly a century). Without sneezing at current environmental crises, without excusing the misguided policies that in many cases precipitated them, without denying the rapacity of an iron triad of bureaucrats-contractors-loggers, and without forgetting the Himalayan scale of basic needs for food, fuel, and fodder nearly everywhere in rural South Asia, it does not follow that every contemporary environmental crisis could have been averted had traditional values and practices been observed. Nor can we conclude that every popular movement to protect a threatened river or forest proves once and for all the bankruptcy of modern science and technology. The worst-case examples that are waved in our faces (the Bhopal fertilizer-plant explosion, the dams on the Narmada River and its tributaries) can be read as instructive warnings rather than as signposts to a fated future.[5] Moreover, the turn toward social learning and resistance in the final lines of the SEN suggests that despair is inappropriate.

Neglected Fauna and the Vegetal Bias

Another drawback of the SEN, and the one most pertinent to this chapter, is that the Indian Eden is wholly vegetal: if we except a few cows, pre-colonial Nature is drawn essentially without animals. Traditional environmental wisdom and popular worship of nature turn out to be the wisdom and worship of flora, not fauna. I should like to ask, what did Indian villagers in the past know about deer, monkeys, and waterfowl, about large and small carnivores, about elephants, raptors, stinging insects, rodents, reptiles, and fish, all of which are more varied on the Indian subcontinent than anywhere else in the Old World except Africa? The evidence from natural history is that vast numbers of wild animals inhabited jungle, lagoon, and savanna, cheek by jowl with fields and settlements, right into the 1950s.[6] Some of these were dangerous and threatened human well-being. Yet the figure of Nature that inhabits the SEN has been defanged and is no longer "red in tooth and claw."

Let me give some examples of the vegetal bias. Vandana Shiva, whose 1988 study of women, ecology, and development is a leading text on traditional environmental knowledge, quotes a passage from the *Devimahatmya,* a Sanskrit ritual text, in which the Indian Mother Goddess says: "O ye gods, I shall support (i.e. nourish) the whole world with life-sustaining vegetables which shall grow out of my body, during a period of heavy rain. I shall gain fame on Earth then as *Shakambari* (goddess who feeds the herbs)."[7] Shiva comments on this passage: "Forests have always been central to Indian civilization. They have been worshipped as *Aranyani,* the Goddess of the Forest, the primary source of life and fertility. . . . The forest as the highest expression of the earth's fertility and productivity is symbolized in yet another form as the Earth Mother, as Vana Durga or the Tree Goddess. . . . In ancient Indian traditions scientific knowledge of the plant kingdom is evident from such terms as *vriksayurveda* which means the science of the treatment of plant diseases, and *vanaspati vidya* or plant sciences" (pp. 54–56). What Shiva does not report, however, is that in the same chapter of the *Devimahatmya,* the goddess assumes a more ferocious, indeed carnivorous, character: "When I [Durga] shall devour the fierce and great *asuras* (demons) . . . my teeth shall become red like the flower of pomegranate; therefore when *devas* (gods) in heaven and men on earth praise me, [they] shall always talk of me as the 'red-toothed.' "[8]

Durga's essence spills over onto all forms of creation; she is both the ferocity of animals and the fertility of plants. In fact, the representation of Nature in Indian tradition includes the distinctly savage image of ripping open animal flesh alongside the watering of *tulsi*-plants and the plaiting of *darba*-grass. Just as the god Shiva always has beside him his favorite bull, Nandi, and Lord Ganesh has his favorite rodent, so Durga's boon companion (*vahana*) is a full-grown tiger. Lesser deities, such as Waghoba (the "father tiger" of the Konkan), Jambhawati ("bear father" of Himachal Pradesh), and Sona Ray (a tiger god of the Bengal Sunderbans), complement the Durga myth, and such rural deities as Manasa attract a huge human following by commanding the loyalty of poisonous snakes.[9] Folktales and folk epics as well as refined *sastras* offer rich pre-modern discourses on hunting, veterinary medicine, elephant capturing, animal mythography, and shape-shifters in the form of animals—all testifying to a vivid consciousness of wildlife.

Another example: in a masterful survey of the woodfuel crisis in the Third World, Bina Agarwal emphasizes that a conservation ethic already exists in India among rural and tribal peoples, but in this conservation ethic, wild animals find no place. She quotes M. Swaminathan's judgment "There is nothing we can teach [Rajasthani] tribals about trees" and adds that "rural people usually, and forest communities in particular, have traditionally shown a deep sensitivity to the need to maintain the ecological balance and to preserve rather than destroy [forests]."[10] No doubt this is so, but the forests in Agarwal's study are never seen as a habitat within which wild animals shelter, only as a source of fodder and fuel for humans.

The tribal Bishnois of Rajasthan's Thar Desert, who protect with passion not only their *khejdi* groves but also the wild blackbuck antelope that shelter among them, seem to have escaped her notice.[11]

A final example: in a well-known account of resistance movements by Uttarakhand hill people against Forest Department *raj,* Ramachandra Guha draws attention to popular protection of sacred groves: "Through religion, folklore and tradition the village communities had drawn a protective ring around the forests. . . . Often hilltops were dedicated to local deities and the trees around the spot regarded with great respect. . . . In fact, the planting of a [sacred] grove was regarded as 'a work of great religious merit.' In parts of Tehri, even today, leaves are offered to a goddess known as *Patna Devi* (Goddess of Leaves), this being only one of several examples of the association of plants with gods."[12]

From the accounts cited, one might conclude that Indians in the past were skilled agronomists, knowledgeable folk-botanists, and wise plant conservators —but ignoramuses where wild animals were concerned. That this is not the case is proved by numerous texts and by an abundance of sculptures and paintings that show wild animals to have been studied and documented over many centuries across the entire subcontinent. Many examples can be cited (including works by H. Srinivasa Rao, Durga Bhagavat, and Kailash Sankhala),[13] but a particularly accessible source is Francis Zimmermann's study of Ayurvedic texts, *The Jungle and the Aroma of Meats: An Ecological Theme in Hindu Medicine* (1982). Zimmermann not only provides a list of several hundred animals whose flesh, broth, and even blood were prescribed by Ayurvedic practitioners as nutritious foods for the sick, but his Sanskrit sources place these animals in biogeographic zones *(jangala,* or arid terrain; *anupa,* or marshy terrain; and *sadharana,* or middling terrain) that can be superimposed on modern ecosystem complexes.[14] Although some of these animals may be fabulous, most were found in the wild as recently as the 1950s. Among the carnivores, Ayurvedic authors distinguished *guhasayas* (animals that "have a lair" and "sleep in caves"), a category that included the lion, tiger, wolf, hyena, bear, panther, wildcat, and jackal.[15] These are among the animals that were acknowledged in the colonial era as most dangerous and therefore most worthy of practical concern.

After about 1950, consciousness of the animal threat in South Asia faded as wild species were eradicated or driven into retreat. Large carnivores, game fowl, bears, elephants, rhinoceroses, and wild ruminants virtually disappeared from the wild and are now found mainly in zoos, sanctuaries, and reserves. According to the naturalist George Schaller, political independence in 1947 ushered in a period of mass destruction when Indian hunters mowed down game animals everywhere, even in nominal sanctuaries and in public forests. The result was a vast kill-off "that could almost be compared to the slaughter [of buffalo] on the American prairies in the 1880s."[16] Interestingly, Schaller attributes this holocaust to an essentially political impulse: gun licenses and shooting regulations, rigidly enforced under foreign rule, were rejected after Independence as "a form of colonial repression." Further, "as a result of food shortages, the government initiated a national drive [during the 1950s] to protect crops from the depredations of wild animals, and guns were freely issued to farmers, an action which literally doomed almost all animals near cultivation." Finally, the cutting down of nearly half of India's forests since 1947, the drive to bring marginal lands under cultivation, and a vast increase in the number of "undernourished, diseased and unproductive" domestic animals, which have been allowed to graze almost without restriction in almost all forests, have destroyed many former wildlife habitats while establishing a rival population of cattle, buffalo, and goats on the straitened habitats that remain.[17] An actual erasure, then, is one reason why historians' accounts of traditional environmental practices leave wild animals out: the present generation of South Asian and foreign scholars were still in school or came to maturity after South Asia's wildlife had nearly disappeared. The current imbalance in South Asian environmental consciousness, then, reflects a comparatively recent ecological score: mankind and domestic animals one, wild animals zero.

Mayhem by the Wild Beasts

Wild animals once gravely threatened India's rural population; tigers, bears, elephants, leopards, snakes, alligators, wild pigs, and so on not only rubbished field crops and depleted herds of domestic animals but also killed a considerable number of people. The extent of this threat in the pre-colonial period can be only guessed at, but it seems clear that long before European rule, villagers were accustomed to a chronic struggle. If large animals like tigers and elephants could not be repelled by tactical measures, villagers necessarily abandoned their holdings. East India Company records from the eighteenth century testify to the pressure of wildlife on cultivation, and the Revenue Department regarded the presence of wild animals inside village precincts as an economic barometer: as cultivation expanded, wild animals retreated, and when cultivation contracted, wild animals advanced. Officials adopted novel measures like paying cash bounties for dead predators, justifying the costs by noting that marauding beasts kept down revenues. In fact, the bounties paid for the extermination of tigers, leopards, pigs, cobras, and other species are our only measure of the animal hazard for almost a century. Beginning in 1875, however, district officers were asked to keep statistics on losses due to *naturae ferae* ("wild beasts"). Ironically, at almost the moment that such data first began to be collected, sportsmen and naturalists had begun to decry the effects of sport shooting and habitat destruction on the animal population, most especially on game animals like tigers, lions, elephants, rhinos, and panthers. Drawing mainly on colonial sources, I sketch in the rest of this chapter the problem of wild animal management in South Asia since the end of the eighteenth century.

A classic statement of the animal-human seesaw is found in W. W. Hunter's description of lowland Bengal (Birbhum) in the aftermath of the 1769–70 famine:

> As the little rural communities relinquished their hamlets, and drew closer together towards the centre of the district, the wild beasts pressed hungrily on their rear. In vain the Company offered a reward for each tiger's head, sufficient to maintain a peasant's family in comfort for three months; an item of expenditure deemed so necessary, that when, under extraordinary [fiscal] pressure, it had to suspend all payments, the tiger-money and the diet-allowance for prisoners were the sole exceptions to the rule. A belt of jungle, filled with wild beasts, formed round each village; the official records frequently speak of the mail-bag being carried off by wild beasts; and after fruitless injunctions to the landholders to clear the forests, Lord Cornwallis was at length compelled to sanction a public grant to keep open the new military road that passed through Beerbhoom. . . . In two parishes alone, during the last few years of native administration, 56 villages with their communal lands had all been destroyed and gone to jungle, caused by the depredations of the wild elephants, and an official return states that 40 market towns throughout the district had been deserted from the same cause.[18]

Village desertion represented a low point in the contest between jungle and arable. No doubt other species were involved—in lower Bengal one thinks of wild pigs—but because dangerous animals like tigers, leopards, and elephants were held by Europeans to be especially "noble," they were the likeliest to be reported in official records. In fact, reports about tigers and elephants can be followed as sentinels right through nineteenth- and twentieth-century accounts of Indian wildlife.[19]

In general, naturae ferae retreated before the expansion of cultivation, and by the 1860s and 1870s officials were reporting a marked decline in animal attacks. The collector of Puri (in Orissa), for example, stated "In the open part of the country the larger wild beasts have been nearly exterminated," and another observed that "the wild beasts which formerly infested Birbhum (Bengal) have now almost disappeared, with the exception of an occasional tiger or bear which wanders into the cultivated tracts from the jungles of the Santal Parganas."[20] A similar story was recounted in Saran (Bihar), and Dhaka also reported a decline.[21] From Faridpur (Bengal), the decline of predators led to a rise in wild pigs: "The larger sorts of wild animals found in the district consist of buffaloes, leopards and pigs. The latter swarm in almost all the villages in the north-west and south of the country and do considerable damage. Their numbers of late years have increased to such an extent that the Collector expresses his opinion, that unless prompt measures are taken for their destruction, many of the villages of the district will be given back to the jungle. In some villages the outlying lands cannot find cultivators, owing to the depredations of wild hogs. The villagers have not yet learnt self-help, and seldom destroy these animals, which are allowed to breed and multiply undisturbed."[22]

In Singhbhum (Bihar), also, the animal menace had not abated: "The wild animals . . . are very numerous, and their ravages are at present the great obstacle to the spread of cultivation. The inhabitants of villages bordering on the jungles complain not of the personal dangers to themselves, but of the wholesale destruction of their crops, and say that they have to raise grain for the wild beasts as well as for their own families. This cause alone is said to have prevented the people from growing cotton, for which the soil is admirably suited. Of late years some herds of elephants, in the hills between Bonai and the Saranda pir of Singhbhum, did such damage to the crops, that the villages at the foot of the range were abandoned by their inhabitants."[23]

Clearly, the pressure of wild animals constrained agriculture in some parts of eastern India in the nineteenth century, but the overall trend in this period is uncertain. An additional complexity is that elimination through hunting of carnivores high on the food chain, such as tigers and leopards, may simply have liberated other animals, such as wild pigs, to breed and thus wreak greater havoc on crops. This suggestion may be ecologically plausible, but the data are insufficient to demonstrate it.

Wild animals attacked not only crops and domesticated animals but also villagers and their families. W. W. Hunter sent a questionnaire to district officers and collected data on mortality due to animal attacks during the period 1860–76 in the forty districts of present-day Bengal, Bihar, and Orissa. Attentive to the distinction between poisonous reptiles and fatal attacks from carnivores, he asked them to record the number of lives lost to each annually. Of the 12,870 deaths reported over the sixteen-year period, 4,426 were attributed to wild animals and 8,444 to snakes.[24] These bald figures, whose completeness must be doubted (peasants had nothing to gain by reporting such deaths), confirm that the number of deaths from snakebites and animal attacks was substantial. The inquiries begun by Hunter in Bengal were taken up in other provinces. The Bombay gazetteers, for example, reported that between 1856 and 1911 (a fifty-five-year period), a total of 19,657 persons died from wild animal attacks—1,176 from carnivores and 18,481 from snakes.[25] Again, the figures are incomplete, and the compilers acknowledged that their data represented only a portion of actual fatalities, many or even most of which were never reported. Nonetheless, we can infer that fatal attacks from carnivores were less common in the Bombay Presidency than in the Bengal Presidency (averaging 357 and 804 deaths per year, respectively) and that the ratio of reported fatal snakebites to reported fatalities from other animals was eight times greater in Bombay (16 to 1) than in Bengal (2 to 1). Further, we see that at least 32,529 persons in approximately one-half the territory of British India were reported to have met their deaths prematurely from wild animals during the stated periods. The actual number of deaths throughout the subcontinent was, I believe, much larger.

Since the earliest years of British rule, officials laid out money to restrain wild animals. Whether by diverting troops to clean out notorious sites of infestation by wolves, by paying bounties for killing lions, snakes, and leopards, or by dispatching hunters to exterminate notorious man-eaters, the authorities incurred real costs. Compassion, however, was never the motive. In India a competent ruler was required to "protect" his subjects, and the East India Company accepted this duty in part by taking control of wild animals seriously. If the intrusion of wild animals can be regarded as a barometer of the village economy, then official efforts to destroy marauding wildlife can be likened to a thermometer of administrative concern. In these terms, both pressure and temperature fell over the course of the nineteenth century as cultivation expanded and as the system of bounties succeeded, thereby destroying both animals and animal habitats. By the mid-twentieth century wild animals had become an exotic challenge for most officials.[26] As early as the middle Victorian period, one can detect something like astonishment at the decline of the animal threat. Note, for example, this passage from a Bombay district gazetteer: "In 1860, when the district [Panch Mahals] came under British management, the forests were full of big game, and during the next eight seasons 40 to 70 head were killed yearly. [In] 1865 the results of the year's shooting included 22 tigers, 10 panthers, and 38 bears. Besides this destruction, two causes—the clearing of their former haunts and the shortening of their former food supplies—have been at work to reduce the number of big game. . . . Tigers are gradually withdrawing from their old haunts. Even in the thickest and safest covers a stray animal is only occasionally found. Panthers wanting less food and shelter give ground slower. But on them too the spread of tillage presses hard, and their numbers slowly drop off."[27]

Although this succinct account of the disappearance of tigers, bears, and leopards in the Panch Mahals district—the result of habitat destruction, decline of prey species, and overhunting with firearms—is more than a century old, it corresponds precisely to modern explanations for the disappearance of large animals in nearly all parts of India over the past few decades.

Countermeasures against the Beasts

Animal attacks elicited defensive measures long before British rule was consolidated. Some of these measures were taken by individual actors, others required the cooperation of entire villages, and still others depended on landlords or the intervention of regional government. The examples given below, drawn primarily from colonial sources, suggest the inability of unaided villagers to do much more than flee or try to blunt the worst of wild animal attacks. As long as India's forests and savannas were abundant in relation to human settlement, the reservoir of animals could not be disciplined by essentially Neolithic technologies.

Abandonment, toleration, propitiation. Under the severe pressure of animal attacks, cultivators fled their settlements. Short of abandonment, cultivation could be confined to daylight hours or carried on from behind barriers of thorn and wood or conducted at a short distance from safe bases. A kind of commuter cultivation was common along the northern edge of the Sunderbans, the 3,600-square-mile belt of low-lying mangroves around the Bay of Bengal. The Sunderbans were historically feared because of man-eating tigers and destructive tropical storms, both of which still plague the area.[28] Nonetheless, the edges of the Sunderbans have been colonized since the eighteenth century. Paddy cultivation as well as foraging, fishing, and woodcutting have taken place steadily in patches inside the mangrove forest.[29] In the Victorian era, woodcutters and cultivators journeyed into the Sunderbans in country boats accompanied by ritualists called *fakirs,* who offered invocatory rituals (*puja*) to the tiger god, Sona Ray, and marked for use forest areas where men ordinarily dared not intrude. These measures "worked" in the sense that forest cultivation and woodcutting were carried out regularly despite occasional loss of life to tigers. The fakirs were, in effect, tiger specialists, whose hard-won knowledge enabled villagers to pursue a risky calling with confidence.[30]

The ritual practices of the Sunderbans fakirs should remind us that popular attitudes toward wild animals are not always fearful. Animals that are lithe and gorgeous as well as powerful and dangerous are viewed with ambivalence. In a cultural context where power and grace are worshipped as expressions of divine power (*sakti*), it is understandable that "the claw of the tiger is tied to the neck of a frightened child to ward off evil spirits" and that "some shaman and folk-practitioners in rural Nepal use sugar-coated tiger-hairs against hydrophobia."[31] Further, South Asian villagers came to know animals in the wild, not in zoos, and they read into the wars between the species allegories of order that resembled social relations.[32] The possibility that some species might be destined to rule over others for human benefit was thus taken for granted. From Bengal, we read: "So long as [tigers] refrain from the habit of attacking men, their presence is desired rather than dreaded by the cultivators. Dr. Buchanan-Hamilton expresses the opinion when writing of this very region [c. 1810], 'that a few tigers in any part of the country that is overgrown with jungle or long grass are extremely useful in keeping down the number of wild hogs and deer, which are infinitely more mischievous than themselves.' Mr. Pemberton, the Revenue Surveyor in 1848, also states that 'the inhabitants of Gaur are rather partial than otherwise to the tigers, and are unwilling to point out their lairs to sportsmen. They call them their *chaukidars* [watchmen], as being useful to them in destroying the deer and wild hog, with which the place abounds, and which make sad havoc of their crops.' "[33]

Even more positive attitudes toward animals have been reported. Madhav Gadgil, for example, reviews reverential practices that in earlier decades protected tigers, leopards, blackbuck deer, monkeys, storks, turtles, cobras, and other small birds and animals from destruction, and other authors draw attention to the remarkable sect of the Bishnois, founded by the visionary Jamboji in the fifteenth century, which continues to this day to protect both wild ungulates and their forest cover in the Thar desert of Rajasthan.[34] A key technical measure—the Bishnois ring their pastures with thorny barriers of *khejdi* trees, which keep wild carnivores from their domestic cattle—allows them to shelter wild blackbuck antelope as well. The inversion of usual values goes even further among the Bishnois, who have been known both to sacrifice their lives to project their khejdi groves and to kill poachers for daring to hunt the protected wild deer.[35] Ecologists have only begun to plumb the extent of these positive attitudes toward animals, which may offer a base of values upon which modern conservation measures can be erected.[36]

Night watches and noisemaking. Another common village response to wild animal attacks has been to mount weaponless guards. At the beginning of the nineteenth century, for example, Buchanan-Hamilton observed in Rangpur district that "when the rice approaches maturity, the cultivators in the parts which elephants frequent have to keep a watch on the crop every night. Stages are erected on posts 12 or 14 feet high; on one side of the stage a small shed is erected for the watchmen, who keep watch in pairs, one man feeding a fire which is kept constantly burning in the open part of the stage, while the other sleeps. In the event of elephants, deer, or hogs coming to the field, the sleeper is roused, and both men unite in attempting to frighten away the intruders by shouting and beating drums. They never attack the animals."[37]

Seventy years later G. P. Sanderson described a similar practice in Karnataka, where Shologa tribesmen were hired to watch for wild elephants from atop platforms at night in the fields; when crops were threatened, the Sholagas would wield large bamboo torches, eight feet long and eight feet in diameter, which they ignited and whirled to scare the elephants away.[38] Interestingly, modern recommendations for driving off wild elephants include "using spotlights, torches, loud-speakers, firecrackers, drums, and carbide noise-makers . . . [which] can usually drive out a marauding herd of elephants before they cause too much damage."[39]

In ticking off these various methods—retreat, propitiation, watch, and ward—I am not attempting to judge their efficacy in controlling wild marauders; rather, I am trying to show that cultivators had a host of defensive practices premised on the constancy of animal attacks. Their measures against crop predators were coordinated with the matter-of-fact enlargement of their acreage to satisfy the deer, wild pigs, and even elephants who also relished ripe paddy and cane; in effect, cultivators paid a tax to wild animals as much as to the king. Singhbhum farmers complained to the collector, for example, that "[we] have to raise grain for the wild beasts as well as for [our] own families."[40] The defensive methods described so far were all local and nonviolent, but when they failed, indigenous hunt professionals could be found to attack the animals.

Tribal hunters. Sanskrit sources as far back as Manu (around the second century C.E.) describe tribal hunters, and it seems likely that *shikaris* (huntsmen) once existed everywhere on the subcontinent. The absence of a single monograph on the history of Indian hunting, while works on cereal agriculture abound, is a further sign of the vegetarian bias that has overtaken scholarship.[41] Nonetheless, tribal hunters appear in the margins of every ancient epic and medieval chronicle as providers of game for the tables of rulers, and it seems clear that most hunting in India was motivated by the desire for food and not for sport. Although epic, sastric, medical, and modern authors all focus on the game animals that appealed to elite palates—deer, peacock, geese, antelope—there seems always to have been a humbler, unrecorded demand among the poor for wild pig, hares, snakes, and rodents as well as every kind of bird and fish. This demand was met by nearly invisible bands of huntsmen characteristically described as forest-dwelling tribals who were willing to provide their hunting skills and the wild animal flesh needed for food and medications in exchange for grain, cloth, and cash. The customs and values of tribal hunters, marginalized by the state and despised by caste society as violent and degraded, have long been at odds with mainstream Indian culture.

These same tribal hunters appeared in another role, however, when wild animals pressed too hard on peasant cultivation. Then their services were eagerly solicited, and local officials and elites offered them honors, land, and money. Preeminent among them were the *baghmaras,* or tiger killers, who were called upon by landlords and even kings to rid their territories of ferocious beasts. Evidence suggests that pre-colonial governments employed them for this purpose; in Dhaka district, for example, the late Mughals "assigned a rent-free tract of land (*jagir*) to a class of men called *baghmarias* [*sic*] or 'tiger-slayers.' "[42] The same title was employed beyond Bengal as an honorific: "In ancient Nepal the killer of the tiger was regarded with high respect [and] used to be called *baghmere* [*sic*]. Not only this, the brave hero used to be exhibited around local markets with a dignified turban on his head and *gulal* (colored powder) sprinkled all over the body."[43]

The source of the baghmaras' prestige was their ability to kill tigers without the use of firearms. The earliest British writers on "oriental field sports" mention their prowess with a poison-dart machine in use in different parts of eastern India.[44] All the elements of this crossbow-like device—its bamboo mount, bow and armature, multiple strings, metal arrowhead, aconite-derived poison, and safety release to prevent accidental discharge at a blundering cow or human—were assembled from forest materials, and the range of mechanical, faunal, and botanical knowledge required to deploy it with effect suggest that it was one of the most remarkable Neolithic hunting devices ever invented. Only the widespread advent of rifles in rural India after 1950 dispatched it to the scrap heap.

Elephant catchers constituted another remarkable group of specialists. Elephants were not bred but had to be captured to be tamed, trained, and sold as work animals, and the hunting of elephants was in some states a royal monopoly that continued into the colonial era. For example, the Rajas of Susang (Maimansingh), Sarguja (Chota Nagpur), and Hill Tippera (all in eastern India) raised significant revenue from elephant sales. In northern and eastern India the *khedda* (or *kheddah*) technique, which involved 6,000–8,000 beaters who drove wild herds toward a funnel-like palisade of timbers surrounded by a deep ditch that emptied into a high corral, was the specialty of local elephant catchers. A great elephant fair was traditionally held each year at Sonepur on the Ganges, where princely rulers and contractors requiring elephant labor purchased recently tamed beasts. In British India elephant catching was a lucrative government monopoly. In 1866–68, for example, 230 elephants with a market value of at least seventeen thousand rupees were captured by the Chittagong khedda establishment.[45] The shikaris (hunters) and *mahouts* (drivers, handlers) employed in these activities failed to share in the profits, but it was their intimate knowledge of the habits of elephants that justified large government investments. The khedda technique was introduced into princely Mysore and other parts of South India during the Victorian era, when it displaced an older method of pit capture.[46] The value of elephants was so great, in fact, that in 1879 a conservationist bill, the Elephant Preservation Act (no. VI) of 1879, following the lead of Madras earlier in the decade, outlawed the shooting of all but proscribed rogues.[47]

In addition to elephant catchers and tiger hunters, Johnson observed in 1827 that "in almost every district of [eastern] India there are great numbers [of hunters] whose profession or business is solely to catch animals or game" and that "these men (whose forefathers have followed the same profession) are brought up on it from infancy [and] become surprisingly expert."[48] A composite picture of these local hunters can be sketched from the early gazetteers: they were highly specialized and employed numerous techniques (netting, hounding, deadfalls, noosing, shooting with arrows, spearing, hooking, birdliming, poisoning, driving into stockades, setting fires, decoying from ambushes, and more) to pursue particular species in varied terrains: buffalo, crocodile, and rhinoceros in Rangpur and Jalpaiguri; cobras and pythons in Patna and Monghyr; wolves in Hazaribagh; and so on.[49] This variety of expertise and effort—not to mention the widespread collection of honey, eggs, rodents, snakes, and insects, which were important sources of food—drew on an archaic knowledge of animal life that also needs to be inventoried as environmental historians re-create the South Asian Eden.

Among the most skilled of the hunters in colonial accounts were the Doms and Ganrars of Rangpur, the Kols of Orissa, the Kukis or Lushais in Hill Tippera, and the Santals in Burdwan and Monghyr—all tribal or low-caste groups. Unskilled in plow cultivation or lacking the capital to undertake it, they were nonetheless essential to the success of peasant agriculture when ripening crops attracted hordes of wild pig, elephants, and deer. Long-term connections with the hunters were offered by some landlords to assure that protection would always be at hand; otherwise, loose contracts could be established when the pressure on settled tenants became unbearable. We read that "the proprietors of Sunderbans lots regularly employ *shikaris* on their estates," that "some [Rangpur] zamindars keep *shikaris* or huntsmen for the purpose of keeping down wild hogs, which would otherwise overrun the cultivation and drive away the tenants," and that "in the state of Bod [Orissa] 86 persons were devoured [by tigers] and the raja was obliged to engage a huntsman from Sambalpur."[50] The fugitive quality of these remarks should not be taken to mean that destroying predators was not essential to the success of agriculture over large parts of India.

A market in carcasses. British rule made its own special contribution to the control of animal predation. In the earliest days of the raj, colonial authorities introduced bounties—the system of offering rewards for killing vermin species. The essence of the system was its purely financial nature: killing animals was not a matter of honor or caste tradition but solely a question of cash for carcasses. Tangible proof was demanded—a severed head, a flayed skin, or amputated ears, paws, or tail—in order to collect the bounty. An official (usually the district collector or his agent) inspected the proofs. At first glance this may look like a market transaction, but "prices" were set administratively, and the "products" (such as skins or skulls) were not subject to further processing. The result was a very odd kind of market, a monopsony in animal remains.

In theory, a system of bounties should have made animal killing attractive to others besides low-caste and tribal hunters: anyone could claim the reward. To this extent, colonial practice could be said to have been liberal. Nonetheless, trafficking in blood and skins was not to the taste of most peasants, and certainly not those of the higher castes. It was the traditional hunters who exploited the new source of cash. At the same time, the prospect of getting cash for kills fostered a new calculation: "the more animals I kill, the larger my income." Hunters appear to have intensified their pursuit of species most rewarded by the state, and district officials began to detect fraud (the chronic anxiety of an administrative system based on attributing cash values to land, labor, and the "products" of nature). Officials took to smashing the skulls and slashing the skins of bounty animals to prevent their being resubmitted elsewhere for more than one reward. The collector of Purnea, for example, recorded his suspicion in 1788 that some among the *six hundred* tigers for which he had paid one pound sterling each that year were not local products and that wily Marang baghmaras had been bringing in the skins from Nepal.[51] Nonetheless, the system of cash rewards, which began in Bengal as early as 1770, was applied in subsequent decades to tigers, leopards, cobras, elephants, rhinos, hyenas, wolves, bears, civets, crocodiles, and other animals in eastern India.[52]

Bounties varied between districts and according to species, and one can see in these differences a regional pattern of depredations. In the late 1860s and 1870s in eastern India, for example, only the districts of Patna, Tirhut, and Gaya paid a bounty for wolves; only Noakhali, Tippera, Tirhut, and Monghyr paid a bounty for vipers; and only Cuttack and Balasore thought a crocodile worth anything at all. The rise and fall of the bounty for the same animal in the same district over time registered the changed perception of the pressure the animal exerted on cultivation. The bounty for tigers in Darjeeling, for instance, quadrupled from five rupees to twenty rupees in 1869, and in the same year rhinos and elephants, previously ignored, had prices put on their heads for the first time (five and twenty rupees, respectively). Whereas the Board of Revenue in Calcutta gave ultimate sanction to the bounty system, district collectors could redirect attention to newly perceived threats and could even draw distant hunters to their districts by sweetening the reward.

Manipulation of the bounties seems to have had the desired effect in some cases. The six hundred tigers killed in Purnea district in one year have been mentioned. In Malda district, 147 hyena heads were presented on one occasion in 1847 for a total bounty of twenty-four pounds sterling (240 rupees); a subsequent collector, writing in 1871, asserted that no hyenas had been seen in the district during the twenty-five-year interim. In Monghyr district no less than a thousand cobras, "alleged to have been killed within the precincts of the town," were produced for bounties between 1871 and 1873; it was also reported that "a nearly equal number" were turned in at the same time from the nearby Jamalpur municipality. Do these sizable numbers mean that the bounty system worked in the sense of eliminating the feared species? The data are ambiguous. To take the example just given, Hunter reported contradictorily that the number of cobras in Monghyr and Jamalpur "do not appear to have sensibly decreased, but it is satisfactory to learn that deaths from snake-bite are now rare within the towns."[53] It is also difficult to separate out the deterrent effects of the bounty system from the parallel effects of habitat destruction that accompanied expanded cultivation. In other countries well-organized bounty systems, accompanied by poisoning and trapping, have eliminated whole species; the best attested example is the destruction of wolves in North America.[54]

The data are too sketchy for us to determine the accomplishments of the diffuse, decentralized, and poorly reported colonial bounty system. I prefer to see in it not a successful means for eliminating all animal attacks but strong evidence of a continuing official concern, dating to the time of the Mughals, to protect peasant agriculture—which was, after all, at the base of the rural economy as well as a chief source of government revenue. The distinctiveness of the colonial approach was that—in contrast to pre-colonial means, which rewarded hunters for taking up residence near cultivation and enjoined them to be resourceful in repressing wild animals—it rewarded only those hunters who demonstrated kills. What this system also did, subtly and gradually, was to affix a value, a price, directly on animals in the wild. At some point in the nineteenth century it became possible for Indians hunters to say, as a tiger or hyena or cobra glided into the brush, "*Ram!*—there went fifty rupees."[55] This economistic logic—attributing monetary values to objects in nature—came to be applied to such floral species as teak, *chir,* and *sal* trees only much later, after British appetite turned to India's forests.

The Problem of Man-Eaters

A century ago tigers were abundant in most parts of South Asia; the standard estimate by E. P. Gee is that 40,000 tigers roamed India in 1900. During the colonial era, formal knowledge about tigers (and about panthers and lions, India's other large cats) was confined mostly to anecdotes published by European sportsmen.[56] Human encounters with tigers, while never common, occurred across large tracts of the country and sometimes resulted in fatalities.

South Asian tigers normally feed on deer and other ungulates and, of necessity, on smaller mammals, reptiles, and rodents—not on humans. Fatalities occur when incautious humans threaten tigers by approaching too closely, especially when female adults are nurturing cubs or when humans get between tigers and their prey. Under these circumstances tigers will kill humans, but the offending tigers are more appropriately called man-killers than man-eaters because they rarely eat their victims. Very occasionally, less agile tigers (and leopards), especially those with broken teeth or injured limbs, deliberately stalk humans and become habituated to human flesh; these are true man-eaters.[57] Actually, both terms—"man-killer" and "man-eater"—are off the mark, because it is women and children who are most vulnerable and thus disproportionately attacked. They are often taken singly while bent over collecting fodder or firewood or pasturing cattle or relieving themselves at the edge of the forest.

The large number of tigers in India in the early twentieth century made possible a veritable cult of royal and colonial tiger shooting, an undisguised expression of wealth and power that reached its apogee between the world wars. Despite the tendency of authors from this period to glorify tigers as "honorable" and "noble" opponents, unarmed villagers knew better than to believe that tigers followed a feudal code of conduct and gave them a wide berth.[58] Yet avoidance was little help when an established man-eater had determined to feed on human prey: man-eaters were known to shatter the locked doors of huts and punch paws through mud and bamboo walls to get to their victims. They also patrolled large territories and thus might appear without warning twenty-five miles away from a previous night's slaughter. In these circumstances European hunters with good rifles, like Jim Corbett and Kenneth Anderson, were called in by local officials. Corbett presents data on a tiger in the Chowgarh region of Kumaon that roamed over 1,500 square miles between December 1926 and March 1927 and killed sixty-four humans in twenty-seven villages. Similarly, a single man-eating leopard, whose area of operation centered for nearly a decade on Rudraprayag village in Gahrwal, killed 125 people in 1918–19 alone, always stalking the same ninety-four settlements.[59] Corbett was called in by local officials to kill both animals after baghmaras and other more ordinary remedies had failed.

What was the impact of such monsters on village life? Corbett makes clear that paralyzing fear was common:

> No curfew order has ever been more strictly enforced or more implicitly obeyed than the curfew imposed by the man-eating leopard of Rudraprayag. During the hours of sunlight, life in that area carried on in a normal way. . . . As the sun approached the western horizon and the shadows lengthened, the behavior of the entire population of the area underwent a very sudden and a very noticeable change. Men who had sauntered to the bazaars or to outlying villages were hurrying home; women carrying great bundles of grass were stumbling down the steep mountainsides; children who had loitered on their way to school, or who were late in bringing their flocks of goats or the dry sticks they had been sent out to collect, were being called by anxious mothers; and the weary pilgrims [en route to Kedarnath and Badrinath] were being urged by any local inhabitant who passed them to hurry to shelter. When night came an ominous silence brooded over the whole area; no movement, no sound anywhere. The entire local population was behind fast-closed doors—in many cases, for further protection, with additional doors to the existing outer ones. . . . Whether in house or shelter all were silent, for fear of attracting the dreaded man-eater. This is what terror meant to the people of Gahrwal, and to the pilgrims, for eight long years.[60]

Man-eaters invariably carried off their human prey to feed in a protected spot; this habit obliged the victims' families to go in search of the remains. Trackers accompanied grieving relatives along trails marked by shredded clothes, blood, and hair only to arrive at grisly scenes of cracked human bones.[61] Corbett leaves no doubt that when he killed an infamous man-eater, this produced local exaltation and relief, especially among village women.[62]

Anderson, referring to the impact of man-eaters on the populace of rural Karnataka, echoes Corbett:

> A man-eating tiger, or panther [leopard], where it exists, is a scourge and terror to the neighbourhood. The villagers are defenceless and appear to resign themselves to their fate. Victims are killed regularly, both by day and night if the killer is a tiger, and by night only if a panther, the former often repeatedly following a particular circuit over the same area. While the death roll increases, superstition and demoralization play a very considerable part in preventing the villagers from taking any concerted,

> planned action against their adversary. Roads are deserted, village traffic comes to a stop, forest operations, wood-cutting and cattle-grazing cease completely, fields are left uncultivated, and sometimes whole villages are abandoned for safer areas. The greatest difficulty experienced in attempting to shoot such animals is the extraordinary lack of co-operation evinced by the surrounding villagers, actuated as they are by a superstitious fear of retribution by the man-eater, whom they believe will mysteriously come to learn of the part they have attempted to play against it.[63]

Anderson's final sentence, despite the dismissive *s*-word, alludes to the conviction that man-eaters were not always considered animals, strictly speaking, but could be seen as embodied spirits. This belief, which shifts the problem of collective action into a realm of supernatural reality, is recounted by Corbett, who describes an instance in which Gahrwal villagers actually identified the alleged malefactor, a solitary *sadhu.* This shape-shifter was burned alive in his hut after the burden of tiger attacks had become unbearable. Corbett gives a second instance in which the execution of a similarly identified sadhu was foiled by an official.[64] Anderson adduces yet a third case, in which a "voodoo man" was killed by fellow villagers because of his presumed relationship to a man-eater.[65] Thus, despite the scarcity and brevity of these accounts (by authors who may be reluctant to seem too well informed about "superstition"), it seems evident that in some places local beliefs endorsed the identification and execution of presumed malefactors as a means to end the terror caused by man-eaters.[66] That quasi-judicial measures of this sort were ever common seems doubtful, but that tiger attacks on humans would be interpreted as part of a meditated design should not surprise us: the relationship between *animals* and humans in South Asia is no less permeated by ideas of spiritual exchange and moral causation than is the spiritual and moral relationship between *plants* and humans as explicated by Vandana Shiva, Ram Guha, Madhav Gadgil, and other advocates of the Standard Environmental Narrative.

Forest Regulation and the Postwar Blitz

The disastrous consequences of closing India's forests to the populace in the last quarter of the nineteenth century has been well discussed.[67] In these accounts the Forest Act of 1878 is treated as the turning point in the environmental history of the subcontinent. The act was motivated by an "imperial" need to secure timber for railway construction and fuel, but it was justified publicly on the "scientific" ground that Indians had shown themselves incapable of managing forest resources prudently. The act placed vast forest regions off limits to nearby peasant communities and ushered in a new era of forest policing: henceforward it became a crime to cut grass, hunt game, graze cattle, or collect wood and fodder where generations had done so before. The impact on village life in India may have been as devastating as that of enclosures on rural England in the sixteenth and seventeenth centuries. Arbitrarily enacted, driven by security fears remote from a public in whose name the country was ruled, and breathtaking in their absence of legal justification, the 1878 Forest Act and the provincial regulations that stemmed from it are perfect examples of the arbitrary exercise of colonial power.[68] Nonetheless, wildlife conservationists regard the act with mixed feelings because it introduced into South Asia closed hunting seasons, bag limits, and shooting permits on the model of European hunting laws and thus laid the groundwork for wildlife protection.[69]

Pre-Mughal and Mughal rulers had sometimes set aside forest reserves where, accompanied by cavalcades of adjutants and beaters, they indulged in *shikar* from the backs of horses and elephants; the aim seems to have been to kill as many tiger, lion, deer, or buffalo as possible—a kind of warfare against the beasts—as well as to secure animal products of value.[70] Despite these traditions of the Indian royal hunt, Victorian jurists found no precedent in law for controlling and punishing Indian villagers who lived in or entered royal reserves to slay animals; this was dismaying, because in Europe the management of forests had been intimately linked since medieval times with prohibitions against poaching.[71] If Indian villagers exercised forest rights by ancient custom, then revocation of these rights had to be justified by other than legal arguments. Science was enlisted to prove that a rabble of foragers, woodsmen, and herders—so the argument went—were damaging the most valuable stands of teak and pine with wasteful techniques of slash-and-burn; that they overgrazed the forest floor with their herds; and that their gathering of thatching and fodder knew no restraints. The survival of the forests, it was said, was at stake. Further, native hunters used cruel techniques like poisoning and deadfalls, and their puny guns lacked modern rifling and explosive charges; hence the "noblest" animals like lions and tigers were often only wounded and then had to be tracked as they slowly bled to death. Such cruelty must not continue; it was unsporting. Hence indigenous hunters, foragers, herders, and woodsmen were all incompetent and required expulsion. On these grounds the forests were closed except by permit, and ancient means of subsistence were criminalized. A new crime, poaching, was invented so that railway ties (sleepers) could be steadily supplied to the burgeoning railways.[72]

Paradoxically, under the new forest rules Victorian officials, military officers, civilian officials, and their native favorites obtained almost unbridled access to the forests for sport. John M. MacKenzie has described the "obsessive cult" of the hunt in colonial India, which, he argues, performed important psychological functions for the sahibs, such as dramatizing their manly virtues and ceremonializing the social hierarchy.[73] Part and parcel of the cult was an unwritten code insisting that sportsmen must court danger in maintenance of honor and must recognize the essential nobility of the hunted animal; this feudal twist made it unacceptable to fire blindly into a herd, to kill pregnant females or their cubs, to take too many animals of the same species on a single hunt, to shoot from a *machan* (blind), or to take game with nets, deadfalls, or poison—all practices doubly denigrated as ignoble and native. The code itself thus had modest conservationist effects.[74] Yet overt animal conservation developed relatively late in India, perhaps because the threat *from* wild animals had always seemed more obvious than the threat *to* them. The first such measure was the 1877 Act for the Preservation of Wild Birds and Game (no. XX), which established sanctuaries inside the reserved forests where wardens protected larger fauna and game birds from both European and indigenous hunters. Elephants, whose decline in numbers was first mentioned in print around mid-century, were taken off the hunting list by the Elephant Preservation Act of 1879.[75] By the turn of the century nostalgia for an earlier "sportsman's paradise" of vast herds and flocks of deer, wild pig, buffalo, elephant, game birds, and so on became a characteristic feature of the European hunting literature in India.[76]

It is unclear whether Victorian hunting regulations, which in theory limited the slaughter of game, complemented or negated the system of bounties, which tended toward species extinction. Because the two policies responded to different bureaucratic cues—the game laws were formulated in the Forest Department, whereas the Revenue Department oversaw the system of bounties—there was a potential for contradiction. But the distinct micro-geographies involved made the contradiction more apparent than real: bounties were intended to protect lives, crops, and cattle in densely settled lowland agrarian tracts, whereas hunting regulations excluded villagers and sportsmen from forests in the more lightly populated hill areas. When animals caused problems at the intersection of these physical and policy terrains, special arrangements could be made. For example, if cattle grazing by permit in reserved forests threatened to infect wild ungulates with disease, forest and civil officers together would kill the infected herds, sometimes on a vast scale.[77] Similarly, the conservationist Elephant Act of 1879, which took wild elephants off the sporting list, nonetheless allowed district officials to "proclaim" the destruction of musth-maddened bulls. "Proclamation" offered a onetime bounty for the extermination of a specific rogue bull, while neither relaxing the shooting laws nor exposing elephants in general to capture.[78]

When wildlife conservation as such became an overt policy concern, around

1900, it took a form that reflected the habits and outlook of the colonial rulers. According to MacKenzie:

> By the end of the century the hunting and natural history elites were beginning to sound a note of alarm. The combined ravages of over-hunting and rinderpest had produced such a marked diminution of game [in Africa and India] that conservation measures seemed necessary. Pressure groups became active in promoting legislation, the creation of reserves, and the funding of societies dedicated to the preservation of game. Since these pressure groups included many governors and other senior colonial officials, aristocrats and "sporting" hunters, and leading landowners in colonies of settlement, suggestions for preservation were swiftly translated into practice. Inevitably the form that preservation took was shaped by the social and economic realities of Empire. . . . Access to animals was to be progressively restricted to the elite; animals were to be categorised according to sporting rather than utilitarian characteristics; some were to be specially protected for their rarity, others shot indiscriminately as vermin; separation was to be attempted between areas of human settlement and those appropriate to animal occupation.[79]

It is these imperatives that lay behind the Wild Birds and Wild Animals Protection Act of 1912, which on an all-India basis specified seasons and listed animals whose hunt required a permit; the wildlife provisions of the Indian Forest Act of 1927, which authorized appointment of forest officers who could restrict hunting; the Bengal Rhinoceros Act of 1932; the Punjab Wild Birds and Wild Animals Act of 1933; and the National Park Act of 1934.[80] Accompanying these legal enactments came the founding of the Kaziranga game sanctuary in Assam in 1926 and then, ten years later, Hailey National Park in Uttar Pradesh (named after the provincial governor, Sir Malcolm Hailey), which protected both wild animals and their habitats.[81] Two Sri Lankan wildlife parks—Wilpattu and Yala/Ruhuna—were also founded in the 1930s.[82] During the interwar years numerous game sanctuaries—zones of refuge that regulated shooting but lacked the national parks' resources to check poaching, enhance herds, and encourage tourism—were established throughout the subcontinent. Hence the groundwork for a wildlife conservation system was laid in the last decades of colonial rule.[83]

In parallel with the official trend toward wildlife conservation, the interwar period saw a steady rise in South Asia's rural population and a corresponding expansion of cultivation at the expense of remaining forested areas. Given the absence of urban employment opportunities, this expansion was implacable and had a destructive impact on habitats of wild animals. Indiscriminate hunting during World War II tipped the balance against wild animals outside the game parks and reserves, and the fifteen years after Independence saw an outright blitzkrieg. As a result, domestic cattle and goats, no longer in danger from tigers and leopards, entered the remaining forests to challenge wild ungulates in their own habitats. I have already cited Schaller's remarks on this situation, but Balakrishna Seshadri is perhaps the most perceptive witness:

> Real, large-scale destruction of wild life began during the years of war. Forests were felled for timber and to make room for vast army camps. Troops entered them for training in jungle warfare, and the slaughter of the animals and birds began. Animals were machine-gunned in fun, and gregarious herbivores like the *chital* or spotted deer merely stood bewildered and stared in the direction of fire and were mowed down. . . . When the war ended and independence came, enormous quantities of guns and ammunition became cheaply available. The period of political transition was one of many uncertainties. Villagers who had lived, in the main, within the game laws, both from fear of punishment and lack of lethal weapons, assumed that the change of authority meant freedom from control. The weapons could now be procured. Forests were freely cut down for timber and fuel, and the animals and birds were slaughtered for food or sale of skin and feather. . . . [The village hunter] came to consider poaching as a democratic right. None was apprehended when the forest ranger who reported an incident was required to produce a third-party eyewitness for the conviction of the offender. Investigating rangers were threatened with a gun or with physical violence if they persisted. . . . There were cases of rangers being shot dead by poachers. So confident did the latter become that piles of game meat began to appear in open markets. . . . In the last two decades [c. 1947–67] we have seen the emergence and then inordinate increase in the numbers of gentlemen-poachers, casual sportsmen who ride in jeeps and roam the jungle at night dispensing with the trifling formality of shooting permits. The time of year, sex, and age of the animals are all the same to them. But even so, the poacher, tribal or genteel, must be given second place when pinning responsibility for the murder of wild life, followed closely by overgrazing of domestic livestock. . . . I do not seek to minimize the destructive role of either [poaching or overgrazing], but habitat destruction through project work and what follows it is by far the most pertinent [cause] and the one most needing urgent attention. Development projects, among them river valley projects the most immense, pound into nature and leave it breathless and exhausted.[84]

It should be pointed out that the postwar blitz, in which newly armed "sportsmen" (a category begging for ethnographic clarification) turned guns on anything that moved in the forest, was a characteristic feature of decolonization throughout most of the Indo-Malaysian realm.[85] But the larger point in Seshadri's remarks is that after 1950 the micro-technology of unrestrained small-arms fire, when combined with the macro-technologies of logging, river damming, road building, and so on, began at last to reverse the terms of terror between wild animals and humans in India.[86]

For wild animals, the consequences of unrestrained shooting and forest grazing were disastrous. India's tiger population, which had been estimated at 40,000 in 1900, declined to 1,800 by 1972.[87] Wild lions in the Gir forests of Gujarat numbered only 180 in the same year, and all three species of Indian crocodile were on the verge of extinction.[88] Of twenty-nine species of deer worldwide identified in 1977 as threatened with extinction, nine were in South Asia.[89] These trends promised the disappearance not only of all the larger wild animals in South Asia but also the multitude of smaller ones that sheltered alongside them in the same habitats.

The Return of the Man-Eaters

Most colonial hunters and even conservationists lacked interest in the mundane lives of animals, never more so than in their indifference to the natural history of tigers. This gap has been filled in recent decades by ecologists and professional conservation managers, who have made the study of *Panthera tigris* a focused object of research. Experts, however, have been unable to isolate the scientific study of the Indian tiger from the tiger's tangled relations with people in South Asia.[90] An issue that has caused more and more concern since the 1970s is the increase in man-killing tigers at a time when tigers as a species are under threat of extinction.

The colonial rationale for game parks and sanctuaries was that only a limited number of species required protection; these were mostly the very large animals, such as elephants and rhinos, that in an earlier era had been hunted for trophies.[91] Supported after independence with advice from United Nations agencies (for example, U.N. Food and Agriculture Organization and U.N. Environmental Programme), and later by the International Union for the Conservation of Nature (IUCN) and the Worldwide Fund for Nature (WWF), South Asian governments established hundreds of national parks, reserves, and sanctuaries in the period from the 1950s through the 1980s.[92] The political leadership in India was particularly supportive, and the Indian constitution was amended in 1976 to include an article stating "It shall be the duty of every citizen . . . to protect and improve the natural environment, including forests, lakes, rivers and wildlife and to have compassion for living creatures."[93] Conservation officers became familiar figures in and around the parks and were authorized to use strong measures to control poachers, herders, woodcutters, minor forest-product collectors (such as honey gatherers), and other intruders. Despite the marked trend toward habitat destruction, which began as early as the 1930s and has only accelerated, a substantial network of protected areas for wildlife was created.[94]

By the mid-1970s regional governments claimed success in rebuilding wild animal populations.[95] In nearly every case, however, success depended on segregating animals from humans. In most cases national parks, sanctuaries, and biosphere reserves were created by central and local governments "gazetting" large tracts of forest adjacent to established villages, that is, by seizing the commons that had been vital to the life and livelihood of the poorest villagers. Whole settlements, no matter how long-standing, were sometimes forced to relocate outside the parks.[96] Conservationist writings in this period show a distrust of villagers and a disdain for politicians, who had been bypassed in the establishment of the protected-areas network.[97] Democratic process had never been an important part of wildlife protection, and it was not uncommon for naturalists to parody democracy by claiming that they "speak for the animals." By the late 1970s and early 1980s little common cause existed between wildlife professionals, who prioritized improving habitats in and around the reserves, and the local people and their political representatives, who prioritized using forest resources to satisfy basic needs.

Then, just as conservationists' labors on behalf of tigers, elephants, rhinos, blackbuck, and crocodiles began to come to fruition, charismatic animals lost their privileged place in global conservationist thinking. A much broader agenda emerged in the 1980s: to preserve *all* threatened species—birds, insects, fish, plants, even mosses and algae—within carefully inventoried ecosystems. This more expansive "ecosystem" or "biodiversity" approach reflected scientific preoccupation with species extinctions.[98] This approach—global in conception, quantitative in its emphasis on inventorying, and above all elite (the initiative was almost wholly in the hands of conservation biologists)—seemed destined to push aside the large-species approach to conservation. After some jockeying, however, both approaches came to be deemed necessary and appropriate: "Ecosystems are made up of individual species and ultimately it is only through monitoring species that we can tell if ecosystems are healthy or not. . . . [Further] only by tackling conservation at the species level can we address the azonal and non-habitat related threats of hunting, poaching, levels of utilisation, competition with domestic animals, indirect impact of other human developments, pollution and other factors affecting the status of wildlife in the [Indo-Malaysian] realm."[99]

In other words, a species-based approach to biodiversity assessment returns attention to "azonal" and "non-habitat related" on-the-ground issues being forced on reserve managers by villagers. The local populations gnawed constantly at the edge of parks and sanctuaries, suborning the forest guards and all too often simply taking what they needed. The urgency of these threats to the enclave system gradually affected scientific and official thinking.[100] Nowhere was the need to understand and accommodate peasant claims more evident than in and around the Project Tiger reserves in India and Nepal.[101]

Project Tiger in India expanded from nine to twenty-three sites between 1973 and 1993. The tiger reserves are widely distributed in different states and ecological zones and are managed by state-level forest officers. The reserves all have the same structure: a "core" area (varying from 100 to 1,800 square miles) is completely closed to human use and reserved for tigers and their prey; the core is enveloped by a "buffer" zone (varying from 150 to 2,300 square miles), which allows for limited, closely regulated human use. Both core and buffer in most cases are surrounded by a "reserved" or state-controlled forest. At least half the tigers in India reside in the lightly guarded forests outside the reserves, where they are at the mercy of villagers with guns and subject to relentless habitat depletion. Nonetheless, a number of reviews have pronounced Project Tiger successful in keeping Indian and Nepali tigers from extinction.[102] Census data show an increase in tigers in India from around 1,800 in 1973 to more than 3,000 in 1993.[103]

Most conservationists know the upward slope of India's tiger population curve since 1973. What is less well known is that in at least two parks—Corbett and Dudhwa, both in the state of Uttar Pradesh—the tiger population grew to such an extent that it exceeded the parks' capacity around 1978.[104] According to B. N. Upreti, the same happened in Royal Chitawan Park in the Nepal Terai. Evidence that the limits had been breached was concrete: tigers attacked park staff and nearby villagers. Ramesh and Rajesh Bedi, who were present at Dudhwa off and on during the difficult years of 1978–81, suggest that the park staff was unprepared for this development. Tigers of an unviable mix of ages and sexes were crowded together in too small a range, where they were frequently provoked by villagers entering the park to collect wood and grass. As noted, half of India's tigers have never been confined to the Project Tiger reserves; in the late 1970s and early 1980s these "outside" tigers were joined by "insiders," and both kinds roamed through the district of Lakshimpur-kheri, near Dudhwa. More than one hundred people died from tiger attacks. Bedi and Bedi comment, somewhat dramatically, that in Lakshimpur-kheri "the district was infested with tigers relishing human flesh. No one seemed to be safe. As soon as one man-eater was killed another tiger would soon replace it. . . . Hardly had the *sansis* [peasants] settled down with relief when a fresh wave of panic and terror spread through them. Sometimes as many as five man-eaters take their toll of human lives. It is like an epidemic."[105]

When villagers demanded that the man-killers be shot, Project Tiger personnel instead tried to relocate offenders into the core area. Thus tigers were being captured, tranquilized, and transported at the same time that obsequies were being performed for human victims. Although the project did eventually resort to shooting tigers, the capture-and-wean approach was never entirely abandoned, and a confused set of claims and counterclaims in different social and institutional settings followed:

> Baits were provided regularly and the man-eaters started taking them. But the experiment [of capturing and releasing in the cores] suffered a set-back. It was possible to control the movements of local people inside the park; but the reserved forest outside the park, where grazing and a limited felling of trees is permitted, continued to attract the villagers. . . . People ignored the warnings of the officials attempting reformation of the killer tigers; and, after every [human] kill, the pressure to eliminate the offenders rose to such heights that it became a problem for conservationists. . . . Local political leaders deprecated the critical situation and fanned the agitation for getting compensation from the government. They pressed for the arrest of officials, whom they accused of being heartless. In one of the public meetings they incited the people to promise that they would tie up the forest guards and put them up as bait for the man-eaters. . . . It was a golden opportunity for the big farmers to demand the lifting of the ban on big-game hunting. They said they were unable to harvest the sugarcane crop because of the active marauders in the fields. The cold hard fact was that they were unable to find a market for the bumper crop that had already flooded the markets.[106]

As in all democracies, local crises were magnified through the press and in the legislature. Whether one advocated shooting them or hoped to confine them to the cores became a test of how one felt about conservation goals—goals that had never been subjected to popular debate in the early 1970s. It did no good to point out that villagers put themselves in harm's way by illegally entering parks to collect wood and thatch, thus provoking tigers; such arguments simply blamed the victims and underlined the fact that human needs for forest products were not being met. Many of the villagers threatened by tigers had been relocated in order to establish the reserves in the first place; understandably, they and their families were unwilling to forgive the man-killers who now emerged to attack them and their domestic animals directly.

The tiger story in this period is full of irony: although tigers were the object of intense conservation, no one had really expected the work to succeed; yet now there was an excess of them. The man-killers had reappeared as a result of careful park management and the application of ecological science.

Similar stories about unexpected animal violence by other wild species would later surface. Between April 1993 and April 1995 news reports about feral wolves appeared in the Indian press: at least five packs carried off eighty children between the ages of three and eleven from sixty-three villages in the Koderma and Latehar divisions of Bihar state. Only twenty of these children were rescued; the rest were killed and eaten.[107] Because the attacks occurred in remote parts of the country, they failed to attract sustained national or international attention. During the summer and fall of 1996, however, journalists reported the deaths of at least two dozen children from wolf attacks in the densely populated districts of Jaunpur, Sultanpur, and Pratapgarh in Uttar Pradesh. "It's the worst wolf menace anywhere in the world in at least 100 years," said Ram Lakhan Singh, the animal conservationist chosen to lead an official effort to hunt down and kill the wolves. Sounding just like Corbett's and Anderson's accounts from sixty and seventy years earlier, Singh noted that "fear is pervasive. Men stay awake all night, keeping vigil with antique rifles and staves. Mothers keep children from the fields, and infants are kept inside all day."[108]

The deaths of children elicited strange vengeance. The police and other officials revealed that twenty to thirty adults had been killed after being accused by co-villagers. "Villagers have turned against strangers, and sometimes against one another, in lynchings that have killed at least 20 people and prompted the authorities to arrest 150 people."[109] Further, "an alert police officer suggested, as have others, that the lynching of humans accused of being 'werewolves' may in fact have been arranged by criminals who manipulate villagers' belief in demons in human form who change their shape at night to become wolves; by this device criminals manage to get others to kill local officials and other enemies."[110] In response to the threat of vigilante chaos as well as the threat to children's lives, Singh organized hundreds of policemen and thousands of villagers, armed with shotguns and bamboo staves (*lathis*), to patrol hundreds of miles of riverbank in areas known to be favored by wolves. Simultaneously, state revenue authorities offered a ten-thousand-rupee bounty for every dead wolf. The hunt and the bounties were only partly successful, however, and wolves continued to plague rural Uttar Pradesh over the next two years.[111] To add even greater uncertainty to the situation, a former cabinet minister for Forests and the Environment, Maneka Gandhi, asserted in a news conference in mid-1996 that wolves were *not* responsible for children's deaths and that talk of wolves covered up dark deeds. She even initiated a lawsuit against Singh and other officials on the grounds that wolves were a protected species that could not legally be hunted.[112] The eventual consensus was that there indeed had been feral wolves in Uttar Pradesh; like other animals deeply disoriented by relentless loss of habitat to agriculture, they had altered their modus operandi to survive. Wolves, unlike tigers, can shift their prey, their habitat, and their breeding patterns in response to human pressures.[113]

The Standard Environmental Narrative, outlined earlier, is an emerging consensus view of the depletion of nature in India and the onset of environmental disasters during the past 125 years. While not a wholly untrue story, the SEN is seriously incomplete, especially in its failure to incorporate the wide variations in animal-human relationships recognized in the past. The decimation of wild animals in the past two generations under the triple pressures of uncontrolled hunting, advancing cultivation, and rural development projects needs also to be acknowledged. By emphasizing only the vegetal (horticultural, arboreal, and silvicultural) trajectory of Indian links to nature, the SEN sidesteps certain violent realities (as well as remarkable stratagems to ward off these realities) that were omnipresent in rural India before 1950. The fact that wild animal threats nowadays take bizarre forms and come mostly from the margins—an occasional cobra emerges in the stock exchange, a random flock of crows jams a jet aircraft's turbines, a befuddled elephant tramples the caretaker of a suburban garden—might be read as proofs that the animal-human struggle is over, that wildness has come to an end, and that the final score is animals zero, humans one hundred. Another reading, however, is that a new kind of wildness is emerging in parks, reserves, and sanctuaries and in the interstices between shrinking forests and expanding arables: tigers deprived of the space they require to breed lash out at any human who wanders within range; wolf packs turn to preying on infants and children when their habitats go under the plow. This new wildness can be seen as the return of the repressed: what was once thought banished now resurrects itself in particularly ugly and unexpected ways. Scientific studies suggest that these new kinds of attacks have two explanations: on the one hand, they may be only dying spikes in an inevitable process of species extinction; on the other hand, they may be viable adaptations that develop only when animals come under extreme stress. These themes, too, deserve a place in the Standard Environmental Narrative because the last word of the narrative has not yet been pronounced.

A revised account of India's environmental history might go beyond its preoccupation with landlords, artisans, and peasants to include a wider cast, some of whom were remarkably skilled: baghmaras and tribal hunters, mahouts and khedda-drivers, snake catchers, honey gatherers, fisherfolk, game beaters, and many others who for centuries managed the traffic in animals for the benefit of rural communities. Their detailed knowledge of the habits of insects, birds, fish, and forest animals; their expertise in methods to transform wood, stone, fiber, and metal into weapons and snares; their sustaining myths, songs, and cosmologies—all these matters, too, need to find a place at the environmental historian's feast.

Finally, as Richard Grove is wont to remind us, the role of colonial rulers and of European scientists since the eighteenth century in articulating conservation anxieties in India (and elsewhere), and in introducing measures like reserved forests, closed seasons, hunting licenses, protected animal species, game sanctuaries, and national parks, cannot be ignored.[114] The overall environmental record of colonial rule *is* dismal: colonialism's deliberate rapacity and inadvertent blundering were perhaps equally destructive. But the great *shaitan* has to be given his due; there were more and less progressive elements in the colonial bureaucracy, and some saw more clearly than others that laying waste to natural India in the name of progress was a terrible outcome of policy. Grove's striking evidence for "green imperialism," a term once thought an oxymoron, has entered Indian environmental scholarship to stay. Yet Grove, too, because of his preoccupation with the roots of global environmentalism in the long colonial encounter with tropical forests, belongs to the vegetal camp. If we grant there was a doctrine of "green imperialism" under colonial rule, we should recognize that there was also "red imperialism," based on an emerging knowledge of the variety and vulnerability of wild animals.

Acknowledgments

This chapter was originally drafted for a conference titled "Common Property, Collective Action and Ecology in South Asia," sponsored by the Joint Committee on South Asia of the Social Science Research Council (New York) and the Ford Foundation (New Delhi) and held at the Center for Ecological Studies, Indian Institute of Science, Bangalore, India, on 19–21 August 1991. The paper was subsequently presented at seminars of the Program in Agrarian Studies at Yale University in September 1991 and of the Department of History at the University of Edinburgh in November 1992. I gratefully acknowledge the constructive suggestions of scholars in these meetings. Ronald Herring gave the earliest version a very close reading, for which I'm grateful, and I extend thanks also to John Gardner, Michael Dove, T. N. Srinivasan, Deepak Bajracharyya, Keshav Gautam, S. R. Jnawali, Ravi Rajan, Raman Mehta, and Wolfgang Werner, each of whom was helpful in suggesting materials or refining the argument. All errors are my own. I acknowledge the facilities of the Center for International and Comparative Studies of the University of Iowa, the International Center for Integrated Mountain Development in Kathmandu, the Worldwide Fund for Nature in New Delhi, and the South Asia Institute of Heidelberg University, Germany, for their support while I was completing the chapter.

Notes

1. Since this chapter was first written in 1991, a considerable amount of South Asian environmental history work on animals has appeared in print, some of which I engage in "Pathogens and Pugmarks on the Edge of the Emergency" and "The Fractured Forest."
2. With apologies to Bina Agarwal (*Cold Hearths and Barren Slopes*), Madhav Gadgil ("Social Restraints"), Ramachandra Guha ("An Early Environmental Debate"), Madhav Gadgil and Romila Thapar ("Human Ecology in India"), Vandana Shiva (*Staying Alive*), Helena Norberg-Hodge (*Ancient Futures*), Madhav Gadgil and Ramachandra Guha (*This Fissured Land*) and others.
3. The Chipko movement of the 1970s was an anti-commercial logging movement in Uttarakhand, North India; organized nonviolent resistance took the form of women and

children in the area hugging trees, which they regarded as living beings. The movement has been instrumental in the struggle against social and ecological disintegration of Indian hill society and in foregrounding the redefinition of gender roles. See Guha, *The Unquiet Woods,* and S. Sinha et al., "The 'New Traditionalist' Discourse." The Narmada Bachao Andolan (NBA, or Save the Narmada Movement) has been struggling since the 1980s to prevent the completion of a network of more than three thousand large and small dams of the Narmada River and its tributaries as it flows west through the states of Madhya Pradesh, Maharashtra, and Gujarat. The NBA's determined civil disobedience efforts, forging unusual combinations between peasant and tribal adherents in three states, have periodically succeeded in halting dam construction by holding up environmental, human rights, and legal issues before the conscience of India and foreign countries. See Drèze et al., *The Dam and the Nation.*

4. The Puranas are medieval Indian texts, generally in Sanskrit, that recount the past in a mythological idiom; hence the adjective *Puranic* implies a richly fanciful but not wholly untrue narrative.
5. Bhopal, the capital of Madhya Pradesh, was the site of an explosion of a Union Carbide fertilizer-component factory inside the city limits in 1984. More than 4,000 people were killed outright, and another 200,000–400,000 suffered serious injuries. See Hazarika, *Bhopal, Lessons of Tragedy.*
6. I use the terms *forest* and *jungle* interchangeably but am aware of the important conceptual and historical distinctions. See Dove, "Dialectical History."
7. Shiva, *Staying Alive,* 56. The next quotation is also from this source. Shakambari (*sakambari*) is more properly translated, according to T. N. Srinivasan of Yale University, as "one who has vegetables and herbs as her dress." I am grateful to Professor Srinivasan for the clarification of this and other Sanskrit terms.
8. Jagadishswarananda, *Devi Mahatmyam,* 147–48.
9. For Sona Ray, see De, *The Sundarbans,* 46–49; for Manasa, see Maity, *Historical Studies in the Cult of the Goddess Manasa;* Jash, "The Cult of Manasa in Bengal"; and Rao, *The Legend of Manasa Devi.* Information on Waghoba and Jambhawati is from a personal communication with Hugh van Skyhawk, South Asian Institute, Heidelberg University, March 1992.
10. Agarwal, *Cold Hearths and Barren Slopes,* 108, 109.
11. Sankhala and Jackson, "People, Trees and Antelopes."
12. Guha, *The Unquiet Woods,* 29–31.
13. Rao, "History"; Bhagavat, "Bear in Indian Culture"; Sankhala, *Wild Beauty;* Sankhala, *Tiger!*
14. Zimmermann, *The Jungle,* 103–11 and *passim;* Dove, "Dialectical History."
15. Zimmermann, *The Jungle,* 174; an alternative archaic classification refers to *prasaha*—"carnivorous land quadrupeds and birds such as fall on their prey with force." See Rao, "History," 255.
16. Schaller, *The Deer and Tiger,* 5. Professional hunters whose experience spans both sides of Independence (1947) are unanimous about the disastrous effects of overhunting since 1950. See, for example, Anderson, *The Call of the Man-Eater,* 11; Burton, "A History of Shikar," 862; Corbett, *Jungle Lore,* 54. An extensive account drawn from Seshadri appears later in this chapter.

17. Schaller, *The Deer and Tiger,* 3, 6–7.
18. Hunter, *Annals,* 67–68.
19. See, for example, Williamson, *Oriental Field Sports;* Johnson, *Sketches of Indian Field Sports;* Tennent, *Ceylon;* Sanderson, "The Asiatic Elephant"; and the useful bibliographies provided on tigers by Sankhala (*Tiger!*) and on elephants by Sukumar (*The Asian Elephant*).
20. For these and subsequent references see Hunter, *Statistical Account,* s.v. *naturae ferae* in relevant volumes.
21. For Saran: "Formerly, both leopards and tigers were very common in the District, but they have now completely disappeared. Wolves and pigs are still found. . . . Wolves carry off a good many children, and sometimes attack sheep." For Dhaka: "Tigers and leopards infest the jungles of the northern tract, but their numbers have decreased of late years owing to the clearing of jungle and the spread of cultivation."
22. Hunter, *Statistical Account,* "Faridpur."
23. Ibid., "Singhbhum."
24. Ibid, *passim* (the sum of all snakebites listed in *Naturae ferae* sections).
25. *Gazetteers of the Bombay Presidency.* Not all districts in the presidency provided data on human mortality from snakebite and wild animal attacks.
26. See, for example, George Orwell, "Shooting an Elephant," in *Collected Essays.*
27. *Gazetteers of the Bombay Presidency, Kaira and Panch Mahals,* 210–11. Other sources in this reference work reported similar observations: "Year by year the tiger is becoming scarcer" (Surat 1877), "The spread of tillage and the efforts of European sportsmen . . . have so reduced [tiger] numbers that they are now only occasionally met with" (Kaira 1879), "Of late years tigers and lions have almost entirely disappeared" (Cutch 1880), "Tigers and hill panthers, though yearly becoming fewer, are still found in considerable numbers" (Rewa Kantha 1880), "Of late years the extensive felling of forest has greatly reduced [tigers'] number" (Janjira 1883), "The spread of tillage and the increase of population constantly reduce the number of wild animals; the tiger, panther, leopard and the bear are found only in the Sahyadris, and even there in small numbers" (Poona 1885), and "The increase of population and the spread of tillage have reduced their numbers, but tigers and panthers still find shelter in western Kolhapur" (Kolhapur 1886).
28. Annual mortality due to tiger attacks in the Sunderbans in Bangladesh in the mid-1980s was estimated to be around 100 (Jackson, "Man-Eaters!"; Ward, "India's Intensifying Dilemma"). For Victorian representation of the threat emanating from the Sunderbans, see Greenough, "Hunter's Drowned Land."
29. Herring, *Agriculture and Human Values.*
30. For more detail regarding Sunderbans fakirs, see Greenough, "Hunter's Drowned Land."
31. Shrestha, *Wildlife of Nepal,* 184.
32. It is this "social" aspect of interspecies relationships that Rudyard Kipling discovered in Indian folktales and incorporated so effectively into his *Jungle Book.*
33. Hunter, *Statistical Account,* "Malda."
34. Gadgil, "Social Restraints"; Sankhala and Jackson, "People, Trees and Antelopes."
35. Sankhala and Jackson, "People, Trees and Antelopes."

36. Gadgil, "Indian Heritage."
37. Hunter, *Statistical Account,* "Rangpur."
38. MacKenzie, *Empire of Nature,* 184, citing G. P. Sanderson.
39. MacKinnon and MacKinnon, *Review of the Protected Areas System,* 235.
40. Hunter, *Statistical Account,* "Singhbhum," 168.
41. MacKenzie, *Empire of Nature,* 167–99, marks a decent start in resurrecting the human-animal links, including hunting. MacKenzie's focus, however, is principally on colonial methods and motives, not indigenous hunting. See also Rangarajan, *Fencing the Forest,* 138–97, on hunting in the Central Provinces.
42. Hunter, *Statistical Account,* "Dhaka."
43. Shrestha, *Wildlife of Nepal,* 183.
44. Williamson, *Oriental Field Sports;* Johnson, *Sketches of Indian Field Sports;* and Hunter, *Statistical Account* (Purnea and Lohardaga districts) confirm the existence of this machine.
45. Hunter, *Statistical Account,* "Chittagong."
46. Sanderson, "The Asiatic Elephant" and *Thirteen Years;* Sukumar, *Asian Elephant.*
47. MacKenzie, *Empire of Nature,* 185.
48. Johnson, *Sketches of Indian Field Sports,* 1–2.
49. Hunter, *Statistical Account,* s.v. relevant vols.
50. Ibid., *passim.*
51. Ibid., "Purnea," 236–37.
52. Ibid., *passim.*
53. Ibid., "Monghyr," 45.
54. Lopez, *Of Wolves and Men.*
55. Determining the exact monetary amount sufficient to attract hunters' interest in killing a proscribed animal was the subject of an official discourse conducted in memoranda and printed articles. In 1890 in Central Provinces, for example, "in the case of man-eating panthers and wolves, and of man-eating tigers, for the destruction of which the ordinary reward of Rs. 100 is found insufficient, special rewards are sanctioned, as occasion requires, and these have sometimes been as much as Rs. 500. The general opinion is that in most respects this scale is sufficiently liberal, and the Chief Commissioner would prefer not to alter it at present or until he has had further experience of its operation" ("Extermination of Wild Beasts," 410).
56. The 1900 estimate is from Indian Board for Wildlife, *Task Force.* For colonial era accounts of hunting man-eaters, see Corbett, *Man-Eaters of India;* Burton, *A Book of Man-Eaters;* Turner, *Man-Eaters and Memories;* and Anderson, *Man-Eaters and Jungle Killers.* For similar post-Independence accounts see Shakoor Khan, *Wild Life and Hunting;* and Singh, *Hints on Tiger Shooting.* A curious memoir by Dean Witter, founder of a powerful Wall Street brokerage firm, links itself unselfconsciously to the exhausted tradition of royal and colonial tiger hunting (Witter, *Shikar*); I am grateful to Terry Burke for this reference.
57. Corbett, *Jungle Lore;* Corbett, *Man-Eaters of India;* Anderson, *Nine Man-Eaters;* Anderson, *Man-Eaters and Jungle Killers;* Anderson, *The Call of the Man-Eater;* Schaller, *The Deer and Tiger;* Sankhala, *Tiger!;* Jackson, "Man Eaters!"; Ward, "India's Intensifying Dilemma"; Ward, "The People and the Tiger."

58. The behavior of tigers was relentlessly anthropomorphized. Note, for example, the agreement between Kenneth Anderson, based in South India, and Jim Corbett, based in far North India, about this matter: "The Man-Eating Tiger is an abnormality, for under normal circumstances, the King of the Indian jungles is a gentleman and of noble nature" (Anderson, *Nine Man-Eaters,* 7) and "There is, however, one point on which I am convinced that all sportsmen—no matter whether their point of view has been platform or a tree, the back of an elephant, or their own feet—will agree with me, and that is, that the tiger is a large-hearted gentleman with boundless courage" (Corbett, *Man-Eaters of India,* xii).
59. Corbett, *Man-Eaters of India,* "Kumaon," 45–47, and "Rudraprayag," 34.
60. Ibid., "Rudraprayag," 10–12.
61. Ibid., "Champawat," 11.
62. Ibid., 28–31.
63. Anderson, *Nine Man-Eaters,* 8–9.
64. Corbett, *Man-Eaters of India,* "Rudraprayag," 19–21, 22–23.
65. Contemporary reports of popular executions of men who are presumed to be baby-snatching werewolves are discussed below.
66. Anderson, *The Call of the Man-Eater,* 56–58.
67. Guha, "An Early Environmental Debate" and *The Unquiet Woods;* Singh, *Common Property;* Rangarajan, *Fencing the Forest.*
68. Guha ("An Early Environmental Debate") describes the policy debates surrounding the act's enactment.
69. For an example of provincial hunting regulations derived from the 1878 act, see "Extermination of Wild Beasts," 258–60. For a specimen of a hunting permit from Madras in 1882, see "The Madras Game Law," 114–16.
70. Kangle, *The Kautiliya Arthasastra,* 129–30, 157–58; MacKenzie, *Empire of Nature,* 174–75.
71. Baden-Powell, *Forest Law,* 71, 338, 364.
72. Among other authorities, Stebbings, *The Forests of India,* was written long ago but is very informative; see also Tucker, "The Depletion of India's Forests," and Guha, "An Early Environmental Debate."
73. MacKenzie, *Empire of Nature.* Richard Tucker has noted the connections between forest law and privileged sport shooting in closed forests. See Tucker, " 'Have You Shot an Indian Tiger?' The History of Wildlife Preservation in India," paper presented at South Asian Studies Conference, Madison, Wis., 20 October 1995.
74. The emergence of this hunter's code is particularly evident in James E. Tennent's strictures on the cruel and pointless slaughter of elephants in Ceylon and Africa. Tennent, a follower of phrenology, believed that Europeans had larger cranial "organs of destructiveness" than Sinhalas (Tennent, *Ceylon,* vol. 2, 812–19).
75. Ibid., 770; MacKenzie, *Empire of Nature,* 182–86).
76. See, e.g., Storey, *Hunting and Shooting in Ceylon,* xv–xix.
77. A remarkable account of such measures to control rinderpest in Edwardian Ceylon is given by Leonard Woolf in his autobiography (*Growing*).
78. MacKenzie, *Empire of Nature,* 185.
79. Ibid., 201.

80. Kothari et al., *Management of National Parks,* 106–7.
81. Hailey Park originally comprised 158 square kilometers (about 100 square miles) but was expanded to more than 520 (about 300 square miles) after Independence and renamed Ramganga Park and then Corbett National Park. Because of its priority, its proximity to New Delhi, and its superb mix of wild elephants, tigers, *gavial,* and mugger crocodiles, Corbett Park became the flagship of India's park system. (Uttar Pradesh, *Corbett National Park,* 1)
82. IUCN/UNEP/UNESCO, *1980 United Nations List.*
83. Gee, "Management of India's Wild Life"; Singh, "National Parks." The organization of national parks and reserves in Pakistan and Nepal lagged behind that of India and followed decolonization. See United Nations World Conservation Monitoring Centre website (www.unep-wcmc.org) and Israel and Sinclair, *Indian Wildlife.*
84. Seshadri, *India's Wildlife,* 14–15, 19–21.
85. MacKinnon and MacKinnon, *Review of the Protected Areas System,* 52.
86. This analysis is concurred in by Putnam ("India Struggles").
87. Indian Board for Wild Life, *Task Force;* Kothari et al., *Management of National Parks,* 108.
88. Kothari et al., *Management of National Parks,* 112–13.
89. Cowan and Holloway 1978: 11.
90. See, for example, Schaller, *The Deer and Tiger;* Sunquist, "The Social Organization of Tigers," Sunquist and Sunquist, *Tiger Moon;* Karanth, "Ecology and Management"; Seidensticker et al., *Riding the Tiger.*
91. The transition from big game hunting to conservation is clearly registered in the professional reorientation of hunters and naturalists—professional and amateur—who exchanged their rifles for cameras in the period 1950–60. Examples include Jim Corbett, D. Chaturvedi, E. P. Gee, P. D. Stracey, Z. Futehally, and M. Krishnan.
92. Kothari et al., *Management of National Parks;* World Conservation Monitoring Centre online country reports (www.unep-wcmc.org). In 1992–93 India alone had 62 national parks and 316 sanctuaries; the corresponding figures for Pakistan were 10 parks and 45 sanctuaries; for Sri Lanka, 21 national parks and 36 reserves, wildernesses, and sanctuaries; for Nepal, 8 national parks and 4 reserves; and for Bangladesh, 2 national parks and 6 sanctuaries (WCMC, online country reports).
93. Article 51-A, sec. g, of the Indian constitution was enacted as part of the 42nd amendment. This was an emergency measure, and thus the depth of political support for it even at the time is questionable. The amendment entered into effect in January 1977, just before Indira Gandhi was voted from office. The "fundamental duties" in question are admitted to be unenforceable except by public opinion (Venkataramiah, *Citizenship,* 52–62). In Pakistan the Pakistan Environmental Protection Ordinance (1983) outlined a comprehensive administrative regime for conservation and assessments of industrial impacts on the environment (WCMC, online country reports). This ordinance has not had much effect, according to Shah ("Environment and the Role of the Judiciary"). I am grateful to Dieter Conrad for these references.
94. By 1999 India's conservation system had increased to 84 National Parks and 447 Wildlife Sanctuaries covering 150,000 square miles or 5 percent of the country's surface area according to the Ministry of Environment and Forests website (s.v. "protected areas": http://envfor.delhi.nic.in/search/search.html).

95. Laurie, *Indian Rhinoceros Study;* Singh, *Tiger Haven;* Singh, "The Status of the Swamp Deer"; Ranjitsinh, "The Manipur Brow-Antlered Deer"; Panwar, "Decline and Restoration Success"; Indian Board for Wild Life, *Task Force.*
96. For an example in Nepal, see Sunquist and Sunquist, *Tiger Moon,* 44.
97. For example, "The compromise which the politician seeks between an increasing and voting public and a decreasing and inarticulate section of wildlife, in the shape of human intrusion into animal habitats must be resisted . . . . Unless conservationists can successfully combat this dénouement of the democratic processes of government, all hope is ended" (Singh, "National Parks," 134).
98. IUCN, *The Biosphere Reserve;* Tolba and El-Kholy, "Loss of Biological Diversity." "Biodiversity" is a relatively recent scientific term for the entire community of species associated with a particular setting; it draws attention to zones and sites that have a high species diversity. Preservation of biodiversity, an initiative of the 1980s, differs from the "Man in the Biosphere" initiative of the 1970s, which was advocated by UNESCO and which emphasized a balance between conservation and human use of resources in demarcated, legally protected, and globally networked ecological zones called "biosphere reserves."
99. MacKinnon and MacKinnon, *Review of the Protected Areas System,* xi.
100. Indian Board for Wild Life, "Eliciting Public Support for Wildlife Conservation"; McNeely and Pitt, *Culture and Conservation;* Kothari et al., *Management of National Parks;* Berwick and Saharia, *Wildlife Research and Management.*
101. Indian Board for Wildlife, *Task Force;* Project Tiger, *International Symposium;* Sunquist and Sunquist, *Tiger Moon.*
102. Singh, "National Parks"; Kothari et al., *Management of National Parks;* Project Tiger, *International Symposium.*
103. Project Tiger, *International Symposium.* A more recent estimate suggests that in 1993–94 there were between 2,500 and 3,750 tigers in India (Seidensticker et al., *Riding the Tiger,* xvii, table 0.1). About this estimate, however, the Ministry of Environment and Forests states, "The figures indicate that while the overall tiger population has not fallen in tiger reserves, the population has indeed declined outside reserve areas between 1989–93." In addition to tigers in India, there are an estimated 600–800 in the rest of the subcontinent, which is home to 60–65 percent of the world's total tigers. Nepal's tigers are found in a network of reserves in the Terai region similar to that of Corbett Park in India. Bangladesh's tigers occupy a very different terrain—the mangrove forests of the 3,600-square-mile Sunderbans region—and their habitat has never been as seriously threatened by human use or required so elaborate a conservation effort as has been mounted in India.
104. Bedi and Bedi, *Indian Wildlife,* 64–79.
105. Ibid., 78.
106. Ibid.
107. Rajpurohit, "Child Lifting."
108. Burns, "Attacks by Wolves"; also see Goldenberg, "Wolf Packs Take Indian Children"; Balaji, "Battle of Wits." Contrary to Singh's suggestion, murderous wolf attacks on children were not rare: 624 children had been killed a century earlier in 1878 (Burns, "Attacks by Wolves"), and during 1974–75 wolves and hyenas carried off numerous infants in Gulbarga district and parts of Uttar Pradesh (Krishnan, "Crying Wolf?").

109. Burns, "Attacks by Wolves."
110. Ibid.; see also "Deaths Exploited to Settle Scores." Allegations of murder disguised as animal attacks are a staple of reports on animal violence in India.
111. Majumdar, "Wolves Target Kids."
112. "Animal Rights Activist to Sue State."
113. Some predator populations cannot be exterminated. The 400-person Animal Damage Control Program of the U.S. Department of Agriculture, with an annual budget of $38 million, killed 86,502 coyotes in the United States in 1989 using guns, gas, and helicopters. Yet biologists find that coyotes have adapted their breeding habits to the fact of relentless predation by humans. See Schneider, "Mediating the Federal War on Wildlife."
114. Grove, *Green Imperialism.*

References

Agarwal, Bina. 1986. *Cold Hearths and Barren Slopes: The Woodfuel Crisis in the Third World.* Riverdale, Md.: Riverdale.

Anderson, Kenneth. 1955. *Nine Man-Eaters and One Rogue.* New York: E. P. Dutton.

———. 1957. *Man-Eaters and Jungle Killers.* New York: T. Nelson.

———. 1961a. "The Man-Eating Tiger Problem." *Oryx* (London Fauna Preservation Society) 11, no. 4: 231.

———. 1961b. *The Call of the Man-Eater.* Philadelphia: Chilton Books.

"Animal Rights Activist to Sue State for Killing Wolves." 1996. *Deutsche-Presse Agentur* (2 July).

Baden-Powell, B. H. 1892. "Forest Settlements in India." *Indian Forester* 18: 132–47.

———. 1893. *Forest Law: A Course of Lectures on the Principles of Civil and Criminal Law and on the Law of the Forest.* London: Bradbury, Agnew.

Balaji, M. V. 1996. "Battle of Wits between Man and Beast in Uttar Pradesh." *Deutsche Presse-Argentur* (10 August).

Bedi, Ramesh, and Rajesh Bedi. 1984. *Indian Wildlife.* New Delhi: Brijbashi.

Beinart, William. 1991. "Empire, Hunting and Ecological Change in Southern and Central Africa: Review Article." *Past and Present* 128: 162–86.

Bennett, S. 1984. "Shikar and the Raj." *South Asia,* new series, 7.

Berwick, Stephen H., and V. B. Saharia, eds. 1995. *Wildlife Research and Management: Asian and American Approaches.* New Delhi: Oxford University Press.

Bhagavat, Durga. 1966–67. "Bear in Indian Culture." *Journal of the Bombay Branch of the Royal Asiatic Society* 31–93.

Bhattacharjee, Tarun Kumar. 1987. *Alluring Frontiers.* Guwahati, Assam, India: Omsons Publications.

Blandford, W. T. 1888. *The Fauna of British India.* Vol. 1, *Mammalia.* London: Taylor and Francis.

Brander, A. A. 1923. *Wild Animals in Central India.* London: Dunbar.

Brightman, Robert A. 1990. "Conservation and Resource Depletion." In *The Question of the Commons: The Culture and Ecology of Communal Resources,* ed. B. J. McCay and J. M. Acheson. Tucson: University of Arizona Press.

Bryant, H. B., and D. Hadfield. 1895. "Elephant-Catching Operations on the Anaimalai [*sic*] Hills." *Indian Forester* 21: 118–20, 200–201, 275–76.

Burns, John F. 1996. "In India Attacks by Wolves Spark Old Fears and Hatreds." *New York Times* (1 September).

Burton, R. G. 1931. *A Book of Man-Eaters.* London: Hutchinson.

Burton, R. W. 1952. "A History of Shikar in India." *Journal of the Bombay Natural History Society* 50, no. 4: 845–69.

Causey, Ann S. 1989. "On the Morality of Hunting." *Environmental Ethics* 11: 327–43.

Corbett, Jim. 1953. *Jungle Lore.* London: Oxford University Press.

———. *Man-Eaters of India.* 1957. Including "Man-Eaters of Kumaon," "The Man-Eating Leopard of Rudraprayog," "The Temple Tiger and More Man-Eaters of Kumaon." 1944. Reprint, New York: Oxford University Press.

Daniel, J. C., ed. 1983. *A Century of Natural History.* Bombay: Bombay Natural History Society.

De, Rathindranath. 1990. *The Sundarbans.* Calcutta: Oxford University Press.

"Deaths Exploited to Settle Scores in Indian State." 1996. *Deutsche Presse-Argentur* (24 July).

Dewar, D. 1923. *Beasts of an Indian Village.* London: Oxford University Press.

Dove, Michael R. 1992. "The Dialectical History of 'Jungle' in Pakistan: An Examination of the Relationship between Nature and Culture." *Journal of Anthropological Research* 48: 231–53.

Drèze, Jean, Meera Sampson, and Satyajit Singh, eds. 1997. *The Dam and the Nation: Displacement and Resettlement in the Narmada Valley.* Delhi: Oxford University Press.

Elliott, J. G. 1973. *Field Sports in India, 1800–1947.* London: Gentry Books.

"Extermination of Wild Beasts in the Central Provinces." *Indian Forester* 17 (1891): 409–13; 18: 255–60.

Feeny, David, Fikret Berkes, Bonnie J. McCay, and James M. Acheson. 1990. "The Tragedy of the Commons Twenty-Two Years Later." *Human Ecology* 18, no. 1: 1–19.

Gadgil, Madhav. 1985. "Social Restraints on Resource Utilization: The Indian Experience." In *Culture and Conservation: The Human Dimension in Environmental Planning,* ed. J. A. McNeely and D. Pitt. London: Croom Helm.

———. 1989. "The Indian Heritage of a Conservation Ethic." In *Conservation of the Indian Heritage,* ed. B. Allchin, E. R. Allchin, and B. K. Thapar. New Delhi: Cosmo Publications.

Gadgil, Madhav, and Ramachandra Guha. 1992. *This Fissured Land: An Ecological History of India.* Delhi: Oxford University Press.

Gadgil, Madhav, and Romila Thapar. 1990. "Human Ecology in India: Some Historical Perspectives." *Interdisciplinary Science Reviews* 15, no. 3: 209–23.

Gamekeeper (pseud.). 1887. "Destruction of Game in Government Reserves during the Rains." *Indian Forester* 13 (1887).

Gazetteers of the Bombay Presidency. 1877–1904. 27 vols. Bombay.

Gee, E. P. 1962. "The Management of India's Wild Life Sanctuaries and National Parks" (pt. 4). *Journal of the Bombay Natural History Society* 59, no. 2: 453–85.

———. 1964. *The Wild Life of India.* London: Collins.

Gilbert, Reginald. 1889. "Notes on Man-Eating Tigers." *Journal of the Bombay Natural History Society* 4, no. 3: 195–206.

Goldenberg, S. 1996. "Wolf Packs Take Indian Children." *Observer* (8 September).

Greenough, Paul. 1998. "Hunter's Drowned Land: A Science Fantasy of the Bengal Sunderbans." In *Nature and the Orient: Essays on the Environmental History of South and Southeast Asia,* ed. Richard Grove, Vinita Damodaran, and Satpal Sangwan. Delhi: Oxford University Press.

———. In press. "Pathogens and Pugmarks on the Edge of the Emergency: Smallpox Eradication and Tiger Conservation in South Asia, 1972–79." In *Imagination and Distress in Southern Environmental Projects,* ed. Paul Greenough and Anna L. Tsing. Durham, N.C.: Duke University Press.

———. In press. "Ironies of the Fractured Forest: Chaos in South Asia's Tiger Reserves." In *In Search of the Rainforest,* ed. Candace Slater. Berkeley: University of California Press.

Grove, Richard. 1995. *Green Imperialism: Colonial Expansion, Tropical Island Edens and the Origins of Environmentalism, 1600–1860.* Cambridge: Cambridge University Press.

Guha, Ramachandra. 1990a. "An Early Environmental Debate: The Making of the 1878 Forest Act." *Indian Economic and Social History Review* 27, no. 1: 65–84.

———. 1990b. *The Unquiet Woods: Ecological Change and Peasant Resistance in the Himalaya.* Berkeley: University of California Press.

Hazarika, Sanjoy. *Bhopal: The Lessons of a Tragedy.* New Delhi: Penguin, 1987.

Herring, Ronald, ed. 1990. *Agriculture and Human Values* 7, no. 2 (Spring). Special issue on Bangladesh and the Commons.

Hunter, William W. 1897. *Annals of Rural Bengal.* 7th ed. London: Smith, Elder.

———. 1976. *Statistical Account of Bengal.* 1875–77. 20 vols. Reprint, Delhi: D. K. Publishing House.

India. 1987. *Eco-Degradation at Kuttanad.* New Delhi: Government of India, Ministry of Environment and Forests.

Indian Board for Wildlife. 1972. "Task Force, Project Tiger. A Planning Proposal for Preservation of Tiger (*Panthera tigris tigris* Linn.) in India." New Delhi: Ministry of Agriculture.

———. 1983. "Eliciting Public Support for Wildlife Conservation." Report of the Task Force, Indian Board for Wildlife. New Delhi: Department for Environment.

Inveriarty, J. D. 1888. "Unscientific Notes on the Tiger." *Journal of the Bombay Natural History Society* 3, no. 3: 143–54.

Israel, Samuel, and Toby Sinclair, eds. 1988. *Indian Wildlife.* An APA Insight Guide. Singapore: APA Publications.

IUCN (International Union for the Conservation of Nature). 1978. *Threatened Deer.* Proceedings of a Working Meeting of the Deer Specialist Group . . . 26 September–1 October 1977. Morges, Switzerland: International Union for the Conservation of Nature and Natural Resources.

———. 1979. *The Biosphere Reserve and Its Relationship to Other Protected Areas.* Gland, Switzerland: IUCN.

IUCN/CNPPA (International Union for the Conservation of Nature and Commission on Natural Parks and Protected Reserves). 1978. *Categories, Objectives and Criteria for Protected Areas.* Gland, Switzerland: IUCN.

IUCN/UNEP/UNESCO (International Union for the Conservation of Nature, United Nations Environmental Programme, and United Nations Economic, Social, and Cultural Organization). 1980. *1980 United Nations List of National Parks and Equivalent Reserves.* Gland, Switzerland: IUCN.

Jackson, Peter. 1985. "Man-Eaters!" *International Wildlife* 15 (November): 4–11.

Jash, Pranabananda. 1986. "The Cult of Manasa in Bengal." In *47th Indian History Congress, Proceedings.* Delhi.

Jagadishwarananda, Swami. 1982. *Devi Mahatmyam* [Glory of the Divine Mother]. *Markendeyapurana* (English and Sanskrit). Mylapore, Madras, India: Sri Ramakrishna Math.

Johnson, Daniel. 1827. *Sketches of Indian Field Sports with Observations on the Animals [and] also an Account of Some of the Customs of the Inhabitants with a Description of the Art of Catching Serpents as Practised by the Conjoors, and Their Method of Curing Themselves When Bitten, with Remarks on Hydrophobia and Rabid Animals.* London: privately published.

Kangle, R. P. 1969. *The Kautiliya Arthasastra,* pt. 2. 2nd ed., ed. and trans. R. P. Kangle. University of Bombay Studies; Sanskrit, Prakrit and Pali, No. 2. Bombay: University of Bombay.

Karanth, K. U. 1991. "Ecology and Management of the Tiger in Tropical Asia." In *Wildlife Conservation: Present Trends and Perspectives,* ed. N. Maruyama et al. Tokyo: Japan Wildlife Research Centre.

Khan, Tahawar Ali. 1961. *Man-Eaters of the Sunderbans.* Lahore, Pakistan: International Publishers.

Kothari, A., P. Pande, S. Singh, and D. Variava. 1989. *Management of National Parks and Sanctuaries in India: A Status Report.* New Delhi: Indian Institute of Public Administration.

Krishnan, M. 1975. "Crying Wolf?" *Times of India* (28 September).

Laurie, Andrew. 1973–74. "Indian Rhinoceros Study: Ecology and Behavior of the [One-Horned] Indian Rhinoceros." Progress Reports 1–4. Kathmandu, Nepal.

Lopez, Barry H. 1978. *Of Wolves and Men.* New York: Macmillan.

Lydekker, R. 1907. *Game Animals of India, Burma, Malaya and Tibet.* London: Rowland and Ward.

MacKenzie, John M. 1988. *The Empire of Nature: Hunting, Conservation and British Imperialism.* Manchester: Manchester University Press, 1988.

MacKinnon, John, and Kathy MacKinnon. 1986. *Review of the Protected Areas System in the Indo-Malayan Realm.* Gland, Switzerland: International Union for Conservation of Nature and United Nations Environmental Programme.

"The Madras Game Law." 1889. *Indian Forester* 15: 113–16.

Maity, Pradyot Kumar. 1966. *Historical Studies in the Cult of the Goddess Manasa: A Socio-Cultural Study.* With a foreword by A. L. Basham. Calcutta: Punthi Pustak.

Majumdar, Jaideep. 1998. "Wolves Target Kids in Indian State." Associated Press. 14 August.

McCay, B. J., and J. M. Acheson, eds. 1990. *The Question of the Commons: The Culture and Ecology of Communal Resources.* Tucson: University of Arizona Press.

McNeely, J. A., and D. Pitt, eds. 1985. *Culture and Conservation: The Human Dimension in Environmental Planning.* London: Croom Helm.

Milton, J. P., and G. A. Binney. 1980. *Ecological Planning in the Nepalese Terai: A Report on Resolving Resource Conflicts between Wildlife Conservation and Agricultural Land Use in Padampur Panchayat.* Washington, D.C.: Threshold International Center for Environmental Renewal.

Mishra, Hemanta R. 1982. "Balancing Human Needs and Conservation in Nepal's Royal Chitwan Park." *Ambio* 2, no. 5: 246–51.

———. n.d. *Nature Conservation in Nepal: An Introduction to the National Parks and Wildlife Conservation Programme of His Majesty's Government.* Kathmandu: National Parks and Wildlife Office.

Nelson, J. G., R. D. Needham, and D. L. Mann, eds. 1978. *International Experience with National Parks and Related Reserves.* Department of Geography Publication Series No. 12. Waterloo, Canada: University of Waterloo.

Nilakantha (of Rajamangalam). 1985. *The Elephant Lore of the Hindus (Matangalila).* Trans. Franklin Edgerton. 1931, New Haven: Yale University Press; reprint, Delhi: Motilal Banarsidass.

Norberg-Hodge, Helena. 1991. *Ancient Futures: Learning from Ladakh.* San Francisco: Serra Books.

Orwell, George. 1961. *Collected Essays.* London: Secker and Warburg.

Panwar, H. 1978. "Decline and Restoration Success of the Central India Barasingha (*Cervus duvauceli branderi*)." In IUCN 1978.

Prater, S. 1980. *The Book of Indian Animals.* Bombay: Bombay Natural History Society.

Project Tiger. 1993. *International Symposium on the Tiger, New Delhi, India, February 22–24, 1993: Proceedings and Resolutions.* New Delhi: Ministry of Environment and Forests, Government of India.

Putnam, John J. 1976. "India Struggles to Save Her Wildlife." *National Geographic* 150, no. 3 (September): 299–342.

Rajpurohit, Kishan Singh. 1999. "Child Lifting: Wolves in Hazaribagh, India." *Ambio* 28, no. 2 (March): 162–66.

Rameshwar Rao, Shanta. 1977. *The Legend of Manasa Devi.* Bombay: Orient Longman.

Rangarajan, Mahesh. 1996. *Fencing the Forest: Conservation and Ecological Change in India's Central Provinces, 1860–1914.* Delhi: Oxford University Press.

Ranjitsinh, M. K. 1978. "The Manipur Brow-Antlered Deer (*Cervus eldi eldi*) —A Case History." In IUCN 1978.

Rao, H. Srinivasa. 1957. "History of Our Knowledge of the Indian Fauna through the Ages." *Journal of the Bombay Natural History Society* 54, no. 2: 251–80.

Rawat, A. J. 1985. "Forest Movement in Uttar Pradesh Himalaya, 1906–1947." In *Environmental Regeneration in Himalaya: Concepts and Strategies,* ed. J. S. Singh. Nainital, India: Central Himalayan Environment Association and Gyanodaya Prakashan.

Saharia, V. B., ed. 1981. *Wildlife in India.* New Delhi: Department of Agriculture, Government of India.

Sanderson, G. P. 1878. *Thirteen Years among the Wild Beasts of India.* London.

———. 1884–85. "The Asiatic Elephant in Freedom and Captivity." *Indian Forester* 10 (1884): 533–39, 576–82; 11 (1885): 32–38, 75–81.

Sankhala, Kailash. 1973. *Wild Beauty: A Study of Indian Wild Life.* New Delhi: National Book Trust of India.

———. 1977. *Tiger! The Story of the Indian Tiger.* New York: Simon and Schuster.

———. n.d. *National Parks.* Dehra Dun, India: Wild Life Preservation Society of India.

Sankhala, K. S., and Peter Jackson. 1985. "People, Trees and Antelopes in the Indian Desert." In *Culture and Conservation: The Human Dimension in Environmental Planning,* ed. J. A. McNeely and D. Pitt. London: Croom Helm.

Schaller, George. 1967. *The Deer and Tiger.* Chicago: University of Chicago Press.

Schneider, Keith. 1991. "Mediating the Federal War on Wildlife." *New York Times* (9 June).

Seidensticker, John, Sarah Christie, and Peter Jackson, eds. 1999. *Riding the Tiger: Tiger Conservation in Human Dominated Landscapes.* Cambridge: Cambridge University Press.

Seshadri, Balakrisna. 1986. *India's Wildlife and Wildlife Reserves.* New Delhi: Sterling Publishers.

Shah, Nasim Hasan. 1992. "Environment and the Role of the Judiciary." *Journal All Pakistan Legal Decisions* 44: 21–29.

Shakespeare, Capt. Henry. 1860. *From the Wild Sports of India.* London.

Shakoor Khan, Sahibzada Abdul. 1978. *Wild Life and Hunting.* New Delhi: Light and Life.

Shiva, Vandana. 1988. *Staying Alive: Women, Ecology, and Development.* London: Zed Books.

Shrestha, Tej Kumar. 1981. *Wildlife of Nepal: A Study of Renewable Resources of Nepal Himalayas.* Katmandu, Nepal: Tribhuvan University

Singh, A. 1973. *Tiger Haven.* New York: Harper and Row.

———. 1978. "The Status of the Swamp Deer (*Cervus d. duvauceli*) in the Dudhwa National Park." In IUCN 1978.

Singh, Chhatrapati. 1986. *Common Property and Common Poverty: India's Forests, Forest Dwellers and the Law.* Delhi: Oxford University Press.

Singh, Col. Kesri. 1965 [1969]. *Hints on Tiger Shooting.* Paperback ed. Bombay: Jaico Publishing.

Singh, R. B. 1978. "National Parks, Game Sanctuaries and Public Reserves of India." In

International Experience with National Parks and Related Reserves, ed. J. G. Nelson, R. D. Needham, and D. L. Mann. Waterloo, Canada: University of Waterloo.

Singh, Samar. 1984. *Conservation of India's Wildlife Heritage.* New Delhi: Department of Environment, Government of India.

———. 1986. *India's Wildlife Heritage.* Dehra Dun, India: Natraj.

Sinha, S., S. Gururani, and B. Greenberg. 1997. "The 'New Traditionalist' Discourse of Indian Environmentalism." *Journal of Peasant Studies* 24, no. 3 (April): 65–99.

Stebbings, E. P. 1920. *Diary of a Sportsman and Naturalist in India.* London.

———. *The Forests of India.* 1922–62. 4 vols. London: J. Lane. Reprint, New Delhi: A. J. Reprints, 1982.

Sterndala, R. A. 1984. *Natural History of the Mammalia of India and Ceylon.* Calcutta: Thacker and Spink.

Storey, Harry. 1969. *Hunting and Shooting in Ceylon.* 1907. Reprint, Dehiwala, Ceylon: Tisara Prakasakayo.

Sukumar, R. 1989. *The Asian Elephant: Ecology and Management.* Cambridge: Cambridge University Press.

———. 1990. "Ecology of the Asian Elephant in Southern India, II: Feeding Habits and Crop Raiding Patterns." *Journal of Tropical Ecology* 6: 33–53.

Sunquist, Fiona, and Mel Sunquist. 1988. *Tiger Moon.* Chicago: University of Chicago Press.

Sunquist, Mel. 1981. "The Social Organization of Tigers (*Panthera tigris*) in Royal Chitawan Park, Nepal." *Smithsonian Contributions to Biology* 336: 1–98.

Tennent, James E. 1977. *Ceylon: An Account of the Island Physical, Historical and Topographical.* 2 vols. Sixth ed. London, 1860; reprint, Dehiwala, Sri Lanka: Tisara Prakasakayo.

Tikader, B. K. 1983. *Threatened Animals of India.* Calcutta: Zoological Survey of India.

Tober, James A. 1981. *Who Owns the Wildlife? The Political Economy of Conservation in Nineteenth Century America.* Westport, Conn.: Greenwood Press.

Tolba, Mostafa K., and Osama A. El-Kholy, eds. 1992. "Loss of Biological Diversity." In *The World Environment, 1972–1992.* United Nations Environment Programme. London: Chapman and Hall.

Tucker, Richard P. 1988. "The Depletion of India's Forests under British Imperialism: Planters, Foresters and Peasants in Assam and Kerala." In *The Ends of the Earth: Perspectives on Modern Environmental History,* ed. Donald Worster. Cambridge: Cambridge University Press.

Turner, J. E. C. 1959. *Man-Eaters and Memories.* London: R. Hale.

Uttar Pradesh. n.d. *Corbett National Park.* Lucknow, Uttar Pradesh, India: Wildlife Preservation Organisation, Forest Department.

Venkataramiah, E. S. 1988. *Citizenship—Rights and Duties: R. K. Tankha Memorial Lecture.* Bangalore, India: Naga Publishers.

Ward, Geoffrey C. 1987. "India's Intensifying Dilemma: Can Tigers and People Coexist?" *Smithsonian* 18 (November): 53–65.

———. 1992. "India's Wildlife Dilemma." *National Geographic* (May): 2–28.

———. 1994. "The People and the Tiger." *Audubon* (July–August): 65–69.

Watt, Sir George. 1989. *Dictionary of the Economic Products of India,* s.v. "The Indian Elephant." Calcutta: Superintendent of Government Printing.

Williamson, Captain Thomas. 1820. *Oriental Field Sports, Being a Complete, Detailed and Accurate Description of the Wild Sports of the East.* 2nd ed. London: Edward Orme.

Witter, Dean. 1961. *Shikar.* San Francisco: Barry.

Woolf, Leonard. 1961. *Growing: An Autobiography of the Years 1904–11.* London: Hogarth Press.

Zimmermann, Francis. 1987. *The Jungle and the Aroma of Meats: An Ecological Theme in Hindu Medicine,* trans. Janet Lloyd. Berkeley: University of California Press.

Zoological Survey of India. 1981. *Rare and Endangered Animals of India.* New Delhi: Government of India.

CHAPTER SIX

Disease, Resistance, and India's Ecological Frontier, 1770–1947

DAVID ARNOLD

It would be perverse any longer to lament, as Madhav Gadgil and Ramachandra Guha did in 1992, "the almost universal neglect of Indian ecological history."[1] Recent years have seen a remarkable volume of writing on the subject, much of it based on extensive archival and oral research. Enormous gaps no doubt remain (not least for pre-colonial India), but a wealth of material and interpretation has been opened up for critical examination and analysis.[2] Environmental history in India has directed particular attention to the fate of the forests (understandably, in view of their former extent and ecological significance) and generated a lively discussion about the nature and imperatives of colonial forestry, its effects on forest-dwelling tribal or *adivasi* peoples, and their subordination, or resistance, to increasing commercial exploitation and state control. Environmental history has thus added an important new dimension to the earlier discussion of tribal protest and rural rebellion pioneered by Sumit Sarkar and others in the late 1970s and early 1980s.[3] At the same time the new environmental history has often been critical of earlier scholarship for failing to take ecological factors into fuller consideration in discussing tribal society and protest movements.[4] There has been growing skepticism, too, in this and related literature about the earlier assumption that tribal societies were, until the colonial and nationalist interventions of the late nineteenth and early twentieth centuries, largely cut off from the rest of Indian society. Accounts of trade, and narratives of political and military relations and cultural contact, now suggest a substantial degree of engagement between forest tribes and the adjacent states and agrarian economies, although the intensified penetration, incorporation, and exploitation of the colonial era is, nonetheless, still widely acknowledged.[5]

In an attempt to revisit earlier interpretations of tribal resistance and to extend the discussion of its ecological dimensions, this chapter focuses on one of the more neglected aspects of tribal/forest ecology, taking up disease, specifically malaria, as a factor ignored (or given only marginal consideration) in most accounts of tribal society and rebellion.[6] Central to this discussion is the duality of the concept of "resistance," with its combination of military and medical meanings suggestive of the wider issue of human agency versus "natural" causes that underlies so much recent environmental history. This is admittedly a tricky field to enter, not just because of the extremely complex nature of India's disease ecology and the limited sources available for its historical reconstruction, but also because emphasis on disease immunities has often given rise to a form of biological determinism, informed by a Darwinian paradigm of the "survival of the fittest," that makes the ascendancy of certain social groups appear inevitable and natural while denying effective agency to supposedly "weaker races." These determinist arguments have most commonly appeared in accounts of the demographic and cultural collapse of Amerindian societies in the face of European invasion in the sixteenth and seventeenth centuries and the importation into the West Indies and Brazil of African slaves, with their greater degree of resistance to yellow fever, malaria, and other "tropical" diseases.[7] While seeking to avoid narrowly deterministic arguments, in this chapter I nonetheless argue that in the period under consideration (the colonial *longue durée* from the Bengal famine of 1770 to the introduction of DDT in 1946–47) and for the tribal triangle of eastern central India (from Chota Nagpur in the west to the Rajmahal Hills in the east and the Madras Agency in the south), malaria is too important a factor to ignore. It is crucial in any attempt to understand the historical division between tribal/forest and plains/agrarian societies in India and to assess the nature of the ecological frontier in sustaining or undermining tribal resistance and in bringing about new patterns of social, economic, and epidemiological integration.

An Ecological Frontier

The idea of an ecological frontier has been less developed for South Asia than it might have been. In its original form the "frontier thesis," as first expounded more than a century ago by Frederick Jackson Turner for North America, seems to hold little relevance for modern Indian history.[8] In India there was no advancing tide of European settlement and wholesale displacement of indigenes, as there was in North America, no comparable interaction between white man and "wilderness," no pretense that a new spirit of democracy and self-reliance arose along *this* Indian frontier. But recent debates over the American frontier, in reappraising Turner, have given a more explicitly environmental twist to the story of European expansion in North America.[9] Taking the Americas as a whole, it could be argued that there existed an advancing ecological (and not merely human) frontier, beginning with the West Indies in the 1490s and continuing in the remoter parts of Alaska and Amazonia to this day. This was a frontier most dramatically marked by forest clearance—whether for European farms in New England or for sugar estates in the West Indies—with all the attendant changes in climate, fauna, flora, and human land use, although the effects on America's grasslands could be no less radical. In A. W. Crosby's writings, the frontier (though never specifically labeled as such) becomes a remorseless tide of Old World biota, in which epidemic diseases—along with hoofed herbivores and imported grasses, weeds, and pests—effectively cleared the way for the advancing whites, elbowing aside "weaker races" of peoples, plants, and animals. The frontier both describes a transforming process and serves as a metaphor for deep and irreversible ecological and human disjuncture, a trauma from which indigenous human, plant, and animal communities can never fully recover. Disease holds an exemplary place in this transformation: swept aside by "virgin soil" epidemics of smallpox, measles, and other Old World diseases, "inferior" peoples, marked by extreme susceptibility and severe mortality, were unable to resist biological as well as human incursions. But for Crosby "ecological imperialism" is confined to those temperate lands he dubs "neo-Europes," captured territory suited to white settlement and European-style agricultural systems. The tropics, by contrast, are seen as biological no-go areas for Europeans and the accompanying biota. Even when they hit the beaches of the "torrid zone," they are repulsed by more potent and entrenched forces, with disease once more to the fore.[10]

Clearly, India never became a "neo-Europe" in Crosby's sense, nor did it experience so radical a break in its ecological history as the Americas, although the degree of ecological transformation that did occur in the centuries following the arrival of the Portuguese—new diseases, plants, and animals, and new patterns of land use, water management, and forest exploitation—should not be underestimated. But historians are also aware of how extensively other, earlier agents modified the South Asian landscape through fire, axe, and plow, through pastoralism and *shikar,* through centuries of deforestation and the slow spread of rice cultivation. The process of ecological change was clearly a protracted one, but at the start of the colonial era there remained many regions of India that were still heavily forested, were inhabited by tribal societies, and lay outside, or only partially integrated into, the main areas of political and military control. This included a large part of eastern central India. Here, between the forests on the one hand and the cultivated plains on the other, lay an Indian version of an ecological frontier. In times of famine, war, and pestilence the forest/tribal area might creep into neighboring plains, while at other times vigorous government, expanding cultivation, and the pressures of land hunger might cause its partial contraction. Taken together, topography, vegetation, disease, climate, wild animals, and tribal inhabitants made up a kind of ecological package that helped over the centuries to preserve a degree of distinction between hills and plains, forests and cleared land, tribal and agrarian societies.

Where the "frontier" has entered Indian historiography it has generally been in a more restricted sense. Following Paul Wittek's *Rise of the Ottoman Empire* in the 1930s, some South Asia scholars have invoked an Islamic frontier of conquest and proselytization. Stephen Dale made passing reference to Turner's frontier thesis in his account of the Mappilas but failed to explore the environmental dynamics of Islamic expansion in Malabar from the late fifteenth century onward. Richard Eaton, by contrast, illuminatingly linked the spread of Islam and environmental change in his seminal study of the "Bengal frontier" from the thirteenth to the mid-eighteenth centuries, while Michael Adas, in a more secular vein, has described the development of an "Asian rice frontier" in colonial Burma, with immigrants from Burma's dry zone (and from India) overcoming a series of "prodigious" environmental hazards (not least malaria) to transform the delta jungles and swamps into productive paddy fields.[11] Elements of an internal frontier thesis have also long been implicit in historical accounts of tribal India, at least as far back as W. W. Hunter's description of an advancing tide of Aryan settlement pushing back the "aboriginal tribes of Bengal" into the forested fastnesses of the "highlands of Beerbhoom." A more refined version of this tribal frontier appeared in F. G. Bailey's account of the "economic and administrative frontier" in highland Orissa.[12] The idea of an Indian ecological frontier may not have been clearly spelled out, but it is certainly implicit in a great deal of writing linking India's internal and external boundaries with both human and environmental factors.

As the English East India Company extended its hold over the subcontinent in the late eighteenth and early nineteenth centuries, it acquired (or laid claim to) a number of environmentally less accessible and politically marginal territories. Many of these were inhabited by tribal peoples following a variety of agricultural practices, from swidden to settled cultivation, as well as hunting, fishing, and the collection of diverse forest products for consumption, sale, and exchange. In the course of their own expansionary careers, the Mughals and Marathas had also come into contact with these "forest polities": they encountered many of the problems the British themselves were to face and, to some extent, employed similar solutions. But a period of warfare, famine, and epidemic disease from the mid-eighteenth to early nineteenth centuries allowed the tribal/forest domains to regain something of their former extent. Mountstuart Elphinstone's description of Khandesh, taken from the Marathas in 1818—"The greater part of Candeish is covered with thick jungle, full of tigers and other wild beasts, but scattered with the ruin of former villages"—could have applied to many other frontier regions at the time.[13] To the problem of establishing political control and effective systems of revenue extraction in such areas was added the determination of forest- and hill-dwelling peoples not to submit passively to Company rule. And yet, despite the environmental and administrative hazards, Company officials felt compelled to intervene because they saw the opportunity to tax forest land and produce or because they regarded periodic raiding for cattle and grain as a threat to revenue and to law and order in the plains. As one of Elphinstone's exasperated lieutenants declared of the Bhils of Khandesh, "Nothing but a war of extermination can put an end to these troubles for the race of these marauders is evidently not to be reclaimed by any measures of indulgence."[14]

From about 1770 in the Jungle Mahals on the western borders of Bengal, from 1819 in Khandesh, and from the 1820s and 1830s in the hill tracts of the Northern Sircars in Madras, the British launched punitive raids against hill and jungle tribes, sometimes capturing, imprisoning, or executing a hill raja, or winning the promise of future obedience from "refractory chiefs."[15] But, as soon became evident, hill and forest populations were protected in no small part by the terrain, vegetation, and diseases (or climate) of the regions they inhabited. One can see evidence of this for the hill tracts of Rampa and Gudem in the Madras Presidency. In 1845 the Magistrate of Rajahmundry reported that Rampa was "an immense tract of hill and jungle infested with tigers, and rendered almost uninhabitable by a deadly fever [malaria] that rages from the setting in of the rains in June until the month of February. The people are said to be more uncivilized than those in the plains and to delight in robbery and war. They are armed with matchlocks, spears, hatchets and bows and arrows. The lowlanders seldom venture into the hills being deterred principally by the fever."[16]

When Company troops were sent on a punitive expedition into the hills in 1846, their commander was soon reporting that "severe sickness paralyses every effort, disheartens the men, and fosters the preconceived belief of the superiority and valour of the insurgents."[17] In 1847 the District Magistrate reminded Madras that "the climate of Juddingly [Rampa] is so deadly during some months that to retain it as a Circar Talook [under direct state control] is to ensure the loss of many lives annually."[18] A parallel incursion into the adjacent Gudem hills in 1847 met with similar results. The rebels were said to be "contemptible in numbers and courage"; "the only obstacle [to] their destruction is the denseness of the jungles and the unhealthiness of the climate."[19] It was further indicated that the "insurgents" were "fully aware of the deadly effects of their climate upon our troops, . . . and will do all in their power to protract operations till the setting in of the rains."[20]

The Madras Army and its medical staff sought ways to restrict soldiers' exposure to such a debilitating and demoralizing environment. Elsewhere, in the Jungle Mahals and Khandesh, the Company experimented with recruiting levies from the tribal population in order to police restless and unhealthy areas, and one can see in this a reflection of a wider imperial strategy of protecting European lives by substituting local troops, whose perceived assets included their apparent immunity to prevalent diseases.[21] In Madras the government was persuaded that military expeditions into the malarial hills and jungles were so costly and ineffective that it was more expedient to compromise with local chiefs, where these could be identified and drawn into negotiations. As the Madras Board of Revenue concluded at the close of this particular episode, "Tracts such as that under consideration—wild and unproductive—and from which the character of the country and the climate must be difficult of management by the officers of Government, are always best confided to the administration of their native chiefs."[22] As governor of Madras, Thomas Munro had reached much the same conclusion twenty-five years earlier when he warned, "We are every day liable to be dragged into a petty warfare among unhealthy hills, where an enemy is hardly ever seen, where numbers of valuable lives are lost by the climate, and where we lose but never gain reputation."[23]

In many of these tribal/forest regions the support that disease and other environmental factors gave to political dissent lasted well into the second half of the nineteenth century. Of the 2,400 soldiers sent to suppress the rebellion in Rampa in 1879–80, nearly a quarter were rapidly incapacitated by malaria.[24] As late as the Rampa rising of 1922–24 malaria and blackwater fever (attributed to inadequate quinine prophylaxis against malaria) severely hampered military operations against the rebels. "I do not suppose that there is a single officer or man whose health has not been more or less injured by the climate of this notoriously hot and fever-stricken region," reported the inspector-general of police.[25]

It could be argued that the tribal/forest situation was only an extreme example of a wider linkage between environment and resistance and of the conscious use of deforestation and other forms of environmental change to enhance state power in the Indian countryside. As Richard Grove has remarked, "Control of forests became synonymous with the political control of dissent."[26] In contemplating the annexation of Awadh in 1856, the British were covetous of the region's extensive woodlands and valuable reserves of timber yet simultaneously censorious of the weakness of Nawabi rule, which had allowed unruly "barons" to thrive in jungle forts and forest hideaways at the expense of rural peace and productive agriculture.[27] The difficulties the British experienced in quashing the revolt in Awadh in 1858 and the "dacoit" gangs that continued for years thereafter to haunt the jungly and malarial Terai must have further persuaded them of the political dangers of vast and unregulated forest tracts. Commenting on the consequences of large-scale deforestation in north India, three historians have noted how many rural communities thereby lost their protection against a predatory state: "An open landscape provides little refuge for rebel, fugitive, or bandit. Just as the colonial regime cleared the countryside of firearms following the Mutiny, so also did economic development clear away the natural defences of local society against state power."[28]

There are a number of ways in which we might understand this colonial discourse on lawless forests and "fever-stricken" tracts. It would be possible to examine the process by which the colonial authorities began to distinguish between India's different ecological zones, identifying certain types of landscapes and associated forms of vegetation, wildlife, and disease with certain kinds of human inhabitants, their ways of life, and cultural characteristics. The equation, evident as early as the 1780s, of uncultivated "wastes" (or "jungle") with lawlessness and primitiveness is particularly striking.[29] As other forms of resistance to British rule were erased, India's tribals came to represent a kind of residual "primitiveness" in contrast to the more "civilized" and settled denizens of the cultivated plains. The growth of such perceptions of India's "wildernesses" enhanced the sense (in middle-class Indian as well as colonial minds) of the India of the tribes and the forests as intrinsically different from the rest of the country, and so provided a basis for ideological pronouncements and administrative measures designed to subordinate, incorporate, and remodel tribal societies, preserving, even poeticizing, primitiveness whenever it might be useful or unthreatening but rigorously seeking to curb or expunge it when it ran counter to the dictates of "scientific forestry" or took on more rebellious forms of noncompliance and dissent.[30] But this process of differentiation was surely more than a matter of mentalities alone. One contributing factor was a colonial recognition, increasingly backed by medical science, that India's forests, hill tracts, and tribal peoples *were* ecologically distinct from the neighboring plains, and that this was a phenomenon that had to be investigated, understood, and surmounted before colonial rule could become fully effective and the dual resistance of the tribals overcome. The late eighteenth-

and early nineteenth-century references to "deadly fevers" thus served as the prelude to the later, more scientific investigation of India's disease ecology.

Malaria and Tribal India

Although the cause of malaria and the manner of its transmission remained unknown until the very end of the nineteenth century, colonial medical and topographical texts repeatedly identified "fever" as a primary attribute of the Indian environment. An initial concern for European health quickly merged into discussions of the destructive impact of malarial fever on the indigenous population, especially in Bengal. Following the current belief that malaria was an airborne poison generated by decaying animal and vegetable matter, a miasma that accumulated with particular intensity in dense undergrowth and on marshy ground, James Taylor graphically described the effects produced by floods, swamps, and jungles in the environs of Dhaka. The inhabitants of some localities, he reported, were so afflicted with the disease as to be "in a state of perpetual fever." Among betel-nut groves on the east bank of the Meghna in Tippera conditions were even worse. The inhabitants of this "gloomy region" showed "too plainly by their sallow, cadaverous looks, tumid bodies and shrunk emaciated limbs, the noxious atmosphere they breathe." They rapidly fell victim to the "poison of malaria."[31] Observation and experience emboldened some physicians to try to identify those regions of India where malaria appeared particularly prevalent and severe. These obviously included marshy or water-logged areas and "low jungles" like the Terai but also included almost anyplace where tropical heat and moisture combined, such as lower Bengal, Assam, and the canal tracts of northern India.[32] Even though the precise cause of malaria remained obscure, it was widely believed that forest clearance helped to remove the disease and so "open up" hill, forest, and tribal areas to outside settlement and agricultural colonization. In advocating such a policy for the eastern forests of the Central Provinces in 1871, Captain J. Forsyth argued that the obstacle of malaria would persist "only so long as the country continues to be uncleared." Experience elsewhere in the region had already shown this, and he cited Wynad, Assam, and Cachar as other "standing instances of the successful occupation of malarious countries by the help of European enterprise."[33] Malaria, in other words, was already being seen as an enemy that could be conquered.

When the vital breakthrough in medical understanding came with Ronald Ross's discovery of the role of the anopheles mosquito in the transmission of malaria in 1896–98, the knowledge that malaria was a parasite spread from person to person by certain kinds of mosquitoes sparked off a series of local malaria surveys and detailed investigations of anopheline ecology. Indeed, a considerable share of the technical expertise of the Indian Medical Service over the next thirty years was devoted to this intricate task (one which, ironically, the advent of DDT in the late 1940s rendered almost irrelevant to the anti-malarial crusade). In the course of this investigation the distinctive nature of tribal/forest India became, in certain respects, more clearly delineated than ever before, even though this was a time when that distinctiveness was undergoing unprecedented erosion.

In 1914 Major W. H. Kenrick of the Indian Medical Service published a detailed report on malaria in the Central Provinces. At the time nearly 40 percent of the province was still forested, particularly in the hills, and these were the areas where malaria was most prevalent and intense. Endemic malaria was so closely associated with forests that, in Kenrick's mind, the term "jungle fever," though obsolete in some ways, remained strikingly apposite. The denser the jungle, the higher the endemicity; the greater the extent of cultivation, the lower the incidence of malaria. The wooded hills bordering on Raipur and Bastar in the southeast were one of the main foci of hyperendemic malaria in the provinces, while the cotton tracts of Nagpur and the plains of Chattisgarh were largely malaria-free. Although the "dark-skinned aborigines" were apparently immune themselves to serious infection, they acted as carriers and so helped maintain high rates of infection among other, more vulnerable groups: "These tracts are particularly fatal to the fair-skinned non-aborigine and outsider; such quickly acquire the infection in its most pernicious form, and either rapidly succumb or else become permanently invalided. It is this fact more than anything else which has retarded the colonization of the more backward tracts of the Provinces and left them to the aboriginal tribes. It will be found that a backward hyperendemic tract, which has become habitable to non-aboriginal inhabitants, has been rendered so by the opening out of the forest and undergrowth which, whether it reduces the mosquito population or not by diminishing the amount of shade, restricts the conditions favorable to the life of the malaria parasite."[34]

In addition to his own inquiries, Kenrick was able to cite the evidence of officials in Balaghat district to the effect that in recent years malarial fever had been a "terrible scourge among all except Gonds and Baigas. The new settlers especially suffered, and hardly a Government official, tahsil or police, escaped. Any official sent up from the lower tahsil was generally down with it within a few days of his transfer. It was pitiful to see many adults and children with spleens that quite disfigured them. Everything possible was done to mitigate the evil. In some cases malarious villages with good land were taken up by two or three families in succession, only a few months sufficing for the malaria to clear away each whole family."[35]

It was widely doubted at the time that true "racial immunity" to malaria existed anywhere, with the possible exception of West Africa.[36] But Kenrick showed that, while spleen rates (known since the 1840s as a reliable diagnostic test for malarial infection) were extremely high (81–89 percent) among the Baigas and Gonds, adults of these tribal communities showed a much higher degree of resistance to the disease than non-tribals. He thought it "probable that the jungle tribes acquire immunity sooner than is the case with others, as the degree of splenic enlargement in their children is lower than with other children." Infection began early among tribal children, with large numbers of babies dying as a result. But those who survived early attacks acquired a high level of tolerance of malaria and so enjoyed a greater degree of resistance to the disease than outsiders not exposed since birth to intensive malarial infection.[37] The converse of this apparent ecological equilibrium was that in areas of the Central Provinces where forest clearance had taken place, tribals were displaced by outside immigrants, or at the least, the bulk of recent population growth had occurred among incoming settlers and not the Gonds and Baigas. In his settlement report on the *zamindari* estates in Bilaspur district, C. U. Wills showed that the tribal population had declined by nearly 50 percent over the previous forty-five years, partly because of tribal migration to Assam but more especially because of the immigration of non-tribals and the associated process of forest clearance and settled cultivation. But, Wills added of the remaining forest areas, "as long as the present extensive malarial conditions exist in this tract there need be no fear of the displacement of the aboriginal tribe by immigration from the Khalsa."[38]

In the same year as Kenrick's report on the Central Provinces, E. L. Perry, another officer of the Indian Medical Service, published a study of endemic malaria across the provincial boundary in Jeypore in the Madras Agency. Perry contrasted the relatively healthy and populous plains of coastal Andhra with the jungle-clad, malarial, and thinly settled slopes and interior plateaus of the Eastern Ghats. In passing from the plains of Vizagapatam (Vishakhapatnam) to the hills of Jeypore, Perry observed, "we pass abruptly from a region practically free from malaria into one that is notorious as the most malarious in Southern India. We pass equally suddenly from amongst races with a highly developed civilisation into a country almost exclusively peopled by tribes of primitive savages."[39] Little other than malaria, he believed, could have been responsible for preserving this distinction: "For a mountainous country it is exceedingly easy to enter and to traverse. The truth almost certainly must be that were it not for the great virulence of the malaria which these people harbor and the ample anopheline fauna of their jungles, such tribes would have been absorbed into the great Dravidian or Aryan civilizations of India and practically lost sight of long ago."[40]

Spleen rates in some interior areas were extraordinarily high. On the Malakanagiri Plateau, one thousand feet above sea level, the population density was as low as fifteen people per square mile but the spleen rate stood at a monumental 90.7 percent; elsewhere, as on the two-thousand-foot plateau in the northwest, abutting Bastar, the population density was considerably higher, at seventy-four people per square mile, with a correspondingly lower spleen rate of 69.9 percent. In a detailed analysis Perry matched fifteen anopheline species with their ecologies and the seasonal incidence, distribution, and malignancy of the quartan malaria. Like Kenrick, he saw a close correlation between tribals and high tolerance of malaria. He noted:

> The fact that the aborigines of the Jeypore country have in childhood a spleen rate of 80% to 90% and a parasite rate . . . of 80% has not the slightest effect on the physique to which they subsequently attain. The children do look as if they might be better nourished and more clothed with flesh, but whether their appearance is due to the effects of malaria or is merely the effect of their general conditions of life I cannot say. The physique of the adults both men and women is splendid. If a European were to go and live in one of these villages, amongst these people, without medical treatment and without protection from mosquitoes his health would quickly be shattered and he could scarcely hope to be alive at the end of two years. If we bear this in mind we shall realize at once the reality of immunity in malaria. The aborigine harbors large numbers of the malaria parasite in his blood when in childhood but is in a high degree tolerant of its presence and immune to its effects.[41]

Among tribal groups like the Parajas, Perry speculated, the high level of infection and yet tolerance of malaria was evidence of "harmonious relations" between host and parasite, "which possibly became fully established many thousands of years ago. . . . a compact appears to have been agreed upon, a *modus vivendi* established, and a condition of equilibrium now prevails." By contrast, even those residents who were long-settled immigrants from the plains suffered to a far greater extent.[42] Like Forsyth and Kenrick, Perry concluded that destroying the forest habitat in which malarial anophelines flourished was the key to material improvement and more effective government in such regions. Perry declared that "every furlong of road, every arch of culvert, every span of bridge that is built on the Jeypore plateau will do its share of silent service in the great work of converting vast wastes of malarious jungle into populous, healthy, and fertile land." Until this was achieved, special precautions had to be taken, especially for the benefit of European health, but through personal prophylaxis, he believed this could be done relatively cheaply and efficiently.[43]

Perry's advice had little immediate effect, although in establishing a new

Agency Division for the hills in 1920 the Madras administration closely associated the suppression of malaria with the task of bringing "development" to the area.[44] But when the Rampa Rebellion broke out in August 1922, the government was promptly reminded of the difficulty of trying to regain control over an intensely malarial area. As the remarks of the inspector-general of police demonstrated, malaria (and blackwater fever) had a debilitating effect on police operations and helped to sustain the guerrilla struggle for two years until its leader, Alluri Sitarama Raju, was captured and shot in May 1924.[45] Thereafter, the control of malaria in the agency became, despite the considerable technical problems and expense involved, one of the Madras government's priorities and the sine qua non for other pacificatory measures, such as road building and education.[46]

Ecological Change and Tribal Integration

To what extent, then, did one form of tribals' resistance (that to malaria) help to sustain another form of resistance (political and economic) to growing encroachment on their land and society? It clearly did so in only a limited sense. The hyperendemicity of many forest/tribal areas was undoubtedly an obstacle to outside intervention and control as it had been in the pre-colonial past, and it continued to exact a heavy toll on incoming settlers, soldiers, police, and officials. But once the ecological frontier had been breached and the forests were felled, the ecological package that had helped protect the tribals for so long rapidly fell apart, and further encroachment ensued. It should be noted, too, that a high degree of tolerance of endemic malaria did not help tribals when it came to other diseases, an increasing number of which found their way into the forests and hills. As roads, railways, and markets brought tribal populations into greater contact with the outside world, so were they exposed to diseases against which they had no immunity or effective prophylaxis. Smallpox and cholera took their toll, and so, too, did bubonic plague (which followed the opening of the Bengal-Nagpur Railway into the tribal interior of Orissa and the eastern Central Provinces in 1903–4).[47] Even more destructive was the influenza epidemic of 1918–19 in areas like Rampa.[48] The combination of thinly populated but naturally abundant forests and local tolerance of malaria may have been an asset to forest-dwelling tribes during the famine episodes of the 1860s and 1870s: forests provided one of the few available reserves of food in times of drought and hunger, and the malaria that devastated plains populations in the wake of famine was likely to be less severe among partially immune tribals. But as the forests shrank or restrictions were placed on their use, so famine and attendant diseases began to have more of an impact on tribal populations.[49] Given programs of road building, railway construction, sanitary and prophylactic measures targeted against malaria, and a determined policy of "developing" "backward" regions, it is clear that such ecological advantages as tribals formerly had, and which helped protect them in earlier times from external assault and control, were by the late nineteenth and early twentieth centuries rapidly being eroded.

Yet the situation should not be made to appear too one-sided. It is important to note that, in the context of the emerging colonial economy, the tribals' resistance to malaria could be seen as a factor that favored their wider integration and exploitation. As the destruction of their own habitat proceeded apace, tribals were increasingly recruited from Chota Nagpur, the Santal Parganas, and even the Madras Agency as plantation labor for the tea estates of Assam. Prabhu Mohapatra has questioned whether the reputed capacity of tribal laborers to endure "jungle fever" in the newly cleared plantation tracts, a claim "repeated *ad nauseam* in the successive labour emigration reports and by planters themselves," was as significant a factor as their perceived "tractability" (in other words, their lack of resistance to plantation discipline and work regimes).[50] Clearly immunities, real or imagined, were not the *sole* factor involved in the recruitment strategies of the planters and their agents. Nor was tribal labor free from all ailments: Santals leaving Chota Nagpur in the late nineteenth century suffered severely on their way to Assam through exposure to cholera (against which they had no immunity) and from unfamiliar diseases and deficient diets when they reached the estates. But an ability to survive in highly malarial conditions in Assam was one perceived, and empirically grounded, advantage to the use of "jungly coolies." In the early decades of the twentieth century, given the new knowledge of malaria, a close alliance developed between planters and colonial medicine around a mutual interest in the investigation and suppression of malaria, although uncertainty remained as to how far tribal laborers were genuinely immune or could pass on their immunity from generation to generation.[51]

Malaria, though hyperendemic in many tribal/forest areas, was by no means confined to these regions. Taylor found it widespread in the Dhaka district in 1840. Malaria appears to have been spreading in the course of the nineteenth century into newly irrigated areas of Punjab and the North-Western Provinces, where seepage from canals caused waterlogging. In western and central Bengal the eastward shift of the Ganges and its distributaries left behind abandoned riverbeds and stagnant pools in which mosquitoes bred. There, too, railway and road construction obstructed natural drainage patterns and produced extensive tracts of swampy ground.[52] The outbreak of "Burdwan fever," which devastated the once salubrious central districts of Bengal from the mid-nineteenth century onward, almost depopulating some areas, was evidence of the formidable demographic and economic impact that malaria could have on unprotected populations. Malaria reinforced colonial perceptions of the physical weakness of the Bengalis. By the late nineteenth century, faced with the overwhelming testimony of colonial sanitary and census reports, many middle-class Bengali Hindus were also coming to see malaria as threatening their *jati* (race or community) with ultimate extinction unless the tide could be turned against this fell disease.[53] The fear of becoming a "dying race" was intensified by the growth of the Muslim population in the far less malarial districts of eastern Bengal, and so contributed to growing Hindu-Muslim conflict in Bengal.[54]

In the west, though, malaria seemed to be shifting the demographic and ecological balance in the tribals' favor. As malaria gained an endemic hold over once-cultivated farm land, so did its old ally, the jungle. "The increase of jungle, invading streams, ponds, fields and homesteads alike, is . . . symptomatic of the decline of agriculture, the decay of an old inhabited area and the prevalence of malaria. . . . malaria decimates the rural population, leaves them weak and listless, and leads to wholesale emigration from the villages. The land is . . . allowed to lie fallow or there is an increase of jungly growth, which in turn fosters disease and rural exodus."[55] With the decline of the Bengali peasantry, tribal laborers moved in from Chota Nagpur, eastern Bihar, and the Santal Parganas. There was a "constant stream of migration of the aboriginal races to Burdwan, Hooghly, Jessore and parts of Nadia, where, as concomitants of malaria and depopulation, land has gone out of cultivation. Thus the Munda, the Oraon, the Santal and the Bauri fill the gap left by the Bengalee peasant, who is fighting a losing battle with the moribund rivers, agricultural decay and disease."[56]

The advance of the jungle/malaria/tribal frontier was only temporary, just as it had been after the Bengal famine of 1770. Renewed pressure on the land and a new phase of agricultural colonization and "development," backed up by the first use of DDT in India in the late 1940s, soon reversed this trend. By 1952, as "man's mastery of malaria" made rapid advances, even notoriously malarial areas like the Terai, the graveyard of many thousands of immigrants in the nineteenth and early twentieth centuries, were being purged of the disease. Indeed, it was possible in the early 1950s to believe that DDT and the National Malaria Control Programme were together removing the "ancient barriers to progress" in India.[57] Fifty years later it is difficult to be so optimistic about the conquest of malaria, but in the context of the foregoing discussion one can see the late 1940s, at the tail end of the colonial longue durée, as the moment when the ecological assets, which had once helped to protect tribal societies from outside encroachment, were finally overturned.

Acknowledgments

An earlier version of this chapter appeared in *Issues in Modern Indian History: For Sumit Sarkar,* edited by Biswamoy Pati (Mumbai, India: Popular Prakashan, 2000). An even earlier version was presented at Yale University in February 1996. I am particularly grateful to Jim Scott and K. Sivaramakrishnan for their comments and would also like to thank Prathama Banerjee for her help.

Notes

1. Madhav Gadgil and Ramachandra Guha, "State Forestry and Social Conflict in British India," in David Hardiman (ed.), *Peasant Resistance in India, 1858–1914* (Delhi, 1992), p. 261.
2. Ramachandra Guha, *The Unquiet Woods: Ecological Change and Peasant Resistance in the Western Himalaya* (Delhi, 1989); Madhav Gadgil and Ramachandra Guha, *This Fissured Land: An Ecological History of India* (Delhi, 1992); Richard Grove, "Colonial Conservation, Ecological Hegemony and Popular Resistance: Towards a Global Synthesis," in J. M. Mackenzie (ed.), *Imperialism and the Natural World* (Manchester, 1990), pp. 15–47; Ajay Skaria, "A Forest Polity in Western India: The Dangs, 1880s–1920s" (Ph.D. thesis, Cambridge University, 1992); Mahesh Rangarajan, *Fencing the Forest: Conservation and Ecological Change in India's Central Provinces, 1860–1914* (Delhi, 1996).
3. Sumit Sarkar, "Primitive Rebellion and Modern Nationalism: A Note on Forest Satyagraha in the Non-Cooperation and Civil Disobedience Movements," in K. N. Panikkar (ed.), *National and Left Movements in India* (New Delhi, 1980), pp. 14–26; *"Popular" Movements and "Middle Class" Leadership in Late Colonial India: Perspective and Problems of a "History from Below"* (Calcutta, 1983); *Modern India, 1885–1947* (1983; 2nd edition, London, 1989), pp. 44–48, 153–54, 239–40, 298–301.
4. See, e.g., Ramachandra Guha, "Forestry and Social Protest in British Kumaun, c. 1893–1921," in Ranajit Guha (ed.), *Subaltern Studies IV* (Delhi, 1985), pp. 55–56.
5. "Colonialism's distinctive contribution was not in integrating these regions into some wider system, but in changing the terms of this integration": Nandini Sundar, *Subalterns and Sovereigns: An Anthropological History of Bastar, 1854–1996* (Delhi, 1997), p. 4. See also Rangarajan, *Fencing the Forest,* chap. 1, and D. E. U. Baker, *Colonialism in an Indian Hinterland: The Central Provinces, 1820–1920* (Delhi, 1993), chap. 1.
6. A significant exception is Sumit Guha, "Forest Polities and Agrarian Empires: The Khandesh Bhils, c. 1700–1850," *Indian Economic and Social History Review* 33, no. 2 (1996): 147–50.
7. Alfred W. Crosby, *Ecological Imperialism: The Biological Expansion of Europe, 900–1900* (Cambridge, England, 1986); Kenneth F. Kiple, *The Caribbean Slave: A Biological History* (Cambridge, England, 1984). For a critique, see David Arnold, *The Problem of Nature: Environment, Culture and European Expansion* (Oxford, 1996).
8. Frederick Jackson Turner, "The Significance of the Frontier in American History" (1893), in Turner, *The Frontier in American History* (New York, 1953).
9. See, e.g., William Cronon, George Miles, and Jay Gatlin (eds), *Under an Open Sky: Rethinking America's Western Past* (New York, 1992).

10. Crosby, *Ecological Imperialism,* pp. 134–44.
11. Stephen F. Dale, *Islamic Society on the South Asian Frontier: The Mappilas of Malabar, 1498–1922* (Oxford, 1980); Richard M. Eaton, *The Rise of Islam and the Bengal Frontier, 1204–1760* (Berkeley, 1993); Michael Adas, *The Burma Delta: Economic Development and Social Change on an Asian Rice Frontier, 1852–1941* (Madison, Wis., 1974). Expanding rice cultivation has been seen as a factor in maintaining a "frontier" environment in Bengal well into the twentieth century: Willem van Schendel, *Three Deltas: Accumulation and Poverty in Rural Burma, Bengal and South India* (New Delhi, 1991), pp. 112–15.
12. W. W. Hunter, *Annals of Rural Bengal,* 5th ed. (London, 1872), pp. 88–89; F. G. Bailey, *Caste and the Economic Frontier: A Village in Highland Orissa* (Manchester, 1957), pp. 4–13.
13. Cited in K. Ballhatchet, *Social Policy and Social Change in Western India, 1817–1830* (London, 1957), p. 25. See also Guha, "Forest Polities," pp. 133–53. The impact of the 1770 famine on Bengal, and the attendant human and environmental changes, is described at length in Hunter's *Annals.*
14. Ballhatchet, *Social Policy,* p. 214.
15. For the Jungle Mahals, see B. S. Das, *Civil Rebellion in the Frontier Bengal, 1760–1805* (Calcutta, 1973).
16. T. Prendergast to Secretary, Judicial Department, Madras, 27 June 1845, Madras Judicial Proceedings (hereafter MJP), nos. 20–21, 15 July 1845, Oriental and India Office Collections (hereafter OIOC), London. For a similar situation in western India, see David Hardiman, "Power in the Forest: The Dangs, 1820–1940," in David Arnold and David Hardiman (eds), *Subaltern Studies VIII* (Delhi, 1994), pp. 106–7.
17. Lt.-Col. J. Campbell to Deputy Assistant Adjutant-General, Northern Division, 15 June 1846, MJP, no. 6, 4 August 1846, OIOC. For these and subsequent conflicts, see David Arnold, "Rebellious Hillmen: The Gudem-Rampa Risings, 1839–1924," in Ranajit Guha (ed.), *Subaltern Studies I* (Delhi, 1982), pp. 88–142.
18. Prendergast to Secretary, Judicial Department, Madras, 20 January 1847, MJP, no. 20, 9 February 1847, OIOC.
19. P. B. Smollet, Agent, to Chief Secretary, 5 June 1847, MJP, no. 15, 29 June 1847, OIOC.
20. Brigadier-General Dyce, Officer Commanding Northern Division, to Chief Secretary, 4 December 1847, MJP, no. 64, 24 December 1847, OIOC.
21. Seema Alavi, *The Sepoys and the Company: Tradition and Transition in Northern India, 1770–1830* (Delhi, 1995), chap. 4, "The Military Experiment with the Hill People"; A. H. A. Simcox, *A Memoir of the Khandesh Bhil Corps, 1825–1891* (Bombay, 1912). A similar strategy was at work in the recruitment of black soldiers to fight in the West Indies at a time of high European mortality from yellow fever: Roger Norman Buckley, *Slaves in Red Coats: The British West India Regiments, 1795–1815* (New Haven, 1979).
22. Minute, 24 August 1848, Board of Revenue Proceedings, 24 August 1848, OIOC.
23. Munro, cited in F. R. Hemingway, *Madras District Gazetteers: Godavari* (Madras, India, 1915), p. 36.
24. Ibid., p. 275. See also H. G. Turner, Agent, to Chief Secretary, Madras, 17 October 1886, MJP, no. 3080, 18 November 1886, OIOC.
25. Inspector-general's endorsement, 19 June 1923, Government Order 572, Madras Public Proceedings (MPP), 23 July 1923, OIOC.

26. Grove, "Colonial Conservation," p. 31.
27. "The Physical Capabilities of Oudh," *Calcutta Review,* no. 52 (June 1856): 415–44.
28. J. F. Richards, J. R. Hagen, and E. S. Haynes, "Changing Land Use in Bihar, Punjab and Haryana, 1850–1970," *Modern Asian Studies* 19, no. 3 (1985): 725.
29. J. Browne, *India Tracts* (London, 1788), is especially suggestive in this regard.
30. Ajay Skaria, "Shades of Wildness: Tribe, Caste, and Gender in Western India," *Journal of Asian Studies* 56, no. 3 (1997): 726–45.
31. James Taylor, *A Sketch of the Topography and Statistics of Dacca* (Calcutta, 1840), pp. 330–31.
32. Joseph Fayrer, *On the Climate and Fevers of India* (London, 1882), pp. 23–49.
33. J. Forsyth, *The Highlands of Central India* (London, 1871), p. 373.
34. W. H. Kenrick, *Report upon Malaria in the Central Provinces* (Nagpur, 1914), p. 5.
35. Ibid., p. 6; see C. E. Low, *Central Provinces District Gazetteers: Balaghat District* (Allahabad, India, 1907), p. 76.
36. Patrick Hehir, *Malaria in India* (London, 1927), pp. 34–39.
37. Kenrick, *Report,* p. 23.
38. C. U. Wills, *Final Report of the Zamindari Estates of the Bilaspur District of the Central Provinces* (1912), cited in Kenrick, *Report,* p. 43.
39. E. L. Perry, "Endemic Malaria of the Jeypore Hill Tracts of the Madras Presidency," *Indian Journal of Medical Research* 2, no. 2 (1914): 459.
40. Ibid., p. 460.
41. Ibid., p. 475.
42. Ibid., p. 483.
43. Ibid., p. 488.
44. L. T. Harris, Agency Commissioner, to Secretary, Revenue, 11 May 1921, GO 1933A, Revenue, 25 August 1921, Madras Revenue Proceedings, OIOC.
45. Arnold, "Rebellious Hillmen," pp. 134–40. There are, however, suggestions that the Koyas of this area were not particularly resistant to malaria but suffered terribly from it during the winter season. As quinine was unavailable to them, their only remedies were said to be medicinal roots and herbs from the forests. The large quantity of opium consumed in the hills (and attributed by colonial opinion to tribal indolence) was also explained in terms of its use as a prophylactic against malaria. P. Sundara Sivarow, "The Koyas: The Hill Tribe of the Godavary Agency," *Modern Review* (April 1910): 375.
46. See especially report of the "Agency Conference" at Waltair in February 1928, in GO 1531, Local Self-Government (Public Health), 31 July 1928, Tamil Nadu Archives (hereafter TNA), Madras. This concern with malaria control in the agency continued well into the 1930s: see GO 407, Local Self-Government (Public Health), 26 February 1932, TNA.
47. Low, *Balaghat District,* p. 77.
48. Harris to Secretary, Revenue, 11 May 1921, GO 1933A, Revenue, 25 August 1921, Madras Revenue Proceedings, OIOC. About four thousand Koyas died of influenza in 1918: K. N. Krishnaswami Aiyar, *Madras District Gazetteers: Statistical Appendix for Godavari District* (Madras, India, 1935), p. 203.
49. As in the Madras hill tracts: *Report of the Famine in the Madras Presidency during 1896 and 1897, volume I: General Report* (Madras, India, 1898), pp. 20–21, 37–38.
50. Prabhu Prasad Mohapatra, "Coolies and Colliers: A Study of the Agrarian Context of

Labour Migration from Chotanagpur, 1880–1920," *Studies in History* 1, no. 2 (1985): 264–65.

51. Ralph Shlomowitz and Lance Brennan, "Mortality and Migrant Labour in Assam, 1865–1921," *Indian Economic and Social History Review* 27, no. 1 (1990): 87–110; C. Strickland, *Abridged Report on Malaria in the Assam Tea Gardens* (Calcutta, ca. 1929), pp. 5–7, 20, 41–42.
52. Elizabeth Whitcombe, "The Environmental Costs of Irrigation in British India: Waterlogging, Salinity, Malaria," in David Arnold and Ramachandra Guha (eds), *Nature, Culture, Imperialism: Essays on the Environmental History of South Asia* (Delhi, 1995), pp. 237–59; C. A. Bentley, *Malaria and Agriculture in Bengal* (Calcutta, 1925).
53. David Arnold, "'An Ancient Race Outworn': Malaria and Race in Colonial Bengal, 1860–1930," in Waltraud Ernst and Bernard Harris (eds), *Race, Society and Medicine, 1700–1960* (London, 1999), pp. 123–43.
54. Papia Chakravarty, *Hindu Response to Nationalist Ferment, Bengal, 1909–1935* (Calcutta, 1992), pp. 33–37.
55. Radhakamal Mukerjee, *The Changing Face of Bengal: A Study in Riverine Economy* (Calcutta, 1938), p. 83.
56. Ibid., p. 87.
57. Paul F. Russell, *Man's Mastery of Malaria* (London, 1955), pp. 168–70, 244.

References

Adas, Michael. *The Burma Delta: Economic Development and Social Change on an Asian Rice Frontier, 1852–1941.* Madison, Wis., 1974.

Aiyar, K. N. Krishnaswami. *Madras District Gazetteers: Statistical Appendix for Godavari District.* Madras, India, 1935.

Alavi, Seema. "The Military Experiment with the Hill People." In *The Sepoys and the Company: Tradition and Transition in Northern India, 1770–1830.* Delhi, 1995.

Arnold, David. "Rebellious Hillmen: The Gudem-Rampa Risings, 1839–1924." In Ranajit Guha (ed.), *Subaltern Studies I.* Delhi, 1982.

———. *The Problem of Nature: Environment, Culture and European Expansion.* Oxford, 1996.

———. "'An Ancient Race Outworn': Malaria and Race in Colonial Bengal, 1860–1930." In Waltraud Ernst and Bernard Harris (eds.), *Race, Society and Medicine, 1700–1960.* London, 1999.

Bailey, F. G. *Caste and the Economic Frontier: A Village in Highland Orissa.* Manchester, 1957.

Baker, D. E. U. *Colonialism in an Indian Hinterland: The Central Provinces, 1820–1920.* Delhi, 1993.

Ballhatchet, K. *Social Policy and Social Change in Western India, 1817–1830.* London, 1957.

Bentley, C. A. *Malaria and Agriculture in Bengal.* Calcutta, 1925.

Browne, J. *India Tracts.* London, 1788.

Buckley, Roger Norman. *Slaves in Red Coats: The British West India Regiments, 1795–1815.* New Haven, 1979.

Chakravarty, Papia. *Hindu Response to Nationalist Ferment, Bengal, 1909–1935.* Calcutta, 1992.

Cronon, William, George Miles, and Jay Gatlin (eds.), *Under an Open Sky: Rethinking America's Western Past.* New York, 1992.

Crosby, Alfred W. *Ecological Imperialism: The Biological Expansion of Europe, 900–1900.* Cambridge, England, 1986.

Dale, Stephen F. *Islamic Society on the South Asian Frontier: The Mappilas of Malabar, 1498–1922.* Oxford, 1980.

Das, B. S. *Civil Rebellion in the Frontier Bengal, 1760–1805.* Calcutta, 1973.

Eaton, Richard. *The Rise of Islam and the Bengal Frontier, 1204–1760.* Berkeley, 1993.

Fayrer, Joseph. *On the Climate and Fevers of India.* London, 1882.

Forsyth, Joseph. *The Highlands of Central India.* London, 1871.

Gadgil, Madhav, and Ramachandra Guha. "State Forestry and Social Conflict in British India." In David Hardiman (ed.), *Peasant Resistance in India, 1858–1914.* Delhi, 1992.

———. *This Fissured Land: An Ecological History of India.* Delhi, 1992.

Grove, Richard. "Colonial Conservation, Ecological Hegemony and Popular Resistance: Towards a Global Synthesis." In J. M. Mackenzie (ed.), *Imperialism and the Natural World.* Manchester, 1990.

Guha, Ramachandra. "Forestry and Social Protest in British Kumaun, c. 1893–1921." In Ranajit Guha (ed.), *Subaltern Studies IV.* Delhi, 1985.

———. *The Unquiet Woods: Ecological Change and Peasant Resistance in the Western Himalaya.* New Delhi, 1989.

Guha, Sumit. "Forest Polities and Agrarian Empires: The Khandesh Bhils, c. 1700–1850." *Indian Economic and Social History Review* 33, no. 2 (1996): 147–50.

Hardiman, David. "Power in the Forest: The Dangs, 1820–1940." In David Arnold and David Hardiman (eds.), *Subaltern Studies VIII.* Delhi, 1994.

Hehir, Patrick. *Malaria in India.* London, 1927.

Hemingway, F. R. *Madras District Gazetteers: Godavari.* Madras, India, 1915.

Hunter, W. W. *Annals of Rural Bengal.* 5th ed. London, 1872.

Kenrick, W. H. *Report upon Malaria in the Central Provinces.* Nagpur, India, 1914.

Kiple, Kenneth F. *The Caribbean Slave: A Biological History.* Cambridge, England, 1984.

Low, C. E. *Central Provinces District Gazetteers: Balaghat District.* Allahabad, India, 1907.

Mohapatra, Prabhu Prasad. "Coolies and Colliers: A Study of the Agrarian Context of Labour Migration from Chotanagpur, 1880–1920." *Studies in History* 1, no. 2 (1985).

Mukerjee, Radhakamal. *The Changing Face of Bengal: A Study in Riverine Economy.* Calcutta, 1938.

Perry, E. L. "Endemic Malaria of the Jeypore Hill Tracts of the Madras Presidency." *Indian Journal of Medical Research* 2, no. 2 (1914).

"The Physical Capabilities of Oudh." *Calcutta Review,* no. 52 (June 1856): 415–44.

Rangarajan, Mahesh. *Fencing the Forest: Conservation and Ecological Change in India's Central Provinces, 1860–1914.* Delhi, 1996.

Report of the Famine in the Madras Presidency during 1896 and 1897, volume I: General Report. Madras, India, 1898.

Richards, J. F., J. R. Hagen, and E. S. Haynes. "Changing Land Use in Bihar, Punjab and Haryana, 1850–1970." *Modern Asian Studies* 19, no. 3 (1985): 725.

Russell, Paul F. *Man's Mastery of Malaria.* London, 1955.

Sarkar, Sumit. "Primitive Rebellion and Modern Nationalism: A Note on Forest Satyagraha in the Non-Cooperation and Civil Disobedience Movements." In K. N. Panikkar (ed.), *National and Left Movements in India.* New Delhi, 1980.

———. *"Popular" Movements and "Middle Class" Leadership in Late Colonial India: Perspective and Problems of a "History from Below."* Calcutta, 1983.

———. *Modern India, 1885–1947.* 2nd ed. London, 1989.

Schendel, Willem van. *Three Deltas: Accumulation and Poverty in Rural Burma, Bengal and South India.* New Delhi, 1991.

Shlomowitz, Ralph, and Lance Brennan. "Mortality and Migrant Labour in Assam, 1865–1921." *Indian Economic and Social History Review* 27, no. 1 (1990).

Simcox, A. H. A. *A Memoir of the Khandesh Bhil Corps, 1825–1891.* Bombay, 1912.

Sivarow, Sundara P. "The Koyas: The Hill Tribe of the Godavary Agency." *Modern Review* (April 1910).

Skaria, Ajay. "A Forest Polity in Western India: The Dangs, 1880s-1920s." Ph.D. thesis, Cambridge University, 1992.

———. "Shades of Wildness: Tribe, Caste, and Gender in Western India." *Journal of Asian Studies* 56, no. 3 (1997): 726–45.

Strickland, C. *Abridged Report on Malaria in the Assam Tea Gardens.* Calcutta, ca. 1929.

Sundar, Nandini. *Subalterns and Sovereigns: An Anthropological History of Bastar, 1854–1996.* Delhi, 1997.

Taylor, James. *A Sketch of the Topography and Statistics of Dacca.* Calcutta, 1840.

Turner, Frederick Jackson. "The Significance of the Frontier in American History" (1893). In Frederick Jackson Turner, *The Frontier in American History.* New York, 1953.

Whitcombe, Elizabeth. "The Environmental Costs of Irrigation in British India: Waterlogging, Salinity, Malaria." In David Arnold and Ramachandra Guha (eds.), *Nature, Culture, Imperialism: Essays on the Environmental History of South Asia* (Delhi, 1995).

Wills, C. U. *Final Report of the Zamindari Estates of the Bilaspur District of the Central Provinces* (1912). In W. H. Kenrick, *Report upon Malaria in the Central Provinces.* Nagpur, India, 1914.

CHAPTER SEVEN

Subalterns and Others in the Agrarian History of South Asia

DAVID LUDDEN

Recent trends in history writing about rustic South Asia have been linked by intricate webs of influence—still dimly understood—to geological upheavals in the world of agrarian studies. Until the early 1980s, historians of the countryside were most concerned with economic, political, and social transformation, and their intellectual orientations reflected the contours of state politics and policy.[1] This kind of history informed national efforts to improve living conditions; its role was to measure trends and substantiate critique. Foundations for such agrarian history first emerged with the nationalist critique of imperial policy, following famines in the 1870s. Agrarian studies permeated Indian economics and sociology even before a British agricultural officer, William Moreland, published the first academic monograph on agrarian history in 1929, partly in response to the nationalist critique.[2] In 1930 Nehru became president of the All-India Congress Committee and announced with these words the arrival of one enduring theme in agrarian studies: "The great poverty and misery of the Indian People are due, not only to foreign exploitation in India but also to the economic structure of society, which the alien rulers support so that their exploitation may continue. In order therefore to remove this poverty and misery and to ameliorate the condition of the masses, it is essential to make revolutionary changes in the present economic and social structure of society and to remove the gross inequalities."[3] By 1947 Nehru had etched agrarian history into the Congress platform. Like other renditions of history for national development, his formulation attracted intense academic scrutiny in later years:

> Though poverty is widespread in India, it is essentially a rural problem, caused chiefly by overpressure on land and a lack of other wealth-producing occupations. India, under British rule, has been progressively ruralized, many of her avenues of work and employment closed, a vast mass of the population thrown on the land, which has undergone continuous fragmentation, till a very large number of holdings have become uneconomic. It is essential, therefore, that the problem of the land should be dealt with in all its aspects. Agriculture has to be improved on scientific lines and industry has to be developed rapidly in its various forms . . . so as not only to produce wealth but also to absorb people from the land. . . . Planning must lead to maximum employment, indeed to the employment of every able-bodied person.[4]

Such pronouncements stimulated massive academic attention to agrarian issues, and studies in agrarian history had implications for land reform, planning priorities, local democracy, farm finance, industrialization, and many other topics of hot dispute in struggles between socialism and capitalism.[5] Historians assumed that national states, histories, identities, and destinies formed every nation. After 1975, this assumption died.

In India its demise began with Indira Gandhi's Emergency, in 1975, and during the 1980s the much-discussed "fragmentation of the nation" proceeded along regional, communal, and other lines. In the 1990s India's national polity became a crazy quilt of incoherent coalitions.[6] Globally, national states in general lost their previous intellectual status. The end of central planning and the death of socialist possibilities in the Third World challenged state capacities to control economies and produce a better future for the nation.[7] States lost legitimacy in the culture of late capitalism.[8] Development dilemmas facing scholars came to include the state's seemingly congenital inability to represent the destiny and promise of the nation. In India, state leadership dissipated with debt-driven liberalization, disarray in the Congress Party, and the rise of Hindu chauvinists who claimed to represent a true national essence suppressed by the old party of Nehru and Gandhi.[9] In South Asia, as in most of the world, national governments ceased to represent forward-moving modernity. The market economy and non-governmental organizations sidelined the state in development ideology, as new social movements spawned critical theories that made "seeing like a state" at best inefficient, and at worst oppressive.[10] Structural adjustment and liberalization, forced by the World Bank and the International Monetary Fund, opened South Asia to globalization, as transnational intellectuals challenged the state's role in development everywhere.[11] Indian economic experts promoted India's liberalization at Yale and Columbia Universities. Financial support for Khalistan, Tamil Tigers, and

Hindu militants came from England, Canada, and the United States. Global media became prominent sources for up-to-date images of India *inside* India, and glossy magazines and academics alike declared that venality and rent seeking were at the heart of state development bureaucracies.[12] The national state lost intellectual authority.

In this changing environment, a new school of history evolved.[13] Called Subaltern Studies after a series of anthologies from Oxford University Press in Delhi, its publications first appeared in 1982 and multiplied rapidly; its authors now number forty-two, and its books and articles, more than 160.[14] Its influence is widespread but hard to measure because of the concurrent multiplication of studies "from below" in many disciplines, for which Subaltern terms and texts have been handy. Subaltern Studies itself emerged with the decline of state-oriented histories and the rise of "history from below" in the academy. In the 1970s historians of South Asia broke the grip of nationalist and imperial formulas to explore the activity of countless groups in the political struggles of the colonial period, and Sumit Sarkar's 1983 distillation of this work, *Modern India, 1885–1947,* marks a permanent shift from old histories of political parties and official elites to new histories of popular politics.[15] History from below had arrived before Subaltern Studies, which filled the notion of subalternity with new contents. In 1980 "subaltern" still referred primarily to ranks in the military, but by 1990 it had regained old connotations of categorical subordination and dependency, as among feudal vassals or logical propositions. Today subalternity includes all the elements of being downtrodden, dependent, dominated, and resistant, and it also indicates a generic bottom-up perspective on society, politics, and culture.[16] But Subaltern Studies focused specifically on colonial India, and its goal was to reshape the politics of knowledge that informs historical research, staking its claim to the future on its commitment to subaltern history. Ranajit Guha guided the project in its first decade, editing six Subaltern Studies volumes, and his major monograph on peasant insurgency appeared alongside the first of those volumes.[17] His introduction to volume 1 declared that nationalist history was elite history and that histories from below should have primacy and autonomy. Subaltern history was to be the real people's history, the true heart of the nation.

But Subaltern Studies dug a narrow shaft into the rural past. The study of agrarian history had grown rapidly in the 1970s by drawing ideas and material from the social sciences of agrarian studies—economics, sociology, anthropology, and political science.[18] Subaltern agrarian history ignored most of this literature and constituted agrarian realities that lacked political parties, economic development, class structures, technological change, and social mobility. The Subaltern agrarian past became a separate sphere of tribal and peasant life, filled with expressions, languages, and texts of domination, subordination, and resistance. Urbanites, middle classes, bankers, and other modern elites inhabit other worlds of agency and identity, pressing their power upon subalterns in the country.[19] Subaltern Studies produced a rich body of new knowledge on subaltern forms of historical experience, memory, and consciousness. This opened new frontiers for historians who had managed to miss the fact that most people in South Asia eluded nationalism but still have histories worth studying, and it stimulated more historians to look at rural life outside the teleology of nationalism. Yet for scholars of agrarian history, for whom rustic politics were more important than nationalism to begin with and for whom economic well-being and social change were more important than whether historians concocted collective memories from above or from below, Subaltern Studies was less compelling. As it opened new agrarian frontiers for some historians, Subaltern Studies drained agrarian history of much of its content in order to reconstruct it in Subaltern terms. This diverted attention from central issues, inhibited empirical investigations, and blocked collaboration with the social sciences.

Ranajit Guha cut his project off from the previous research on questions that preoccupy Subaltern Studies: inequality, rebellion, social power, domination, and the many mundane struggles of everyday life in the countryside. His motive seems to have been to launch a new school of history against the legacy of nationalism, which included agrarian studies. Overtly, Subaltern Studies opposed "nationalist historiography," but all histories from below, including agrarian history, had already moved away from the state-centered history. As a distinguishing mark of its intervention, Subaltern Studies covertly jettisoned modes of agrarian studies that focused on the complexities of production and class. One specific indication of this erasure is the absence of any reference by Guha and his colleagues to two prominent scholars who had intensely explored agrarian histories from below: A. R. Desai and Kathleen Gough. Desai wrote the first social history of Indian nationalism, which was published in 1948 and reprinted four times in the next twenty years.[20] As Desai said in the preface to the first edition, it was "an attempt to give a composite picture of the complex and variegated process of the rise of Indian nationalism and its various manifestations" (p. xii). In this effort to explain nationalism as a social process, Desai devotes three chapters to transformations of agrarian India under British rule. This is the first history of nationalism "from below." Its long eleventh chapter, "Rise of New Social Classes in India" (pp. 174–220), describes the evolution of various classes in the city and countryside and presents propositions that still merit inquiry. Despite their diversity, Desai argues, "one striking characteristic of the new social classes was their national character. This was due to the fact that they were integral parts of a single national economy of India and further, they lived under a single state regime. This engendered community of economic, political and other interests of the members of each of the new social classes on an all-India national basis. . . . they felt an urge to organize themselves on an all-India scale and start movement to advance their common interests on a national basis" (p. 214). Its nationalist tone may have rendered this book archaic in the 1980s, but Desai, inspired by Gough's decades of work on oppression and rebelliousness among Dalits in Tanjore (a summary of which was published in 1981), also edited two large collections of essays on agrarian struggles in British India and independent India, to delineate the "roles of specific sections and classes of the rural population which took leadership, provided guidance, raised specific issues and elaborated various forms of mobilization and struggle."[21] Ranajit Guha's monograph on peasant insurgency cites historians of colonial upheavals like Eric Stokes with approval but ignores A. R. Desai, and yet Guha concludes, much like Desai, that in the evidence of peasant upheaval we can see that "a welter of individual instances . . . stands for a generality in which ideas, mentalities, notions, beliefs, attitudes, etc. of many different kinds come together and constitute a whole."[22] For Guha, however, this "whole" was a nation apart from nationalism; his peasant struggles stand outside nationalist history altogether, to express the autonomous mentality of the Indian subaltern.

This radical break between subaltern and elite nationality was a founding idea for Subaltern Studies and underpinned its distinctive opposition to the historiography of "colonialist elitism and bourgeois-nationalist elitism."[23] By 1980, however, scholars had filled agrarian history with the substance of other histories from below, which were simply erased to form a Subaltern agrarian past. In the Subaltern countryside, resistance reigned supreme. "The ideology operative in this domain, taken as a whole, reflected the diversity of its social composition. . . . However, in spite of such diversity one of its invariant features was a notion of resistance to elite domination. This followed from the subalternity common to all the social constituents of this domain and as such distinguished it sharply from that of elite politics." The first principle of Subaltern history thus became to "demarcate the domain of subaltern politics from that of elite politics": "The coexistence of these two domains or streams, which can be sensed by intuition and proved by demonstrations as well, was the index of an important historical truth, that is, the *failure of the Indian bourgeoisie to speak for the nation.* There were vast areas in the life and consciousness of the people which were never integrated into their hegemony. The *structural dichotomy* that arose from this is a datum of Indian history of the colonial period, which no one who sets out to interpret it can ignore without falling into error."[24]

Society as Externality

Subaltern history has relied on and reinforced in the minds of its avid readership a primitive historical sociology. The Subaltern project began with a theoretical division of the social world between elite and subaltern, and Subaltern historians have since described forms, means, events, and activities of social power and resistance without attending to the inadequacy of this initial bipolar division of social life. The result is sophisticated social history in what we can call an American style. Based on deep archival work and thick (anthropological and synchronic) description of events (mostly in the late colonial period), it is literary, self-referential, progressive, populist, righteous, and sensitive. Often heart-rending and uplifting, it concentrates on the underdog who fights for a place in (national) society and in academic consciousness. Its sociology is quite shoddy, however. Guha's own *Elementary Aspects of Peasant Insurgency,* for example, includes hill tribes that farm slash-and-burn plots as well as commercial farmers in the category of "peasant." David Hardiman's *Peasant Nationalists of Gujarat* examines subalterns who were some of the most successful commercial farmers in British India and whose opposition to taxation was much like that of bourgeois activists in many parts of the world, which made them darlings of elite Indian National Congress agitators.[25] The political gap between Congress leaders and peasants in Uttar Pradesh, discussed by Gyanendra Pandey in the research that launched his Subaltern career, turns out to separate urban professional and rural tenant elites. Rudrangshu Mukherjee ignored research by Eric Stokes and others to (re)construct the 1857 rebellion in rural North India as a monolithic outburst of subaltern rage against British rule.[26] These few examples indicate a more general pattern. Repeatedly, data to indicate how a subaltern's world is constructed disappear. A bipolar elite-subaltern model creates a compelling, dramatic motif that obliterates the complex social world of caste, kinship, class, patron-client ties, and above all, production relations. The result is a trademark head-to-head confrontation of power and resistance, modernity and tradition, English and Indian, elite and subaltern, which distinguishes the Subaltern Studies series and many spin-off publications that now constitute the corpus of Subaltern history.

Indications of why Subaltern history treats society as external to elite-subaltern interaction lie buried between the lines, but not deeply. Ranajit Guha opens *Subaltern Studies I* (1982) by attacking the elitism of nationalist writers who document national unity, and he ends *Subaltern Studies VI* (1989) by describing subalternity as the essence of the Indian nation.[27] Subaltern history thus becomes the route to India's authentic national history. Subaltern scholars who work in India, Britain, Australia, and the United States have in effect reinvented the Indian nation for a new, global postcolonial context. The subaltern seems to be a quintessential Indian citizen, who has suffered feudal, colonial, and capitalist oppression; who is alienated from the Indian state and India's nationalist past; who has been denied a voice in history; and whose consciousness and identity Subaltern histories reveal. The growing prominence of cultural studies, literary flourish, deconstruction, and postcolonial theory in Subaltern history may thus be seen as a dual effort to enrich India's new (inter)national *imaginaire* and to obliterate empirical knowledge that threatens to undermine subaltern (Indian) *communitas.* This effort to enrich the national imagination by recounting its subordination and resistance sounds very much like the original nationalist project. But Subaltern nationality expels Indian elites from the real India. English-speaking urban intellectuals, government officials, capitalists, landlords, and patriarchs appear as new elites who carry on the (colonial) subordination of (subaltern) India.

Subaltern history has simply ignored a huge body of research indicating that agrarian India was complexly differentiated and changing dramatically in centuries before British rule. Erased are the evolution of agrarian classes, outlined initially by A. R. Desai, and thus the histories of societies in which subaltern struggles occur. Social change and social mobility disappear in Subaltern history, along with any idea that subalterns regularly form local elites and turn against their own kind. In this respect, odd similarities among Subaltern history, neoclassical economics, and rational choice theory must be noted, because they all depict social actors who are driven by strict rules of behavior and consciousness, established in theory. The complexity and ambiguity of changing social environments become external to historical agency and mentalities. In addition, such theories of agrarian life avoid the task of describing temporal change and spatial variation in social formations. They concentrate rather on the predicaments of individuals within fixed relations of order, inequality, and resistance.

In its current form, Subaltern history cannot be disturbed by empirical evidence concerning agrarian social organization or changing societies. It simply ignores the documentation of complicity with power, communalism, corruption, and clientelism among subalterns, while also ignoring everyday social relations of production and reproduction. Much of agrarian social history disappears as Subaltern historians describe the subjective conditions and tribulations of theoretical underdogs. Everyday life among people who speak for subalternity appears complexly constituted by power and resistance, but social environments in which their struggles occur seem quite simple, being composed of colonial power forged in collaboration with (emerging) native bourgeois elites. The internal world of subaltern subjectivity is elaborately detailed, while the external world of society, especially agrarian society, is a flat background, mere externality.

Problems of Location

Solutions in Subaltern history to three problems of theory undermine its rendition of agrarian South Asia. I call these "problems of location" because they involve the location of (1) theoretical concepts in social worlds, (2) qualities in social entities, and (3) narratives in epistemological landscapes.

The first problem is the etic/emic, or "in itself/for itself," problem. It arises whenever we attribute consciousness (for itself) expressed in "their own" (native, emic) terms (or "voice") to a social entity that has been defined externally (in itself) in the alien (etic) terms of "our own" (that is, the theorists') social theory. Do the Indian subalterns recognize themselves as such? This question cannot be solved theoretically; it must be addressed in practice, and its difficulties must be recognized. Folding "for itself" representations into "in itself" categories has become standard procedure in social history and anthropology. When Subaltern scholars call people or groups "subaltern" and ascribe a "subaltern" consciousness to them, questions remain about how the person or group would respond to such an attribution. Subaltern history avoids this problem by taking definitions of subaltern identity outside the real social world and refusing to test the fit between subaltern identity attributions and evidence about self-identification generated by subalterns themselves.[28]

The second difficulty—that of defining groups by their qualities—is posed by simple questions. Who is a subaltern? When is a person or group subaltern? Does "subaltern" denote a person, a kind of relationship, or an aspect of a relationship? A story in the spirit of Antonio Gramsci will help to illustrate the problem. Our hero is a subordinate army officer, "subaltern" by definition, no matter what he thinks or does. He is in a "subaltern relationship" with superiors, which has many modalities of power within it and many other qualities as well, perhaps even affection and admiration, but not all of these qualities are called subaltern. Only some are called subaltern, and then only some of the time. Imagine our hero in a parade, marching smartly behind his commander, showing clear signs of pride in his rank. After all, his juniors march in ranks below him. His wife and children obey him. His mates respect him. Peasants shrink when he passes by. But all of a sudden, he lunges forward. He bayonets his commanding officer. With this murder, a mutiny begins, or is it a riot, revolution, or coup? During the fracas, our hero kills other subalterns. When the killing ends, a new commander takes charge and our hero marches proudly again.

This story could represent an anti-colonial struggle, which ends with national independence, told from a Subaltern perspective. But where is the subaltern? The hero is the same throughout, but his qualities shift and change. Do pride, obedience, murder, mutiny, anger, risky decisions, and accepting a new status quo equally characterize a subaltern identity? There is no theoretical reason why not. But Subaltern historians do not tell us how to locate "resistance," "domination," and "power"—key analytic terms—in the complex of subaltern modalities and qualities. In Subaltern Studies resistance and domination stand in opposition. So the act of murder would be "the moment of truth" in the above tale. Certainly for Ranajit Guha, this would be the defining moment for the subaltern. Why not his pride of rank or his killing of other subalterns? This empirical possibility must be avoided; it would certainly defy Subaltern theory. If subalterns in colonial India were to express pride in their subordination and exercise power over others through the power vested by their own rank in society—a situation that would be commonly shared by landed peasants and tenant farmers—this would, first of all, imply a kind of imperial hegemony that Ranajit Guha has rejected, and second, it would mean that power was too widely diffused in society to allow for fixed oppositions between subaltern and elite. Subaltern Studies avoids the search for such empirical discoveries.

The third problem derives from the necessary location of historical stories in a framework of description and narration. Neoclassical economics and rational choice use an array of assumptions to establish their context. The same is true for all historical narratives. Subaltern Studies relies on a host of widely shared cultural understandings of how power works in modern societies, and it also relies on the received authority of India's national master narrative. Ranajit Guha's inaugural essay denigrated official *nationalism* and *nationalist* history in order to reclaim the legacy of *nationality* or *national* history. The superb social history in Subaltern Studies depends on a basic assumption of the primacy of the nation. Readers need to accept certain general ideas about the big picture of national history to locate the up-close, local studies in Subaltern history. The British conquered India. Landlords oppress peasants. Peasants constitute a definite social stratum. Peasants live in village societies that constitute their ancient social and cultural environment. Indian culture is different from Western culture. Vernacular texts express vernacular cultures. And so on. Such ideas form a familiar structure within the citadel of world history. But familiarity is deceptive as well as reassuring. Empirical evidence may kick the props out from under national history, but Subaltern history makes the search for such evidence irrelevant.

These problems of location suggest that we need more than Subaltern Studies to understand the agrarian history of subalternity. Subaltern Studies makes its most important contribution by using primary sources inventively to write "up-close" social history. Its broader use of theory and evidence is another matter. Writers in Subaltern Studies often deploy primitive, sloppy, conservative solutions to problems of location. In this and other respects, Ranajit Guha led the way in *Elementary Aspects of Peasant Insurgency*.[29] Three major flaws in that work have had a significant impact. First, Guha locates the subjectivity of "the peasant" within a social typology defined so loosely as to enable anyone who lives outside the city and is not a landlord, moneylender, or state official to be a peasant. It is impossible to imagine that most people he calls peasants would recognize one another as belonging to the same social category. Second, he takes any act of opposition (especially violent acts) against government to be "insurgency" and defines peasant subaltern qualities by acts of resistance to state power. Insurgency seems to be the only effective expression of a peasant's real identity, and it remains disconnected from the rest of peasant social existence. It seems to be virtually an animal drive, and in fact Guha calls insurgency "necessary" (p. 2ff). Finally, Guha situates his stories firmly in the master narrative of Indian nationalism and locates subaltern modes of consciousness within an epistemological framework built by cultural anthropology to define Indian national culture. His peasant insurgency essentially expresses India's national resistance to colonial domination in the heart of agrarian society.

Agrarian Contexts

Other histories of agrarian power and resistance begin where Subaltern Studies will not venture, that is, with the project of locating actors in complex, changing social relations. This project has become increasingly difficult as we learn how subtly and profoundly agrarian societies in South Asia have changed and varied historically and geographically. Subaltern Studies avoids empirical accounts of historical change, at the top, the bottom, or the vast midsection of agrarian society. Instead, it isolates the top-down aspect of social relations and imagines people only in polarized power relations. Outside Subaltern history, however, scholars still work with data to describe agrarian relations and to locate historical actors in context.[30] Escaping this empirical realm may help to explain Subaltern popularity, because it avoids tedious, inconclusive, locally specific research and eludes endless debates about the contextual specificity of agrarian regions in South Asia during, before, and after the colonial period.

To use documents historically, we need to know where they come from, but situating them in the agrarian past has become more difficult because we have lost confidence in our map and compass. At a broad cultural level, it is no longer acceptable merely to assume that South Asian and Western cultures are strictly bounded entities that confront one another hierarchically under colonialism. We cannot accept the assumption clear lines can be drawn between Muslim and Hindu cultures. It no longer makes sense to assume that cultural strategies prescribed by the "market economy" and by "caste society" are incompatible, or that one of these is Western and the other Indian, or that one is new, modern, and foreign while the other is old, traditional, and native. Thus there is no justification for basing historical narratives on moments of interaction between Indian and European cultures or between Islamic and Hindu cultures.[31]

If we can no longer assume an "Indian culture," how can we situate the logic of social activity in India in a cultural setting? My answer would be to propose that a culture should be seen as a repertoire of possibilities whose symbolic bounding and delimitation are themselves cultural activities that define the historical movement of a culture in space and time. Thus, for instance, medieval intellectuals defined the boundaries of their cultural world differently than do modern nationals, and the inscribing of boundaries around a specifically "Indian culture" is a historical activity that is part of the process of forming Indian nationality, rather than being an inherent feature of Indian cultural life. From this perspective, we can find a number of overlapping cultures in pre-modern South Asia, and we can see that a distinctively modern entity, which we can now justifiably call "Indian culture," came into being in the nineteenth century. Its fissures and solidarities developed interactively with colonialism and nationalism. The study of cultures by scholars and the reification of cultural boundaries by states have helped to solidify national identities. The partition of British India in 1947, and subsequent communal, linguistic, ethnic, and regional conflicts in independent states, represent events in a process of modern culture-formation rather than expressions of ancient opposing loyalties.

We need to locate subaltern subjectivity and experience, therefore, in shifting cultural terrain and to locate subaltern actors in changing agrarian environments. To identify any personal or group mentality requires attention to their positioning within cultural *transformations* in which scholars are also embedded. This is doubly important and difficult in agrarian history, first because scholars, being urban intellectuals, are themselves alienated from agrarian life, and second because rural and urban moderns alike carry encrusted ideas about agrarian life that are constituents of our own modernity. We carry, for instance, the idea that urban and rural histories have fundamentally different qualities and that village life is more constrained by its traditional past. Empirically this may not be true, but modernity assumes it is. Agriculture, rural development, famine, villages, folk, folklore, and the like have accumulated rigid meanings in the intellectual formations of modernity that are difficult to break down so as to devise more realistic ideas about agrarian history. The rigidity of inherited ideas is reinforced by the factual authority that is woven into the fabric of academic disciplines.[32]

It is hard to imagine history without images of colonial power "reaching out" and state power "reaching down" into "village society." The expansion of Europe, spread of capitalism, advance of modernity, and other basic historical phenomena depict village India as a frontier, a destination for historical forces moving out and down from centers and pinnacles of progress. Rural India has been pre-constructed by the discipline of history as a stable, isolated, ancient (native) place of village subsistence, forcibly integrated into a British (colonial) empire that spread out from Europe, first (most thickly) into Indian cities and then (most thinly) into the Indian countryside.[33] It is equally hard to imagine anthropology without traditional (village Indian) society, composed of sedentary farmers, operating in its own (indigenous) cultural terms for centuries before (being disrupted by) Western colonial impact. What would anthropology have to say about culture if villages today did not provide data on traditional times, if cultural retention and reproduction did not connect "the ethnographic present" to India's (ancient village) past?[34]

Disciplines of professional practice thus prefigure agrarian South Asia as a particular kind of historical entity. Although history may have followed trajectories other than those built into theories of orientalism and modernization, into disciplines of history and anthropology, and into ideologies of nationality and development,[35] where do we find the motivation to search for such empirical possibilities? It has come from the impulse to break national territory apart theoretically to explore the substance of its local and global components, each with its own variety of subaltern history.

Agrarian history in South Asia is regional and local in its academic construction. This procedure emerged soon after Independence, as people sought their location in the national system. Daniel Thorner was in the first generation of scholars to argue that national economic planning needed to avoid reinforcing local power inequalities when setting to increase national wealth; to this end, he argued for regional agrarian studies by citing "the classic agrarian problem, namely, the inter-relation of the institutional framework on the one hand, with the level of output and the distribution of the product on the other."[36] By focusing on regions, scholars can analyze influences on social relations exerted by the nature of material resources, economic geography, land ownership, political order, labor and credit supplies, technology, demography, and other variables. The character of historical sources also encourages regional studies, because regional languages and institutions shape the nature and distribution of records. Regional research also allows closer collaboration between historians and social scientists within the field of agrarian studies.[37]

With his "classic agrarian problem" in view, Thorner defined thirty-nine agrarian regions in India.[38] By mapping dominant crops, geographers have since defined agricultural regions in more detail, showing that India is divided agriculturally east and west along a line that swerves around the north-south line of eighty degrees east longitude. To the east (and on the western coast of the peninsula) the climate is more humid and the dominant food grain is rice; in the drier west and interior, away from the coast, wheat and millets dominate. This division—inscribed by rainfall and drainage—has been significant in agrarian history for millennia and provides a template for agricultural regions today.[39]

Agrarian historians first concentrated their attention on regional revenue and property systems, but regions defined by dominant crops, ecology, social relations, and technologies have attracted more attention in the past few decades. Such complex agricultural regions had long histories before British rule, and they attained specific colonial identities. Each had traditions of governance that inflected its integration into British India. Their specific histories represent their distinctive modes of participation in the British Empire and the capitalist world economy, whereas comparisons among them describe South Asian patterns and trends. Each region adds something to the agrarian history of South Asia as a whole. What holds these regions together is not so much their inherently national unity or uniformity as their long-term dynamics of mobility and integration.

Focusing on regions, scholars have slowly shifted their perspective on conditions in agrarian South Asia as colonialism took hold, in the late eighteenth and early nineteenth centuries. Their efforts have produced a trend in scholarship toward increasingly detailed appreciation of historical interactions among large-scale political and economic forces and small-scale agrarian societies. Until the 1970s pre-colonial history presented a unitary—national—social formation, in which a subsistence-oriented peasantry inhabited self-sufficient villages for many centuries before the British inflicted dramatic shocks upon them. Today, agrarian regions appear to have been dynamic and differentiated for many centuries before colonialism. So the baseline for studies of change in modern times cannot be established any longer by simply assuming pre-colonial agrarian stability and village uniformity.

Agrarian Space and Time

The most radically revisionist arguments contend that agrarian South Asia was not essentially sedentary at all and that old models of "peasant society"—Marxian, Chayanovian, and others—do not apply. Farmers of various peasantry types were surrounded by a large population whose natural condition was movement rather than sedentarism. The mobility of the agrarian population as a whole has shed new light on the character of South Asian political economy more generally and laid new emphasis on the massive sedentarization that accompanied the privatization of property rights under the colonial regime in the nineteenth century.[40]

Rather than resting on peasant household production, pre-colonial agriculture involved large, complex social formations that organized production and distinguished regions by their modes of resource utilization. As irrigated agriculture in Asia differed in its social organization from dry farming in Europe,[41] India's irrigated, dry, shifting, and pastoral regimes differed from one another. In agricultural regions, land-use regimes competed, expanded, and succeeded one another. Struggles to control land, labor, water, animals, and other resources enmeshed farming families in social units—lineages, clans, tribes, villages, kingdoms, sects, religious networks, and states—that defined communities and social powers in production.

The social organization of settled farming differed from that of forest and pastoral production; likewise, short- and long-fallow farming, irrigated and dry farming, and commercial and subsistence farming all differed from one another. Such differences defined spatially segregated production regimes, which interacted politically and commercially—as they competed for control of the landscape and succeeded one another over time, as more intense forms of land-use expanded, or as jungle invaded farmland and irrigation works declined. The major trends in production entailed changes in population density and composition, in urbanization and economic differentiation, and in forms of land and labor control, and shifts in social relations. The broad trend toward more intense land use progressed as settled farming succeeded shifting cultivation and pastoralism, as farmers invested labor on the land, fought to control investments and output, and conquered pastoral and jungle people. More intense land use and shortening fallow complicated land rights as investments in plows, animals, and irrigation engaged households in hierarchies of rights and powers. In wet rice regimes, the most intense land and labor control entailed the most intricate, ritualized social and political hierarchies.

Pre-colonial agricultural expansion, competition, and succession occurred within political systems that organized power in production and also sustained kingdoms, empires, cities, and markets. Thus the history of agricultural regimes involved transformations in the institutional framework of farming as well as in the volume of output: the nature of subalternity changed and varied as a consequence. Pre-colonial South Asia no longer appears to be a land of self-reproducing villages, barely connected to commercial networks and connected to states only by the extraction of produce in taxes. Articulations among markets, states, and farms now seem complex, variable, and pivotal in the historical dynamics of agricultural expansion. Urban-rural, inter-village, inter-local, and inter-regional integration and differentiation characterized South Asia long before the Delhi Sultanate in the fourteenth century, and expanded during subsequent centuries of migration, agricultural growth, technological diffusion, and political change. In the seventeenth century, many regions larger than European kingdoms generated state revenue in complex networks of commercial exchange that were vital for local peasant reproduction and state revenue alike. In many regions, eighteenth-century agriculture depended on urban demand. In such regions—particularly in wet regions near the coast—European commercial centers laid foundations for colonialism. Near the coast, agrarian regions became deeply involved in worldwide operations of merchant capital. The English East India Company's profits depended on agrarian expansion that sustained overseas trade. Here, colonial power made its initial impact.[42]

In contrast to the older view that Company raj revolutionized agrarian life almost immediately by imposing steep revenue demands, alien forms of property, and commodity production, we now see that external forces emanating from capitalist empire entered agriculture differently and with differing results, depending on place and time. The Company had limited power to alter agrarian life. Colonial wars encouraged the building of alliances to raise revenue and lower administrative expenses. Native state treaties, Permanent Settlement, and Ryotwari settlements punctuated an extended political process in which the Company sought military stability, secure revenue, cheap administration, and allies; revenue settlements had initially modest, conservative effects on many agrarian structures, in a context where resistance and revolt set terms of negotiation. Initially, Company power expressed itself—in sharp contrast to its own self-representation—by increasing the power of the locally powerful social groups on which it conferred entitlements to land.[43] In dry areas of extensive cultivation, the privatization of peasant property cut into long-fallow farming and pastoralism, enabling settled farmers to make new territorial claims and also enabling the best-endowed among them to grab loose land and labor. In more intensely farmed wet regions, new landowners sought rental returns from property and more control over labor. Colonial governance, as it spread to the 1870s, exerted power on systems of agricultural production in a direction that was neither alien nor pressed at revolutionary speed, toward more intense land and labor use and more appropriation by settled farmers of land used by pastoralists and shifting cultivators, thus increasing the labor available to landowners and for agricultural commodity production.

Company policy alone could not revolutionize production. Too many variables intervened. Before 1840, war and declining prices weighed against vast intensification of land use and commercial production. Declining prices after 1820 reflect a slow synchronization of distant agrarian economies, but regional diversity was still so pronounced that agrarian history remained a collection of many regional histories sharing common themes and connections. By 1840 the convergence of regional histories gained a momentum that announced India's passage into the world of industrial capitalism. As Marx argued at the time, industry transformed rural India. As the railway and steamship wove together segments of world commodity production, farmers commercialized British India, whose cotton rushed into world markets with the blockade of the American South in 1860. Though ephemeral in the Deccan, the Civil War cotton boom launched a seventy-year expansion of rural investment that brought new social power into agrarian production relations, based on commodity production. Government propelled expansion with its own investments in infrastructure, to intensify land and labor use. Transportation and transaction costs declined as the railway spread, as Central India was opened up, and as the imperial administration grew. After 1870, interior regions entered the world economy and India became a national economy, as A. R. Desai has said. Cotton was a leading export item that brought the world economy into the dry interior of the peninsula.[44] Workers moved into labor markets inside India and throughout the British Empire. Industrial empire brought all agrarian regions in South Asia into an institutionally unified, national political economy within the capitalist world system.[45]

Yet Indian systems of agricultural production retained regional forms, not only because specific crops thrived in specific regions but because commodity production rested on systems of power with regional dynamics instituted by colonial governance. Colonialism first impressed modes of capitalist accumulation on wet regions near the coast. These areas, dominated by paddy cultivation and rice, had long and deep connections to overseas trade, ample water supplies, dense populations, intricate labor specialization, and cosmopolitan elite strata sustained by rent, commercial capital, and political status. The colonial state forged alliances with these elite classes in revenue settlements. By contrast, in the relatively sparsely populated dry interior tracts, colonialism confronted extensive land-use regimes in which farming concentrated on crops that withstood drought, above all, millets and pulses. In these areas, production relations were organized socially in militarized political systems. Here the British found their "martial races" and "yeoman peasants," in the Deccan, Rajasthan, Central India, the Mughal heartland, and Punjab. Here colonial conquest and revenue settlement weighed in favor of settled farming and farmers, enhancing their powers to effect the expansion of cultivation; it weighed against shifting and long-fallow farming, and above all, against pastoralism. In the dry interior, as opposed to the wet coastal areas, colonialism and capitalist power strove to supplant militarized forms of territorial authority that articulated connections among diverse and conflicting extensive land-use regimes in pre-colonial times.

At the outset of colonialism, in densely populated wet regions near the coast—from eastern Uttar Pradesh and Bihar, throughout Bengal, and along both coasts of the peninsula—a relatively large population of landless and near-landless laborers worked under controls enforced by their subordination to landed elites within intricate hierarchies of social power. To traditional modes of power colonial settlements added property law and debt coercion as means to exploit labor for the expansion of commodity production. By contrast, in the dry interior—from the tip of the peninsula to Rajasthan, Sind, and Punjab, where in 1800 extensive farming mingled with a vast pastoral animal economy, always on the hoof—the labor force moved constantly over short and long distances in the everyday conduct of subsistence, to work land, trade, fight, tend animals, flee drought, seek water, open and defend territory. Here labor was kept under control primarily by the kinship units that did farm work, by tribal and *jati* (caste) alliances that held territory, and by clan and caste alliances that built expansive royal domains, like the Maratha, Rajput, and Jat. Colonial land law reduced the rights of pastoralists and shifting cultivators dramatically, opening new realms for commodity production by settled farmers, who colonized land previously outside their reach, using the labor of groups who had lost traditional entitlements to land. Here the process of labor control was shorn of its military aspect and recast in terms of family, caste, village, market, and private property entitlements to land and labor.

Everywhere, agricultural commodity production entailed a more intensive utilization of land and labor. But in the densely settled wet regions, this effect emerged within domains and forms of established landed class power that had existed before colonialism and evolved to generate more commodity crops. In more sparsely settled interior regions, by contrast, more innovative social powers to produce agricultural commodities were created by the invention of landed property rights and by their allocation to settled farmers, who emerged with a new set of powers for controlling both land and landless and land-poor workers. This contrast explains why dry interior regions have appeared more innovative and progressive to observers of development in the colonial period and ever since. This impression was furthered by irrigation works built mostly in dry regions under the modern state. Broadly speaking (with some notable exceptions being the Krishna-Godavari Delta and the Brahmaputra Valley, where new irrigation created new paddy land, and dry peninsular tracts, where overexploitation destroyed the land), a comparison of India's dry interior with its wet coastal regions after 1800 reveals proportionately greater increase in productivity in the dry interior, and much more radical change in the social framework of agriculture.

Redefining the Nation against the State

I have described elsewhere the institutions of development that arose in the nineteenth century, which were composed of technological and political elements that reorganized production and inflected social power locally throughout the length and breadth of British India. I have also argued that India's development regime worked on the same plane of global engagement with world capitalism as did contemporary regimes in Europe and America. India's development before 1947 did not leave a large zone of agrarian life or agricultural production outside the influence of markets, although most agrarian capitalists were very small operators. In India, Europe, and America alike, family farms and world markets had become very tightly entangled by 1945.[46] India's national development regime was built on a populous but poorly capitalized bourgeoisie (urban and rural), which demanded state protection and investment. Enabling India's large and small, urban and rural, agricultural and industrial, mostly family-owned capital interests to flourish was the central imperative for Nehru's Congress. Although Nehru himself may have sought another kind of socialism, furthering capitalist growth in city and countryside was what his government actually did, because such growth provided the political base of Indian nationalism. In this respect, India is a model democracy and describing Indian nationalism as simply elitist is grossly incorrect.[47]

In all South Asian countries, development institutions covered all regions. States pursued a national development led by an advanced scientific, managerial elite and experts including those from the Ford Foundation and World Bank, but states could never bypass local and regional politics. Agricultural development depended on orderly relations between central planning and regional agrarian polities.[48] Development knowledge continued after 1947, as it did under British rule, to include expertise in how to hold and manipulate political power in agrarian regions. With states reorganization in India in 1956, regional agrarian power became more visible. Development, party formation, vote banks, and social mobility became so entangled that today there is no region where we can talk about local agrarian history without connecting it directly to state political machineries.[49] Modern politics "from below" included powerful movements to secure and expand upward social mobility, and the most powerful of these brought local struggles onto the national stage, as in East Pakistan, where the movement for

Bangladesh was propelled by upwardly mobile middle classes, and in Punjab, where the movement for Khalistan grew out of the Green Revolution. Dominant castes have formed political blocks that wield enough power to shape significantly the course of national events and state investments. It will be interesting to see what happens as the IMF raj tries to reduce agricultural subsidies. Certainly India will find it difficult to tax agricultural profits. International capital will need to accommodate powerful local and regional actors to secure global profiteering locally.

Globalization has brought cries of "capitalist invasion" from new waves of Indian nationalism, as the old-style nationalist state, with its development apparatus designed by Nehru's generation, is being relegated to the past. These complexly related trends derive in significant measure from India's own internal capitalist expansion—represented politically by the urban and rural middle class, which has overflowed the protective bounds of Nehru's development regime and now faces a growing mass of poor people who, if they were given the chance, would vote for massive increases in state expenditures to ameliorate poverty. Casting the old socialist Indian state as the villain of bureaucratic, statist development makes sense in this scenario. Yet the prosperous middle class and overseas Indians who are heavily involved in the new formation of India's economic and cultural identity come disproportionately from agrarian regions which have benefited most from past agrarian development. Economic disparities among Indian states as environments for private investment have increased rapidly in the past few decades, as have disparities between rich and poor in South Asia and internationally. Thus the regions in which local capitalists have gained most from the state have much to gain now from new private investment. Adventurous and well-connected captains of business and farming are taking the next giant step out of the confines of the Indian state and old-style national histories into the world market and transnational culture. Unfortunately, a critique of these dynamics of political culture is not a subject for discussion in Subaltern Studies, which has been at the academic forefront of debunking the historical heritage of nationalist politics.

As in the North American Free Trade Agreement and in the European Union, there is also in South Asia a political split between national and transnational bourgeois mentalities. Governments are deeply torn between globalism and nationalism (*swadeshi*). Nationals seeking state protection and nationals seeking new identities in world markets and "diasporas" are both striving to redefine themselves in Indian terms as nationals, albeit with contrasting views of the past. Media rhetoric about "capitalist invasion" includes cries of cultural chauvinism and calls for protection from multinational capital. This rhetoric does not come from poor people—the absolute subalterns—who lost out under the old regime of state capitalist development and continue to lose under the IMF raj, who still focus their attention and hopes for a better future on the Indian state, despite its failed promises. Poor wage laborers are not analyzing world markets and cultural trends to valorize the Indian state one way or another; they are moving to jobs in villages, cities, towns, in the Gulf, or anywhere, in labor movements that are old but barely visible in Subaltern Studies.[50] The rhetoric of invasion seems to come from aspiring but weak segments of Indian capital, who still need state support and seek to profit safely within India's internal markets.

Today, nationalism, subaltern assertion, and intellectual internationalism in South Asia are, as they have been for a century, simultaneous manifestations of a massive engagement with world capitalism. They represent the many faces of a multicultural, transnational middle class that certainly numbers well over 200 million people, spread over much of the world, from small villages to high-rise apartments in Bombay and New York. Its agrarian segments, the rustic, small-town middle class, are most often forgotten by scholars, particularly by those anthropologically inclined, including Subaltern Studies scholars. There is a vast, diverse rural population of petty, medium, and large bourgeois interests, which has grown steadily since 1850, when it already had a solid base, especially in regions near the great cities of Bombay, Karachi, Lahore, Calcutta, Dhaka, Madras, Delhi, and Colombo. Within a four-hundred-mile radius of those cities we can locate the history of South Asia's most upwardly and outwardly mobile agrarian folk for the past two centuries. Many of their sons and daughters came to the cities, and some went overseas.

From this perspective, agrarian history in South Asia is a continual engagement with world capitalism, and (very partial) national state protections of intranational enterprise from world markets after 1947 represent a brief period of relative isolation. This view is important for understandings of Subaltern Studies. It reveals that in British India, many acts of "peasant resistance" can be interpreted as conflictual, negotiative maneuvers. Many acts of "tribal revolt" turn out to be new versions of old battles among competitors for valuable cropland. So we must wonder why Subaltern Studies has so downplayed the search for empirical evidence on the transformation of agrarian environments, on the dynamics of capitalist development, on regional and class disparities, on the relationship between the state and agrarian social forces, and on the political organization of social conflict—all of which are so fundamental for understanding agrarian South Asia today.

Acknowledgments

This chapter was revised from a paper presented to the Yale Agrarian Studies Program on 16 November 1993. I have included some ideas taken from seminar discussion, for which I thank participants. Dina Siddiqi has also provided constructive critique. I am now also editing a volume of essays, *Reading Subaltern Studies* (Delhi, 2001), which considers the substance of Subaltern Studies more thoroughly. Here my discussion is strictly confined to agrarian issues.

Notes

1. David Ludden, "Agricultural Production and Indian History," in David Ludden, ed., *Agricultural Production and Indian History,* Oxford Themes in Indian History (New Delhi, 1995), pp. 1–35.
2. See David Ludden, *An Agrarian History of South Asia* (Cambridge, England, 1999), pp. 1–17, 180–90.
3. A. Moin Zaidi, ed., *A Tryst with Destiny: A Study of Economic Policy Resolutions of the Indian National Congress Passed During the Last 100 Years* (New Delhi, 1985), p. 54.
4. Ibid., p. 72.
5. The best account of this environment is Francine Frankel, *India's Political Economy, 1947–1977: The Gradual Revolution* (Princeton, 1978).
6. The 1975–76 turning point in India is now the starting point for many discussions of recent trends in political culture: see, for example, Uma Chakravarti, "Saffroning the Past: Of Myths, Histories, and Right-Wing Agendas," *Economic and Political Weekly* (31 January 1998): 225–32.
7. David Ludden, "Agrarian History and Grassroots Development," in Arun Agarwal and K. Sivaramakrishnan, eds., *Agrarian Environments: Resources, Representations, and Rule in India* (Durham, N.C., 2000).
8. See Fredric Jameson, *Postmodernism, or, The Cultural Logic of Late Capitalism* (Durham, N.C., 1991); and Ernest Mandel, *Late Capitalism* (London, 1975).
9. See David Ludden, ed., *Contesting the Nation: Religion, Community, and the Politics of Democracy in India* (Philadelphia, 1996), pp. 1–27 ff.
10. James Scott, *Seeing like a State: How Certain Schemes to Improve the Human Condition Have Failed* (New Haven, 1998).
11. When compared to other South Asian countries, as well as to Malaysia, Indonesia, and South Korea, the Indian state still remains relatively strong in the face of foreign capital.
12. *India Today* (which began publication in 1987 and looks like *Time* magazine), with a large circulation in urban India and North America, typically depicts the ordinary Indian citizen as a middle-class urbanite who is striving for personal fulfillment against a horde of cruelly inefficient government bureaucrats.
13. Burton Stein, "A Decade of Historical Efflorescence," *South Asia Research* 10, no. 2 (November 1990): 125–38.
14. *Subaltern Studies: Writings on South Asian History and Society,* vols. 1–10 (Delhi, 1982–99). See David Ludden, ed., *Reading Subaltern Studies: Critical Histories, Contested Meaning, and the Globalization of South Asia* (New Delhi, 2001).

15. Sumit Sarkar, *Modern India, 1885–1947* (Delhi, 1983).
16. Two examples of recent books that use Subaltern Studies vocabulary without participating in the Subaltern Studies project are Biswamoy Pati, *Resisting Domination: Peasants, Tribals, and the National Movement in Orissa, 1920–1950* (New Delhi, 1993), which is standard political history; and Nandini Sundar, *Subalterns and Sovereigns: An Anthropological History of Bastar, 1854–1996* (Delhi, 1997), which owes more to anthropology than to Subaltern Studies and concludes (p. 232) that Marx's *Eighteenth Brumaire* provides a better way of understanding Bastar rebellions than Ranajit Guha's *Elementary Aspects.*
17. Ranajit Guha, *Elementary Aspects of Peasant Insurgency in Colonial India* (Delhi,1983).
18. See Meghnad Desai, Susanne H. Rudolph, and A. Rudra, eds., *Agrarian Power and Agricultural Productivity in South Asia* (Berkeley, 1984), pp. 51–99.
19. The contrast between urban and rural is most elaborately described in Shahin Amin, *Event, Metaphor, Memory: Chauri Chaura, 1922–1992* (Berkeley, 1995).
20. A. R. Desai, *Social Background of Indian Nationalism* (Bombay, 1948, reprinted in 1954, 1959, and 1966). Page references are to the most recent edition.
21. Kathleen Gough, *Rural Society in Southeast India* (New York, 1981); A. R. Desai, *Peasant Struggles in India* (Delhi, 1979, reprinted 1981 and 1985); A. R. Desai, *Agrarian Struggles in India after Independence* (Delhi, 1986).The quotation is from Desai, *Peasant Struggles,* p. xi.
22. Guha, *Elementary Aspects,* pp. 333–34.
23. Ranajit Guha, "On Some Aspects of the Historiography of Colonial India," in Ranajit Guha and Gayatri Chakravorty Spivak, eds., *Selected Subaltern Studies* (New York, 1988), p. 37 (reprinted from *Subaltern Studies I*).
24. Ibid., pp. 41–42.
25. See S. J. M. Epstein, *The Earthy Soil: Bombay Peasants and the Indian Nationalist Movement, 1919–1947* (Delhi, 1988)
26. David Hardiman, *Peasant Nationalists of Gujarat: Kheda District, 1917–1934* (Delhi, 1981). Guha, *Elementary Aspects.* Gyanendra Pandey, *The Ascendancy of the Congress in Uttar Pradesh, 1926–34: A Study in Imperfect Mobilization* (Delhi, 1978). Rudrangshu Mukherjee, *Awadh in Revolt, 1857–1858: A Study of Popular Resistance* (Delhi, 1984). See also Dipesh Chakrabarty, *Rethinking Working-Class History: Bengal, 1890–1940* (Princeton, 1989).
27. Guha, "On Some Aspects of the Historiography of Colonial India"; Guha, "Dominance without Hegemony and Its Historiography," in Ranajit Guha, ed., *Subaltern Studies VI: Writings on South Asian History and Society* (Delhi, 1989), pp. 210–309. The essay in volume 6 is full of collective possessives—"our people," and so on—to indicate the author's own stance in relation to the subaltern nation.
28. I leave aside an expedient solution, in which an author identifies with and "speaks for" his or her "people."
29. Guha, *Elementary Aspects.*
30. Two important books are Utsa Patnaik, *Peasant Class Differentiation: A Study in Method with Reference to Haryana* (Delhi, 1987); and Atiyur Rahman, *Peasants and Classes: A Study in Differentiation in Bangladesh* (London, 1986). See also David Ludden, *Peasant History in South India* (Princeton, 1985; reprint, Delhi, 1989).

31. Scholars who claim to represent a unique, bounded culture, who claim to represent the integrity and voice of a people, operate within the terms of "imagined communities," and they often reconfigure orientalism as nationalist intellectuals. See David Ludden, "Orientalist Empiricism: Transformations of Colonial Knowledge," in C. A. Breckenridge and P. van der Veer, eds., *Orientalism and the Postcolonial Predicament: Perspectives on South Asia* (Philadelphia, 1993), pp. 250–78.
32. On "factual authority," see ibid. Another reason—the slow advance of empirical research—derives in part from the first. There is no scarcity of sources for this research, but the difficulty of using them to generate far-reaching conclusions escalates the unit cost of knowledge in agrarian historical research.
33. In this context, India's own native colonial elite, as depicted in Subaltern history, arose at the interaction of two worlds and thus became foreign to India, especially rural India.
34. The discipline of Indology analyzes the textual basis for traditional Indian culture; its alliance with anthropology was most brilliantly manifested by Louis Dumont. Development economics and sociology measure village India's modern material and social progress, using baseline measures of traditional (initial) conditions. Each discipline has its chronological limits. History concentrates on modernity, after 1750, and officially ends in 1947 with Independence and Partition. Anthropology and development studies examine post-1947 South Asia. Indologists rarely study any text after 1300 or "the coming of the Muslims."
35. See David Ludden, "India's Development Regime," in Nicholas B. Dirks, ed., *Colonialism and Culture* (Ann Arbor, 1992).
36. Daniel Thorner, "Agrarian Regions," in A. R. Desai, ed., *Rural Sociology* (Bombay, 1959), pp. 152–60.
37. See, for example, Desai et al., *Agrarian Power and Agricultural Productivity.* Also Donald W. Attwood, *Raising Cane: The Political Economy of Sugar in Western India* (Boulder, Colo., 1992).
38. Thorner, "Agrarian Regions." Thorner also compiled data from the 1931 census for Daniel Thorner, ed., *Ecological Regions of South Asia, circa 1930* (Karachi, Pakistan, 1996).
39. See Jasbir Singh, *An Agricultural Atlas of India: A Geographical Analysis* (Kurukshetra, India, 1974), p. 301 ff.
40. David Washbrook, "Progress and Problems: South Asian Economic and Social History," *Modern Asian Studies* 22, no. 1 (1988): 57–96.
41. Francesca Bray, *The Rice Economies: Technology and Development in Asian Societies* (London, 1986).
42. On the above points, see David Ludden, "Agrarian Commercialism in Eighteenth Century South India: Evidence from the 1823 Tirunelveli Census," in Sanjay Subrahmanyam, ed., *Merchants, Markets, and the State in Early Modern India* (Delhi, 1990), 215–41.
43. On these points, see David Ludden, "World Economy and Village India, 1750–1900," in Sugata Bose, ed., *South Asia and World Capitalism* (Delhi, 1990), pp. 159–77.
44. C. Shambu Prasad, "Suicide Deaths and Quality of Indian Cotton: Perspectives from History of Technology and Khadi Movement," *Economic and Political Weekly* (30 January 1999): PE-12–21.
45. See Sumit Guha, "Some Aspects of Agricultural Growth in Nineteenth Century India,"

Studies in History 4, no. 1 (1982): 57–86; and Guha, *The Agrarian Economy of the Bombay Deccan, 1818–1941* (Delhi, 1985). Also see references in Ludden, "World Economy and Village India"; and Ludden, *An Agrarian History of South Asia.*
46. Ludden, "India's Development Regime."
47. See Prem Shankar Jha, *India: A Political Economy of Stagnation* (Bombay, 1980).
48. See Ludden, *An Agrarian History of South Asia.*
49. See Francine Frankel, ed., *Dominance and State Power in Modern India: Decline of a Social Order* (Delhi, 1989).
50. Such movements have now entered the field in the form of a personal memoir from Sudesh Mishra, "Diaspora and the Difficult Art of Dying," in Gautam Bhadra, Gyan Prakash, and Susie Tharu, eds., *Subaltern Studies X: Writings on South Asian History and Society* (Delhi, 1999), pp. 1–7.

References

Amin, Shahin. *Event, Metaphor, Memory: Chauri Chaura, 1922–1992.* Berkeley, 1995.

Attwood, Donald W. *Raising Cane: The Political Economy of Sugar in Western India.* Boulder, Colo., 1992.

Bray, Francesca. *The Rice Economies: Technology and Development in Asian Societies.* London, 1986.

Chakrabarty, Dipesh. *Rethinking Working-Class History: Bengal, 1890–1940.* Princeton, 1989.

Chakravarti, Uma. "Saffroning the Past: Of Myths, Histories, and Right-Wing Agendas." *Economic and Political Weekly* (31 January 1998): 225–32.

Desai, A. R. *Social Background of Indian Nationalism.* Bombay, 1948, reprinted in 1954, 1959, and 1966.

———. *Peasant Struggles in India.* Delhi, 1979, reprinted 1981 and 1985.

———. *Agrarian Struggles in India after Independence.* Delhi, 1986.

Desai, Meghnad, Susanne H. Rudolph, and A. Rudra, eds. *Agrarian Power and Agricultural Productivity in South Asia.* Berkeley, 1984.

Epstein, S. J. M. *The Earthy Soil: Bombay Peasants and the Indian Nationalist Movement, 1919–1947.* Delhi, 1988.

Frankel, Francine. *India's Political Economy, 1947–1977: The Gradual Revolution.* Princeton, 1978.

Frankel, Francine, ed. *Dominance and State Power in Modern India: Decline of a Social Order.* Delhi, 1989.

Gough, Kathleen. *Rural Society in Southeast India.* New York, 1981.

Guha, Ranajit. *Elementary Aspects of Peasant Insurgency in Colonial India.* Delhi, 1983.

———. "On Some Aspects of the Historiography of Colonial India." In Ranajit Guha and Gayatri Chakravorty Spivak, eds., *Selected Subaltern Studies.* New York, 1988.

Guha, Sumit. "Some Aspects of Agricultural Growth in Nineteenth Century India." *Studies in History* 4, no. 1 (1982): 57–86.

———. *The Agrarian Economy of the Bombay Deccan, 1818–1941.* Delhi, 1985.

Hardiman, David. *Peasant Nationalists of Gujarat: Kheda District, 1917–1934.* Delhi, 1981.

Jameson, Fredric. *Postmodernism, or, The Cultural Logic of Late Capitalism.* Durham, N.C., 1991.

Jha, Prem Shankar. *India: A Political Economy of Stagnation.* Bombay, 1980.

Ludden, David. *Peasant History in South India.* Princeton, 1985; reprint, Delhi, 1989.

———. "Agrarian Commercialism in Eighteenth Century South India: Evidence from the 1823 Tirunelveli Census." In Sanjay Subrahmanyam, ed., *Merchants, Markets, and the State in Early Modern India.* Delhi, 1990.

———. "World Economy and Village India, 1750–1900." In Sugata Bose, ed., *South Asia and World Capitalism.* Delhi, 1990.

———. "India's Development Regime." in Nicholas B. Dirks, ed., *Colonialism and Culture.* Ann Arbor, 1992.

———. "Orientalist Empiricism: Transformations of Colonial Knowledge." In C. A. Breckenridge and P. van der Veer, eds., *Orientalism and the Postcolonial Predicament: Perspectives on South Asia.* Philadelphia, 1993.

———. "Agricultural Production and Indian History." In David Ludden, ed., *Agricultural Production and Indian History.* Oxford Themes in Indian History. New Delhi, 1995.

———. *An Agrarian History of South Asia.* Cambridge, England, 1999.

———. "Agrarian History and Grassroots Development." In Arun Agarwal and K. Sivaramakrishnan, eds., *Agrarian Environments: Resources, Representations, and Rule in India.* Durham, N.C., 2000.

Ludden, David, ed. *Contesting the Nation: Religion, Community, and the Politics of Democracy in India.* Philadelphia, 1996.

———. *Reading Subaltern Studies: Critical Histories, Contested Meaning, and the Globalization of South Asia.* New Delhi, 2001.

Mandel, Ernest. *Late Capitalism.* London, 1975.

Mishra, Sudesh. "Diaspora and the Difficult Art of Dying." In Gautam Bhadra, Gyan Prakash, and Susie Tharu, eds., *Subaltern Studies X: Writings on South Asian History and Society.* Delhi, 1999.

Mukherjee, Rudrangshu. *Awadh in Revolt, 1857–1858: A Study of Popular Resistance.* Delhi, 1984.

Pandey, Gyanendra. *The Ascendancy of the Congress in Uttar Pradesh, 1926–34: A Study in Imperfect Mobilization.* Delhi, 1978.

Pati, Biswamoy. *Resisting Domination: Peasants, Tribals, and the National Movement in Orissa, 1920–1950.* New Delhi, 1993.

Patnaik, Utsa. *Peasant Class Differentiation: A Study in Method with Reference to Haryana.* Delhi, 1987.

Prasad, C. Shambu. "Suicide Deaths and Quality of Indian Cotton: Perspectives from History of Technology and Khadi Movement." *Economic and Political Weekly* (30 January 1999).

Rahman, Atiyur. *Peasants and Classes: A Study in Differentiation in Bangladesh.* London, 1986.

Scott, James. *Seeing like a State: How Certain Schemes to Improve the Human Condition Have Failed.* New Haven, 1998.

Singh, Jasbir. *An Agricultural Atlas of India: A Geographical Analysis.* Kurukshetra, India, 1974.

Stein, Burton. "A Decade of Historical Efflorescence." *South Asia Research* 10, no. 2 (November 1990).

Subaltern Studies: Writings on South Asian History and Society. Volumes 1–10. Delhi, 1982–99.

Sundar, Nandini. *Subalterns and Sovereigns: An Anthropological History of Bastar, 1854–1996.* Delhi, 1997.

Thorner, Daniel. "Agrarian Regions." In A. R. Desai, ed., *Rural Sociology.* Bombay, 1959.

Thorner, Daniel, ed. *Ecological Regions of South Asia, circa 1930.* Karachi, Pakistan, 1996.

Washbrook, David. "Progress and Problems: South Asian Economic and Social History." *Modern Asian Studies* 22, no. 1 (1988).

Zaidi, A. Moin, ed. *A Tryst with Destiny: A Study of Economic Policy Resolutions of the Indian National Congress Passed during the Last 100 Years.* New Delhi, 1985.

PART IV *Economic Histories, Local Markets, and Sustainable Development*

CHAPTER EIGHT

Contesting the "Great Transformation": Local Struggles with the Market in South India

RONALD J. HERRING

Anomalies and Puzzles: Interpreting Polanyi

Much social science discourse, implicitly or explicitly, is about the search for central tendencies and structured simplifications: the mean, the generalizable, the patterned. Exceptions to central tendencies are often dismissed as mere "cases," "outliers," or even "error terms." We might learn from the practice of clinicians: anomalies also suggest puzzles.

The state of Kerala in Southwest India is sufficiently anomalous that many Indologists would prefer to ignore it. Like Sri Lanka, it is anomalous for its very high levels of life expectancy and literacy and very low infant mortality rates despite aggregate poverty.[1] Poverty reduction, which has lagged generally in India, has proceeded significantly faster in Kerala than in other Indian states—more than 120 times the rate of Bihar, for example, or 4 times that of Rajasthan.[2] Its agrarian organizations have historically been well developed, in contrast to most of the subcontinent and much of the poor world. Kerala elected what was arguably the first communist government in the world in 1957 and remains—contrary to the global pattern—an electoral stronghold of communism in a nation of decidedly centrist political tendencies.[3] Its agrarian reforms have been radical, abolishing an especially oppressive rentier landlordism integrated with agrestic serfdom, in a period of Indian history dominated by inaction on the agrarian question. Agrarian labor legislation in Kerala establishes entitlements anomalous for the poor world and radical by the standards of rich nations. In a remarkable decade, 1970–80, beginning with the effective date of the land reform and ending with the mobilization of farmers against labor reforms derisively referred to as the "factory acts" and hailed as a magna carta for labor by the government, much of Kerala's agrarian world was turned upside down.

In order to understand the political dynamics of this transformation, I find myself drawn to a reevaluation of the work of Karl Polanyi on responses to transitions to market society. This theoretical turn might itself seem puzzling and anomalous. Polanyi wrote *The Great Transformation* in 1944, when the civilization dominated by market societies seemed to have collapsed into barbarism and disintegration. He explicitly argued that both socialism and fascism were responses to the failures and disintegrative effects of the "self-regulating" market system domestically and internationally. Michael Hechter correctly noted that Polanyi's polemical and problematic tract "appeared to be the kind of book likely to attract hostile notices from nearly all quarters. In fact, it has become something of a classic . . . and interest in Polanyi, if anything, continues to build." Douglass North suggests an explanation: "The stubborn fact of the matter is that Polanyi is correct in his major contention that the nineteenth century was a unique era in which markets played a more important role than at any other time in history. Polanyi not only argued convincingly that economic historians have overplayed the role of markets in ancient economies, but argued with equal force that the market was a declining 'transactional mode' of the twentieth century as well. Even more embarrassing is the failure of economic historians to explain a major phenomenon of the past century—the shift away from the market as the key decision-making unit of economic systems."[4]

Polanyi had a simple answer: market society—a society organized by markets as a central principle—was inherently unstable, always moving away from its putative equilibrium point. At about the time North made this observation, a counterreaction had become observable: a global shift away from non-market allocative devices and toward markets.[5] Polanyi sets this problematic as the core of the modern dilemma: the dialectic of setting the boundary between market and society, between what markets can and should control and what they should not.

Polanyi's conceptualization of the "great transformation" to market society begins with the statement "What we call land is an element of nature inextricably interwoven with man's institutions. To isolate it and form a market out of it was perhaps the weirdest of all undertakings of our ancestors."[6] Simultaneously, market society required the extraction of one element of human existence—the capacity to work—in order to create the complementary commodity of labor. In Polanyi's formulation, pre-market economic relations, norms, and outcomes were "embedded" or "submerged" in social relations generally. The extraction and elevation of market-driven outcomes from their social moorings produce significant social conflicts and centrally involve the state. The creation of commodity markets for land and labor has been accompanied at various times by resistance, coercion, rebellions, imprisonment, and social disorganization of extreme forms; at a minimum, new systems of meaning must emerge in consequence. Forging a new normative order such that bizarre relations (from the perspective of a pre-capitalist normative structure) become normal and legitimate is itself a major social transformation. From Polanyi's perspective, this process was "weird" in the sense of unnatural, and fundamentally wrong in a normative sense.

From the perspective of the late twentieth century, established market society seems far more natural, and the normative complaints less compelling; market society has been more vigorously naturalized through the collapse of its antinomy of state socialism. But the making of market society, Polanyi argues, inevitably produces contestation that shapes and changes not only allocative institutions but also moral economy.

Polanyi's contribution to positive theory is difficult to tease out. He insists on what Block and Somers call "holistic social science" in the study of social change.[7] The standard critique is that he thereby is guilty of reifying "society." But Polanyi in practice illuminates market dynamics as these dynamics affect whole human beings, reembedded analytically in their social moorings. His analysis of class is thus more Weberian than economistic, and it resonates with the thick theory of rationality rather than the thin (which presupposes the universality of the *homo oeconomicus* he deplores). As an institutionalist, he argues that individuals will indeed behave as the interest maximizers of neoclassical economics in market society, but doing so results either in efforts to defeat the market or in social disintegration.

The implication for economic theory is that real markets function through institutions established through political processes reflecting interests, particularly interests in security. The "self-regulating market" society is utopian and ephemeral. Instead, society "protects itself" from the insecurity of "exposure" to the disintegrative effects of atomistic pursuit of material interests; thus, "regulation and market grew up together." Polanyi saw the horror of his time, fascism, as the reunification of polity and market; unbearable tension created by the artificial separation of the two spheres of human existence produced disequilibrating social pressures.[8]

In less apocalyptic terms, public policies and social movements may be usefully analyzed in dialectical relation to market forces. In contrast to invisible-hand notions of integration from individualism, Polanyi argues that disruptions of society from markets produce, inevitably, counterreactions to hem in, bound, and constrain the market. These constraints do not defeat the market but rather produce new dynamics resulting in institutional change. As societies and their institutional bases differ, so will the forms of market regulation, but convergence is to be expected from the very logic of market-society tensions. Whereas powerful nations and individuals may find their interests served by a laissez-faire regime, the periphery of nations and classes contests market dynamics. Stephen Krasner analyzed the international expressions of these interests by poor nations in his book *Structural Conflict* (1985), significantly subtitled *The Third World against Global Liberalism.* Peasant studies deals with social groupings at the periphery of societies; the "defensive-reaction" and "moral-economy" school of peasant studies is deeply indebted to Polanyi's perspective.[9]

Polanyi's polemics may appear picaresque in today's global rush toward markets, but the questions implied in his problematic remain as compelling as ever. Kerala introduces puzzles with regard to these questions because of the vigor of popular reactions to the great transformation, the unusual political expression of these responses, the anomalous character of public policy driven by mobilization, and the notably salubrious human-welfare implications of responses. In the remainder of this chapter I shall illustrate the usefulness of three elements of Polanyi's implicit positive theory: first, the defensive-reaction dynamic as an explanation for anomalous leftist organization that drove Kerala's political exceptionalism; second, on a more micro scale, institutional innovation as a residue from struggles with the market, culminating in the "factory acts" period (1970–80) in Palakkad district; and finally, the necessity of reinserting institutions into economic analysis in order to understand paddy agriculture after the reembedding of markets.

Explaining Public Moral Economies

Karl Marx wrote in the *New York Daily Tribune* of 25 June 1853:

> Now, sickening as it must be to human feeling to witness those myriads of industrious patriarchal and inoffensive social organizations disorganized and dissolved into their units, thrown into a sea of woes, and their individual members losing at the same time their ancient form of civilization and their hereditary means of subsistence, we must not forget that these idyllic village communities . . . had always been the solid foundation of Oriental despotism . . . enslaving [the human mind] beneath traditional rules . . . contaminated by distinctions of caste and by slavery. England, it is true, in causing a social revolution in Hindustan, was actuated only by the vilest of interests. . . . But . . . the question is, can mankind fulfill its destiny without a fundamental revolution in the social state of Asia? If not, whatever may have been

the crimes of England she was the unconscious tool of history in bringing about that revolution.

Marx was enthusiastic about the dissolution of pre-capitalist society because of the effects on material progress engendered thereby; Polanyi romanticized pre-capitalist society because of what he saw as the results of social disintegration engendered by market society. Marx saw new freedom for labor in its liberation from traditional shackles; Polanyi feared that "uprooting" laboring individuals from their "social moorings" through the commoditization of labor created dangers from "exposure." Both argued that societies would attempt to defeat domination by market, whatever the aggregate material advantages it promised. These attempts both reflect and constitute a public moral economy.

In Kerala, the decade 1970-80 compressed a remarkable span of history into a short span of years. New property rights were established, old ones extinguished; novel property-like rights and obligations were created, distributed, and then redistributed. In a sense, the transition was from remnants of an almost "feudal" agrarian system to heavily regulated market capitalism. It began with the replacement of the Left-communist-led coalition government and enforcement of its radical land reforms; it ended with the return of that coalition and the announcement that old-age pensions for agricultural laborers would be paid from general revenues, not from farmers' mandatory contributions as stipulated in a law passed only six years before. It was a decade of struggle and institutional change that incorporated into public law key elements of the pre-market moral economy. Elements of that moral economy were used, as James Scott claims they typically are, as weapons of the weak in their confrontation with the strong.[10]

The pension for farm workers captures this dynamic. The very notion of old-age security was embedded in the ideological (and, to a far lesser extent, material) practice of pre-capitalist landowners.[11] Public law—the Kerala Agricultural Workers Act (KAWA)—imposed on owners of land the responsibility of providing for the sustenance of the living bearer of labor power even after that labor power had ceased to have market value. Farmers mobilized in protest, on grounds of the unprofitability of agriculture under the new regime. Crafting a class compromise to keep production going forced a leftist government attempting to hold together the shards of its historic agrarian alliance to assume the farmers' burden and pay it from a severely strapped treasury. The provision of old-age security then in one decade represented a movement in public moral economy from dyadic paternalism to class redistribution to distributive routines characteristic of social democracy.[12]

Shifts in public moral economy have an irreducibly normative component.

Farmers in Kerala resisted the regulation of agriculture with the telling critique that "a paddy field is not a factory."[13] The original "factory acts" of nineteenth-century England—to which they compared the new legislation—were based in part on the principle that the labor of children was fundamentally different from the labor of adults and should be protected from market allocation. Kerala's agrarian factory acts established security of employment at administered wages whatever the pressures of the market. Farmers considered the "permanency" mandate of the law egregiously inappropriate for agriculture. College professors have long believed that comparable tenure on campus is acceptable and right; state legislatures, mass publics, and some university administrators are not so sure. In both cases the argument is normative and meant to defeat market forces.

Many elements of the public moral economy are likewise disputed along class lines, precisely because the market applies equal logic to unequal situations. Outcomes are driven by success in politics; leftist coalitions have established in Kerala a public moral economy valuing labor over profits at the margin. But other elements of the public moral economy have approached Polanyi's reifying argument that "society" protects itself from the market. Revulsion at slavery, exploitation of child labor, or discarding of the elderly simply because the market has ceased to value their labor power are all legitimately termed a "societal" response to market rule. Nevertheless, the struggles that have been necessary to obtain those outcomes indicate the absence of pre-existing normative consensus. One strong implication of Polanyi's work for positive political theory is that those most vulnerable to market dynamics are most likely to pursue the politics of market rigging. Yet the politics of "farmer movements" and Kerala's Legislative Assembly indicate how broadly the perception of vulnerability may extend—even to sections of society relatively privileged by existing property endowments. The politics of rejecting market vulnerability are practiced by both the weakest and the strongest in Kerala, as in much of the world.

The struggle to abolish landlordism in Kerala pitted bitterly contesting moral economies against one another, but the end result was surprising consensus that the outcome was normatively right: even landlords admitted that "those days have passed." Contrary to the pattern in Latin American societies, no serious movements for reversal of the land reforms have emerged in Kerala, despite changes in regimes. The historical experience of exorbitant power of landowners in a society with so many propertyless people has left a residue of hostility to the normative position that market allocations have legitimacy a priori. The residue of these conflicts is an interventionist and welfarist political economy that attempts to keep the market in its place.

Politics of the Transformation

The great transformation in the northern section of contemporary Kerala—Malabar—was initiated by a colonial state in need of clearly defined property rights and local political allies.[14] Malabar under colonial rule became in many ways the archetypal disintegrating agrarian system; one colonial officer said it attained "the unenviable reputation of being the most rack-rented place on the face of the earth."[15] With the introduction of colonial law, particularly the imposition of a legal system based on the absolute notion of land as private property, traditional overlords were able to evict tenants and raise rents according to market logic enforced by the police powers of a colonial state. Property claims were disentangled from their broader social moorings, and thus from their functions. As courts and administrative law protected the property claims of landlords, the necessity of good patron-client relations diminished; control of assets was guaranteed by higher authority. The Collector and District Magistrate of Malabar commented on the result in 1887: "Fanaticism of this violent type flourishes only upon sterile soil. When the people are poor and discontented, it flourishes apace like other crimes of violence. The grievous insecurity to which the working ryots [peasants] are exposed by the existing system of landed tenures is undoubtedly largely to blame for the impoverished and discontented state of the peasantry."[16]

The transformation was begun in southern Kerala by traditional rulers seeking to consolidate a state from feudal lords and introduce property relations conducive to commercial development. Under both colonial rule and commercializing rajas, slavery was abolished in law in favor of nominally free labor, although societal resistance on the ground prevented the emergence of fully commoditized labor until well into the independent period. Villagers of Padoor and Nallepilli in Palakkad district argue that before "the rules were made" (in the mid-1970s) and especially before "the EMS [first communist] government came to power" (in 1957), the status of field laborers who are the largest class locally was something between serf and slave (*adima*).[17]

Buttressing landlordism with novel property rights under colonialism produced the earliest societal reactions in the form of suicidal *jacqueries* (the "Mappila outrages" of colonial discourse) throughout the nineteenth century and into the twentieth. Although colonial land policy changed one aspect of landlordism—the institutional meaning of landed property—with resulting economic insecurity for the politically important middle peasants, the socially oppressive character of landlordism was left in place, perhaps strengthened.[18] Evictions could be enforced through courts; lands could be bought and sold with no regard for subsidiary traditional claims. Mortgages of smallholders could be and were foreclosed (an especially important phenomenon in the Great Depression, accentuated by Kerala's early and extensive integration into global commerce). The colonial state's revenue demands put additional pressure on lower orders just as its police *thana* became in a real sense the auxiliary arm of landlordism. Simultaneously, severe social degradation along traditional lines remained in force.[19]

Though dramatic, the great transformation was incomplete in one politically important way: land was thoroughly commoditized, but labor remained tied to its social moorings of humiliation and subordination, laying the basis for both moral outrage and militant organization. On a grand scale, these mobilizations created a leftist coalition that was able to attain power by forming governments; these governments legislated away the traditional agrarian system.[20] On a more local scale, consider the history of Padoor, in Alathur Taluk, Palakkad District. Padoor approaches the stereotypical view of rural Kerala: agrarian unions and the communist party are strong on the ground; class conflict has been extensive.

Participants in the agrarian movements in Padoor explain the origins of mobilization as follows. Both tenants and laborers were "slaves" in those days, subordinated thoroughly by landed power of the Brahmin (Namboodiri) and aristocratic Nair communities.[21] Hours of work for the laborers were in principle fixed by the demands of their superiors. The local struggle that precipitated militant organization came in 1948. The issue was primarily one of sexual exploitation of working women, particularly Harijans ("untouchables"), but the conflict spread quickly into other dimensions of inequality and dignity, taking the form of wage and land struggles, reaching a high point in 1952.

The category of field laborer was essentially covariant with caste status of Irava and Cheruma. The Cherumar were subjected to especially humiliating social practices from which the Iravas had incrementally freed themselves. Cherumar were prohibited from coming within a half furlong of the temple. They were not allowed to wear any cloth above the waist or below the knees; wrapping the *torthu* around the head for protection from the sun was forbidden as well.[22] Special language had to be employed by and about Cherumar. They were not to use the first person in addressing superiors but were to construct sentences in the third person (such as "your serf [*adiyan*] seeks permission from the master [*thamburan*] to drink water"). Their food was referred to as *karhikadi* (from charcoal, or ash, connoting black, hence inferior, and a mixture of bran and water [*kadi*] usually fed to buffaloes, connoting animal food), in opposition to the food of superiors, generically referred to as *ari* (hulled rice). Their money was preceded by the implicitly derogatory *chempu* (copper) to distinguish it from superiors' money.

Cherumar were to scrupulously separate themselves from contact with superiors; if a bunch of plantains from their compound was to be delivered to the owner, it had to be left at the gate. The custom that was instrumental in the 1948 strike was a common perversion of the separation injunction: laborers were forced to bring their new brides to the home of the owner for the right of the first night, as it was beneath the dignity of higher castes to go to the hut of the laborer. "Seeing the bride" was an especially galling instance of the general sexual access to lower-caste women claimed by high-caste men and figured prominently in the strike.[23]

The structural fusion of caste and class—of standing and life chances—produced a congruence of aims in the early leftist mobilization. Labor's very embeddedness (in Polanyi's terms) was a source of moral outrage and political action. Polanyi insisted that a material reductionist explanation of behavior was flawed, that individuals in pre-market societies act primarily to "safeguard" their "social standing . . . social aims . . . social assets."[24] All allocative systems are inescapably embedded in a moral order of some sort. During the mobilization to abolish landlordism, there was no contradiction between asserting social standing and material gain. Commoditization of human beings as units of labor is alienating in one sense, as stressed in Marxian analysis, but liberating in another, as the embedding of labor in social relations traditionally produced claims on the person that were unacceptable to the living bearers of labor power. Over time and through local struggles, the only claim of farmers on workers became the specific hours of labor purchased, regulated by law.

Slogans of mobilization in this period were explicitly anti-landlord, but that objective was expressed locally as a necessary condition for wage increases. Poverty exacerbated humiliation that was socially organized against both laborers and tenants. Tenants who hired laborers were understood to be too strapped by rental exactions to pay more, hence the potential for unity between classes with opposed material interests through a promise to share out the rent fund when the landlords were defeated. Major strikes in 1957 coterminous with the first communist ministry escalated to demands for radical land reform, not merely increased wages. As in Kerala generally, mobilization increased after the dismissal of the communist ministry by Delhi, reaching a second high-water mark in 1961. During the 1960s, there was episodically virtual famine in the village; prices for local stocks escalated beyond the reach of laborers. Farmers refused to pay wages in kind as rice prices rose; as a result, workers were unable to afford rice for extended periods. Were it not for ration shops (which worked imperfectly because of the state's difficulty in procuring and distributing grain), laborers believe, many would have starved. The union stood for dignity and wages; ration shops stood between hunger and subsistence. Both were communist projects.

The small numbers, parasitism, and decadence of rentier landlords made them vulnerable symbolic targets in political and normative terms; their claims to fee-simple ownership rights devoid of obligation were as insulting as their continued claim to deference and privilege.[25] The tactic of targeting particularly oppressive and dissolute landlords for local protests associated vile behavior with the institution of landlordism. In Kerala generally, as in Padoor, powerful movements for poverty alleviation and removal of caste indignities were fused under a leftist leadership that understood the integrated character of multiple strands of landed domination and tailored their programs accordingly.[26] Their political analysis was reminiscent more of Polanyi than of the mature Marx, playing on the peculiar form of embedded production relations and their manifestation in severe caste indignities. Their effective political strategies birthed a peculiar form of regulated agrarian capitalism.

Local Origins of the Factory Acts

Polanyi's analysis of a great transformation to market society offers a normative perspective on political mobilization at the bottom, and a positive theory for understanding the politics. Likewise, there are lessons for a positive theory of economics. "Institutionalists" and "substantivists" among economists have a crucial point stressed by Polanyi: the historical origins and institutional structure of any particular economy are vital components of its working logic. In Kerala, these institutions were, somewhat ironically, forged in the struggle against a precapitalist formation almost "feudal" in character as it changed in response to market forces: the local variety of landlordism as a social system. The vector sum of those conflicts imposed a public moral economy as boundaries within which the market can work, recognizing the moral claims of the tenants and then the permanent laborers to the extent politically and fiscally practicable. The resulting regulated capitalism cannot be understood without reference to these new institutions.

Like the land reforms, the factory acts were born in struggle; that struggle was itself a continuation of the mobilization for land reforms. New institutions created in the struggle drew on traditional rights reinterpreted and new demands drawn from the political promise of sharing the rent fund.

The payoff for the laborers for supporting land-to-the-tiller was to be sharing the spoils. Rental exactions on almost half the nonplantation arable in the state ended on 1 January 1970. The United Front government led by the Communist Party of India-Marxist (CPI-M, 1967–69) had formulated an ordinance that took effect after the fall of the ministry: the Kerala Agricultural Workers Payment of Prescribed Wages and Settlement of Agricultural Disputes Ordinance (No. 5, 1970). A district-level Industrial Relations Committee (IRC) was established on 7 July 1970 to deal with agriculture in Palakkad District. By 10 August the *tehsildar* (subdistrict official) of Palakkad Taluk warned that the CPI-M was organizing for immediate implementation of the minimum wage. The Collector—the chief administrative officer of the district, the quaint title reflecting his main function under colonialism—immediately assembled representatives of landowners and laborers to prevent confrontation; the individuals called overlapped considerably with the members of the district-level Industrial Relations Committee.[27]

The Marxist union of agricultural workers (KSKTU) had at this point explicitly decided to avoid confrontation, relying on *dharna* (silent picketing) and sloganeering before representatives of the local state at village offices. Clever and artistic pamphlets were distributed. Demands included increases in the wage rate, cessation of police interference in labor struggles, and withdrawal of criminal cases against activists. The tactics at this stage were called "village-level *satyagraha samaram* (truth struggle)," linking the Gandhian invocation of rightness of cause with the Marxist position on the necessity of class struggle. In response to the implied threat, a prominent CPI-M Member of the Legislative Assembly (MLA) was called back from the capital to meet with the Collector; simultaneously, Revenue Department officials (who are responsible for land issues) began conferences with officers of the Labour Department for implementation of the minimum wage.

Responding to similar threats of confrontation in various parts of Kerala, the state government enhanced the minimum wage by ten percent, by ordinance (No. 11 of 1970). At a meeting in Palakkad town on 9 August, agreement was reached by the IRC on the new wage structure. However, the Deshiya Karshaka Samajam (National Farmers' Association, or DKS) representative dissented vigorously. With peaceful demonstrations continuing, the Kerala Agricultural Workers Bill was introduced in the Assembly. The district MLAs met with representatives of farmers, workers, and government officials in a second conference held in September 1971, where wages were again enhanced. These major conferences were held during harvest season, when strikes usually peak and labor's tactical power is maximized.

The new wage structure formalized the traditional distinction between "season" and "off-season" rates, legitimizing the market forces that made wages higher when demand peaked and lowered them when there was little work in the fields. Formalized into law as well was the wage differential between women and men. In both cases, the difference settled on was one *edangazhi* (750 grams) of paddy. The women's off-season wage was 83 percent of the male rate, the season rate somewhat more. Although some in the villages felt that the gender differential was unjust, both season and off-season gender ratios were better than the aggregate ratio in the United States.

Settling on a wage rate defused one set of conflicts but did not assure implementation. Real power lay with individual farmers; the burden of redress was on unions. Another major conference was held in December 1971 to discuss demands by unions that farmers actually pay the agreed rate. At the same time, the police were brought into taluk-level meetings with representatives of land and labor to formulate means of ensuring enforcement and to create local dispute resolution mechanisms. The DKS sought to minimize harvest conflict by changing the traditional harvest share system (newly fixed at 1/7, previously 1/12) to one in which a gender-differentiated flat rate would be paid. Such a change would allow farmers to capture more of the benefits of technological change; as yields rose, laborers would not share in the increment unless struggles to increase the harvest wage were conducted each season.[28] Moreover, the DKS proposal would have disadvantaged women, who traditionally received the same harvest share as men. The KSKTU flatly rejected the proposal and threatened intensified strikes if any change in the status quo was attempted. Labor's strategy was to defend, selectively, traditional rights while simultaneously enhancing its share of total gross product.

Rigging a rural labor market necessitates surveillance and intervention. New disputes arose over farmers' refusal to comply with a government notification that registers of laborers and records of wage payments must be maintained and wage receipts must be issued. The government saw this as the only means of assuring compliance; farmers saw it as an unacceptable imposition, pleading illiteracy and inexperience with record keeping. Peasants hired accountants. Police, Labour, and Revenue officials finally obtained farmers' consent by agreeing to establish dispute resolution machinery at the *firka* level, thus bringing conciliation closer to the farmers' fields. These registers became central in the ensuing conflicts over permanency rights for individual workers.

As mediation continued, more unions began to appear at meetings; all were nominally independent, but almost all could be easily associated with some political party by looking at their leadership. Most had minimal representation on the ground, where the Left-communist KSKTU dominated, but every major party made symbolic commitments to the most oppressed (and numerous) class in Palakkad by sponsoring unions. It is not clear that proliferation of unions helped labor's cause, because infighting resulted, but farmers became convinced that the entire active political spectrum was deployed against them. Some tried to convert their lands to less labor-intensive crops (such as coconut); others exited agriculture, while still others continued the political struggle. Investment declined.

In spite of the impressive coordination of government departments and representatives of class organizations orchestrated by the Collector, it should not be assumed that the local state was any more unified than the national state. The tehsildar of Mannarkkad, for example, had consistently refused to call the land-labor conferences ordered by the Collector. When responding to a reprimanding communiqué marked "urgent," the tehsildar responded that any move to convene a meeting would only strain the existing cordial relations between land and labor; political parties and vested interests would take advantage of the situation to foment trouble (L.Dis. 7316/71). As local organization was uneven, so, too, were protest and implementation of pacts.

Petitions and letters to the Collector and MLAs proliferated as regional meetings continued to demonstrate that enforcement and local dispute resolution were uneven. The density of organizations was extraordinary. The Malabar Regional Harijan Samajam petitioned the Collector that outcaste laborers in Chittur Taluk were being forced to work excessive hours and were not paid the minimum wage. The Assistant Labour Officer responded that the absence of militant trade unions did mean that the prescribed wage was not being paid "in some areas" of his domain; the state's own machinery on the ground was deemed "inadequate." One small farmer who worked in his own fields wrote to the Collector that Palakkad's wages were already the highest in India; even in Thanjavur (an area of labor militancy in Tamilnadu state) the harvest share was only 1/9, and the local rate had been 1/14 just ten years back. He couldn't pay 1/7 and survive as a farmer. His solution was cottage industries, rather than robbing the farmers. This opinion was widespread.

Divergent moral economies of land and labor emerged quite clearly at a high-level meeting at the labor office in Palakkad town during harvest conflicts of 1972. Farmers' representatives unanimously rejected enforcement of the new minimum wage. Their complaints centered on shrinking profits caused by failure of the monsoon, levy procurement, fertilizer costs, manure shortage, and so on. Agriculture is not amenable to regulation, they contended; "a paddy field is not a factory." The proposal for a twenty-five *paisa* (one-fourth of a rupee) raise was rejected outright; farmers said they were not able to pay even the existing minimum. Their solution was to conduct a study on the costs of cultivation; they presented data for a typical 6.5-acre farm demonstrating that there was no profit in agriculture. Their analysis was supported by officials of the Agriculture Department but opposed by those of the Labour Department. Officials of the two departments consistently represented the views of their respective constituencies.

Union representatives, by contrast, rejected the very idea of a cost-of-cultivation study. The issue to them was not whether farmers could make profit but whether laborers could survive on existing wages. The meeting ended in animosity and no recommendation for the Collector.

Conflicts surrounding harvesting became so severe that the state government eventually invoked the Defense of India Rules to intervene. Writing to the Ministers of Agriculture and Labour on 11 September 1972, the Collector described the harvest conflicts as potentially dangerous and destructive. Combined unions of the CPI and CPI-M had escalated their demand to 1/6 harvest share. He requested the Labour Minister's intervention, fearing violence if the issue were not settled before the union's 15 September deadline. The Labour Minister flew to Palakkad on 14 September. The Labour Department in Thiruvananthapuram immediately promulgated a notification (MS 82/72/LBR) that paddy harvesting in Palakkad (which supplies more than a quarter of all rice in Kerala) was so essential to the community that the Defense of India Rules would be invoked to set the harvest share at 1/7.

Losing control of labor on the ground, farmers went to court. A petition to the High Court by thirteen Palakkad cultivators (associated with the DKS) requested police protection to harvest their crops. The government pleader argued that protection could be given only if the posted wage were paid and laborers engaged in the previous season were allowed to harvest, with which the High Court agreed. This position began the formal recognition of the institution of permanency, which later infuriated farmers. Landowners were at the time attempting to employ additional workers to finish the harvest more quickly, depriving traditionally attached laborers (*steerthozhilali*) of harvest share. Laborers militantly opposed allowing any but attached laborers from participating in the harvest, when wages were double the season rate. The assistant superintendent of police wrote that his practice was then to allow cultivators to employ extra labor only after the tehsildar assessed the labor requirements on the spot. Production decisions were drifting out of the hands of farmers and into those of ministries and police. Sporadic violence occurred as attached laborers refused to let additional laborers enter the fields.

All agricultural work ceased in some areas in 1972. The governor assented to inclusion of paddy-harvesting regulations under the Defense of India Rules (DIR) as an essential service. The DIR inclusion, though originally welcomed by the farmers and sought by the Collector and Labour Department as a means of obtaining legal backing for enforcing agreements, proved to be grounds for court intervention in removing traditional farmer managerial prerogatives from their control.

Conflict shifted to another of the traditional rights of labor the following season. Cultivators led by Ambat Shekera Menon, president of the DKS, refused to pay wages in kind in Chittur Taluk. Union leaders now said they would not press for cash wage increases if in-kind payment could be guaranteed at the posted minimum. Prices of paddy were steadily rising in this period, but the DKS refused on the grounds that the low levy price imposed by the government was making cultivation unprofitable. Additionally, the DKS claimed that laborers were doing such poor work that yields were falling; no one could be disciplined or fired. Cultivators offered higher wages in cash, which was rejected, most adamantly by the KSKTU. Women in particular demanded wages in kind, to assure that the family would always have something to cook, unaffected by the fickle cash nexus. The compromise was half in cash and half in kind until the transplanting season ended, after which another conference would be convened. Confidential reports from tehsildars indicated that 45 percent of the area in Chittur Taluk was affected by strikes; in Alathur, 15–20 percent; in Palakkad, 15 percent. By harvest of 1973, the percentage for Alathur had increased to 50 percent, and for Chittur to 30 percent.

In this process of mediation through virtually every channel of authority, traditional claims of attached laborers were becoming de facto policy, reinforced by High Court rulings. Substitution of nonlocal labor for attached laborers provoked the most violent confrontations. Petitions to the High Court by farmers seeking permission to replace laborers who were on strike (and police protection to do so) were denied; this case law then determined the Collector's practice. The DKS submitted a memorandum to the Collector during the harvest of 1973 calling for official action to break strikes because the production of paddy had been officially declared an essential service under the DIR; leveraging the essential-services logic, they threatened to cease paddy production if no action were taken.

Through resolution of numerous conflicts in various forums, the right of laborers to permanent employment in the field to which they were traditionally attached was becoming policy.[29] Farmers responded by refusing to hire new workers, fearing their eventual claim of tenure. Permanency became the core of conflict around the factory acts; it struck at the heart of the prerogatives of ownership and stood at the center of insecurity fears of labor. But in 1973, the Labour Department published a notification that presaged one of the most radical clauses of the Kerala Agricultural Workers Act: "In cases where the workers are actually in receipt of higher wages than the minimum wage fixed, they shall continue to get the benefit of the higher wage."[30] Wages could only go up, never down. Facing fluctuating yields and fluctuating output prices, and now saddled with a workforce they could not discipline or replace, farmers were frightened by the ratchet of ever-increasing minimum wages.

But what was the minimum wage? Procurement levies on rice complicated the moral economy of paddy values. Farmers complained that after meeting the levy, they had to buy grain in the market in order to pay wages in kind. More galling, they lacked ration cards, unlike the laborers, and had to buy back their own rice at market prices. One compromise suggested by the Collector in a letter to the Food Department on 14 September 1973 was to relax procurement requirements. The procurement levy on farms larger than ten acres then was 9 *quintals;* if the average yield were 120 *paras,* as claimed by the DKS, then the entire surplus would legally be required for sale to the government at a below-market price. Although official estimates of yields on Class A land were then higher than those of farmers (by about 50 percent), the Collector believed, based on his own calculations, that the farmers had a point and recommended levy exemptions for Palakkad rice farmers.

Nothing in the short run could be done about the levy, but the very existence of an administered levy price was appropriated by the DKS as the basis for new tactics. Farmers began to demand conversion of rice to rupees for wages at the levy rate rather than the market rate. The district superintendent of police telephoned the Labour Ministry on 22 September 1973 for help with his conundrum: High Court orders required payment of the minimum wage before police protection of harvesting could be provided, but what was the minimum wage? The Labour Department's official regulation interpreting the minimum wage act said explicitly that a higher prevailing wage superseded the minimum wage; existing wages in much of Palakkad in edangazhi (750-gram measures) were higher than the posted wage in rupees at existing market price conversions. At what rate were the police to convert measures to rupees to determine compliance? The DSP requested legal advice. Officials in the state capital interpreted clause E of the notification to mean that conversion had to be at market prices; it was up to police officials to determine whether minimum wages were being paid, based on the local market price, and to decide on protection of harvesting accordingly.

Requests for police protection overwhelmed the Collector. He ordered the DSP (D3-22105\73) to give protection to farmers only if laborers traditionally attached to the field were being employed, 1/7 harvest share was being given, and wages were being paid at 7 and 6, 6 and 5 edangazhi (season and off-season rates for men and women, respectively) or the market equivalent in cash. If these conditions were met and the workers still refused to work, the farmers could employ other workers. Protection of strikebreakers could be done only after consulting with the striking workers and unions. This resolution placed a large burden on local police; protection of strikebreakers was certain to lead to violence.

Attempts at conciliation by the MLAS, political parties, the Labour Minister, and relevant departments failed. The great strike of 1973 began on 25 September. The tehsildar reported that 90 percent of the laborers were on strike in Alathur, where Padoor is located. The labor crisis totally absorbed district administration, but there were pockets where nothing happened. Proliferation of demands made conciliation at any but the very local level impossible. Disputes in Ootappalam Taluk, for example, varied by union; demand differentiation became a means of capturing membership at a time of high mobilization. The Congress INTUC union led the fight for rehiring of retrenched workers; the CPI-M union led protests for release of laborers arrested by police; the CPI union spearheaded demands for a 25 percent wage increase and uniforms for workers. Variation over time and space in demands and outcomes was remarkable.[31] Farmers desperate to get crops harvested before the grain spoiled made separate deals with individual workers and with local units of unions. No neat curves defined this labor market.

The dominant mode of conflict in the harvest strike of 1973 was police against striking laborers who were preventing harvest on lands determined by the police to be in compliance with the High Court conditions. Other laborers attempted to harvest disputed fields forcefully, physically opposed by farmers who were determined to let the grain rot to protest the new rules. Two DKS leaders were hospitalized after a clash near Nemarra. Violence also occurred between Congress and communist workers. Tamil laborers from Coimbatore were brought in by truck to break the strike. When the Collector learned that labor was being imported to break strikes, he ordered the tehsildars to inform farmers that strikebreaking would not be tolerated; in essence, Malayalee fields were for Malayalees. The DKS rejected the prohibition as having no legal or logical standing ("labor is labor, Tamil or Malayalee").

Police arrested or removed strikers in large numbers but also restrained farmers and their enforcers. Women were numerically prominent in the police reports on the strike; some local actions were carried out entirely by women. Students affiliated with the Marxists boycotted classes to show solidarity with the laborers, resulting in a counterstrike by students affiliated with Congress. Near Kollengode, a landowner shot two laborers; two others were stabbed, leading to a "calling attention" motion in the Assembly and an immediate phone call to the collectorate from the Home Secretary in Thiruvananthapuram. In early October, a district-wide strike was called to protest police handling of the harvest strike. Although people continued to be hospitalized with injuries from the conflict, the end of the harvest season undercut labor's ability to mobilize great numbers. The great tactical strength of labor in the *Onam* season results from the simultaneous occurrence of two peaks of labor demands: harvesting the *virippu* crop and preparing fields for the *mundakkan* crop.[32]

From these details emerges a cumulative picture of a remarkably responsive political and administrative system attuned to the power of the agrarian underclass. The farmer-labor alliance that had carried the land reform was dissolving, although there were KSKTU agitations for removal of the land tax and for supply of fertilizer. The most fundamental structural conflict was over the meaning of ownership of land: the unions rejected farmers' claim to managerial rights to employ whomever they wanted in whatever numbers they wanted.

In response to strikes and violence, the Legislative Assembly passed the Kerala Agricultural Workers Act in 1974. The Industrial Relations Committee legally became the locus of district-level coordination of land and labor, reflecting reality on the ground. But the DKS representative boycotted the first meeting of the IRC in 1974 in protest against permanency and passage of the factory acts. The Collector ordered that all laborers except those against whom a police case was pending had to be employed. The DKS resisted and continued to petition the High Court under section 113(1) of the DIR. The Collector had by then concluded that the DKS leadership was unscrupulous and unreliable, yet they retained the power to disrupt agricultural operations. After their refusal to meet with labor and government, the police intervened in a major conflict in 1974 and arrested 490 activists of the DKS who were preventing permanent workers in Chittur Taluk from entering the fields.

The symbolic end of the period of large-scale conflict was the Olasherri firing. During the harvest of virippu in October 1974, the DKS made a last stand on the estate of Janardharan, one of the most recalcitrant of the large farmers. He refused to allow permanent laborers to harvest; the KSKTU mobilized a force estimated by police to be five thousand to harvest forcibly on Janardharan's land. The police set up camp near Kodumba village. The district magistrate granted their request for an order under the Kerala Police Act prohibiting collective action.[33] On October 15, six hundred farmers and two thousand laborers affiliated with the DKS assembled on Janardharan's land, attempting forcible harvest against resistance of five thousand KSKTU workers. In the ensuing clash, one local laborer was killed and a large number were hospitalized.

The Olasherri firing marked the end of large-scale local opposition to changes in land-labor relations in the half-decade of conflict after passage of the land reform. New strands were later added to labor's share in the bundle of rights: most important, a significant solatium paid to permanent laborers when land was sold or partitioned. Strikes continued, and DKS resistance took other forms, but from Olasherri on, it was clear to the farmers that outright resistance was futile; they would have to live with the factory acts. Labor, too, understood the limits of militant confrontation after the 1974 strike; the CPI-M was beginning to retreat from militance because of splintering of its agrarian coalition, and the laborers were too weak economically and organizationally to contest on their own. The following year, the national Emergency declared by Indira Gandhi quashed labor militance.

In place of multifaceted conflict on the land, the KAWA established a corporatist structure for mediating the conflicting interests of land, labor, and the local state. This stalemated class conflict in paddy agriculture has not been conducive to growth, but it has produced some peace. Response of communists to the resultant dilemma of undercompensated land and destitute labor in Kerala has been twofold: first, attempts have been made to create structures of local corporatism to prevent this contradiction from leading to unmanageable class warfare, and second, there have been efforts to displace the conflict into the distributive and federal arenas—sectoral politics against the market through higher output prices and subsidized costs of production. Rigging the terms of trade in favor of agriculture could blunt the class contradiction between agrarian capital and labor. The model is social democracy: a marriage of regulated capitalism and welfarism mediated through politics. The critical departure from the European model is the size of the sector that needs subsidization in proportion to the aggregate financial resources available. In this period, almost 10 percent of the total population remained on the live employment registers; real unemployment was much higher. About 4.5 million individuals held ration cards for use in "fair-price" shops. Conditions of the laborers were bleak; under- and unemployment continued to rise. Demands for "remunerative prices" for farmers established agrarian scissors politics in the federal arena, in parallel to what Paige calls "commodity reform movements."[34] Simultaneously, the government asked for more state autonomy and resources from Delhi to shoulder the burden of subsistence guarantees.

Extensive market rigging in agriculture is possible, though perhaps "inefficient," in the United States, the European Community, and Japan because of the strength of their economies. Kerala is a poor and food-deficit state in a poor nation. Its locally rigged markets for labor frighten outside investors and prompt local labor-intensive industries to leave the state.[35] Debates in the Legislative Assembly in Kerala in the 1980s frequently revolved around the dilemmas of the periphery. National policy liberalizing imports, while attractive to large Indian manufacturing houses, meant falling prices for Kerala's primary products, a sector already squeezed by high labor costs and perpetually in crisis. The resolution was always "relief" for local capital and labor and pleas to Delhi to take Kerala's plight seriously.[36] Liberalization at the national level in the 1990s deepened this dilemma; states increasingly competed with each other for foreign and domestic capital. Peripheral states lack the levers to control markets because markets extend over a space more extensive than their political authority; exerting authority scares away capital.

As the structural niche for leftist mobilization disappeared with the land reforms, reasons for affiliation with class politics became less compelling. It is not quite true, as the Rudolphs argue, that "no national party, right or left, pursues the politics of class" in India;[37] rather, the CPI-M pursues a politics of class coalition, torn between appeals to interests of objectively opposed classes. On the ground, class identifications remain strong in Palakkad, and the class politics of production relations continues to engender conflicts, damped though these are by the corporatist arrangements hammered out in the 1970s. The dilemma is that the communist party for a time fought market society and could not win; recognizing this, labor simultaneously pursues a politics of particularism and individual survival complementary to its class identification. If this means joining the union of the Congress when Congress is in power, the implication is not the absence of class politics but the absence of an alternative class project as viable political strategy.

Agricultural workers have no illusions about who has been on their side historically, but the communist party reached a strategic cul-de-sac in searching for solutions to their central dilemma.[38] As a consequence, the party's new project became one of balancing class redistribution and a shadow of revolution with strategies for economic growth—a productivist class compromise. Patrick Heller argues that the strength of highly mobilized civil society and the institutions of class compromise augur well for the economic future of the state.[39] The ruling CPI-M, currently in power, has staked a great deal on using the strengths of social democracy to institute a bold project for decentralized planning and resource mobilization, building up from local bodies to state managers rather than the other way around.[40] The limits to potential success remain the vulnerability of a society embedded in much larger market systems beyond anyone's control.

There are, then, clear limits to local struggles with the market; the scale is wrong for the dynamics. In 1999 U.N. Secretary General Kofi Annan appealed to the wealthiest nations to reembed global markets in a decidedly imagined global community: "The spread of markets far outpaces the ability of societies and their political systems to adjust." Without "active commitment and support" of the rich nations for market-endangered "universal values," Annan argued, "I fear we may find it increasingly difficult to make a persuasive case for the open global market."[41]

Reembedded Markets and Human Welfare

The strength of the "moral-economy" model of peasant society lay in its extension of Karl Polanyi's great insight: the making of market society, elevating market outcomes above the social norms in which economic relations had been previously "embedded," produces profound dislocation and propels social forces to reestablish guarantees of economic security and morally acceptable outcomes. Reembedding the market in turn creates its own distortions.

Kerala's public moral economy is embodied in institutions produced by these conflicts; institutional change cumulatively altered and re-created the configuration of public authority. The state's continual reformation was driven not simply by powerful social actors but by historically specific attempts to reintegrate society following ruptures.[42] Although Polanyi is rightly criticized for reifying society and insisting on the priority of "integration," the state as actor clearly has an interest in retaining a degree of social integration conducive to production and administration. The communist party, in particular, has been interested in repairing the ruptures of conflicts because they took place within the class coalition that enabled the Left to come to power.

Contrary to Polanyi's romanticized portrayal of pre-market society, rural Kerala was characterized by extreme subjugation and abuse, grinding poverty for the lower orders, and significant insecurity despite ideational paternalism. These very characteristics propelled communist success, producing a system of fully commoditized and largely depersonalized social relations of production, disembedded from social conditions of servitude, diffuse claims, or extra-economic domination but reembedded in new social institutions in the form of public law. As in James Scott's magisterial work on Malaysia,[43] the ideological origins of these institutions lay in appropriation by the weak of the ideological cover of the strong, with three additions. First, comparative social learning expanded significantly the standards of what subsistence should mean. If former tenants can wear shirts of synthetic cloth and wristwatches, young laborers ask, why can't they? Second, public policy provided crucial cues about rights and social justice, which both revised expectations and were incorporated strategically into the struggle. For many laborers, the just wage became simply "the government wage." Farmers demanded that the translation of minimum cash wages into measures of paddy be done at the administered levy price, not the market price. If the levy price—a mechanism to provide social justice through ration shops—were fair, they argued, it should be the price for all transactions. Finally, the market itself was selectively appropriated ideologically—although no actors in Palakkad's paddy sector wanted exposure to the implications of a thoroughly market-run society.

At the end of the transformative decade in Kerala, the hegemonic policy prescription internationally on politics and markets was for allowing markets more allocative autonomy.[44] This prescription overlooked the extent to which contesting the great transformation had reembedded markets in society. The short-term consequences of removal of producer and consumer subsidies and labor regulations in Palakkad would be painful in human terms; the long term is irrelevant to a household scraping by on one hundred days of annual employment at administered wages. For labor, the factory acts have distinct and recognized advantages over the pre-capitalist configuration; most important is freedom from abuse and harassment, sexual exploitation, and demeaning social observances (of dress, language, movement). Permanency continues one of the few plausible benefits of the pre-capitalist order—some security of employment. The factory acts represent state-mandated agricultural involution, spreading and guaranteeing employment to those with the strongest moral claims—the attached laborers—even if the consequence is less employment per worker in the aggregate and thus a sharing of secure poverty and immobility. At this point, a public moral economy of basic human needs takes over, through subsidized food distribution, extensive social services, and, most important, a pension for agricultural workers.[45]

The Left's struggle with the market thus produced an economic system in which institutional rules resonate with proclaimed normative features of the pre-capitalist structure. In this sense, the system is the outcome of "revolutionary" transformation, where revolution retains its etymological roots connoting turning full circle, absent the landlords. Farmers' demands for protection from the market are accepted; labor's historic claims to subsistence guaranties and old-age security are recognized in the public moral economy. But all protections are limited by the productivity of agriculture and the state's fiscal imperatives. The market itself has no legitimacy other than instrumental for political actors, yet its inescapable logic frames every choice.[46]

Kerala's boundary on the politics-and-markets continuum is far closer to the authority pole than are the boundaries of most states of India, in part because popular responses to unusually severe dislocations were so extensive and militant. A mobilized polity expects extensive intervention of both the regulatory and distributive sorts. Competitive political democracy ensures that preferences matter. People in villages implicitly recognize limits ("the most the government can help is a thousand rupees") but have high expectations nonetheless. The most striking change in Nallepilli in 1989 was the presence of a well-staffed crèche, providing nutritious meals and stimulating educational materials, which made the life of female agricultural laborers considerably easier; how can these perfectly reasonable improvements in the human condition be afforded in poor societies? As much as the "Kerala model" is praised for its accomplishments in quality of life and social justice, a nagging complaint has been that the cost has been economic stagnation, strangling the material base of its robust public moral economy.[47]

Critics of redistributive politics and state intervention in markets reject the enabling potential of the Kerala experience. There is no conclusive answer. But in ethical terms, surely it is the case that a direct and certain means of alleviating poverty and indignity—such as the land and labor reforms—should have preference over the indirect and uncertain—the hope of higher growth rates that trickle down. To be convinced, one need only think of the historical experience of slavery and agrarian oppression in the United States. A promised land reform of "forty acres and a mule" for former slaves during Reconstruction (1865–77) was scuttled by political opposition. Despite impressive growth in aggregate wealth, the descendants of slaves remained disproportionately poor and excluded for generations. Social democracy was disabled on both fronts: economic justice and political participation.

On a small and local scale, in Palakkad district, struggling against the market gave labor more of the strands of the "bundle of rights" we call ownership in land, and public entitlements unusual for a marginal class, but nothing approaching what most told me they wanted: "a factory job." On a grander scale, societies continue to contest the "great transformation," shifting boundaries between politics and markets back and forth for reasons of legitimation and accumulation in response to mobilized social forces. Polanyi seems to argue that the exercise is ultimately futile; the genie of market society is out of the bottle, and attempts to put it back in are counterproductive. No one seems to know how much, or what, has to be sacrificed to the market to win its favor. Claiming to know is the stuff of much political practice.

Notes

The following abbreviations are used in the notes:

KLAG	Kerala Legislative Assembly, *Gleanings from the Question Hour*
KLAS	Kerala Legislative Assembly, *Synopsis of the Proceedings*
KLAR	Kerala Legislative Assembly, *Resume of the Proceedings*

1. For a representative range of positive and critical commentary on the "Kerala model," see Dreze and Sen, *Hunger and Public Action,* pp. 221 ff and passim; Parayil, "The 'Kerala Model' "; Mencher, "Lessons and Non-Lessons"; Herring, "Abolition of Landlordism"; Jeffrey, *Politics;* Heller, *Labor of Development,* chap. 1; Tharamangalam, "Perils." The "model" has become so much a part of public discourse that Vice President Al Gore of the United States has called Kerala a "stunning success story"; it is

not common for U.S. politicians to pay attention at all to India, much less to individual states.

2. Although the percentage of poor in India's population has declined by various measures since Independence, in absolute numbers the long trajectory has been an increase in poverty, with strong regional variation: by World Bank estimates, from 164 million poor in 1951 to 312 million (about 35 percent of the population) in 1993–94. The state comparisons cover the years 1957–58 to 1993–94 (World Bank, *India,* v).
3. The argument for centrism is made by Rudolph and Rudolph (*In Pursuit of Lakshmi*). My characterization in this chapter does not imply the unimportance of class politics, as the Rudolphs' does, nor does it deny the existence of powerful mobilization on the Right and Left. I mean to say only that national political dynamics present a vector sum of politics that is centrist. On Kerala's politics in comparative terms, see Nossiter, *Communism in Kerala;* Nossiter, *Marxist State Governments.*
4. Polanyi, *The Great Transformation,* p. 239 and passim; Hechter, "Karl Polanyi's Social Theory," p. 400; North, "Markets," p. 706.
5. Glade, *State Shrinking,* pp. 2–10.
6. Polanyi, *The Great Transformation,* p. 178.
7. Block and Somers, "Beyond the Economistic Fallacy."
8. Polanyi, *The Great Transformation,* pp. 68 and 223.
9. Scott, *Moral Economy.*
10. Scott, *Weapons of the Weak.*
11. Herring, "Dilemmas."
12. Here *redistributive* and *distributive* have the meanings of Lowi (*The End of Liberalism*), not Polanyi. For a clarifying discussion of Polanyi's classification of allocative systems, see Hechter, "Karl Polanyi's Social Theory."
13. Herring, "Dilemmas."
14. Panikkar, *Against Lord and State.*
15. Varghese, *Agrarian Change,* p. 78.
16. Logan, *A Malabar Manual,* vol. 1, p. 667.
17. *Adima* has the same ambiguity in ordinary usage in Malayalam as the English word "slave," as in "wage slave" or "slave to her husband," retaining some of the core meaning. Real slaves that could be bought and sold have not been known in the area under discussion since Independence (1947). In interview material I found that the dominant meaning of *adima* locally was restriction on labor opportunities: "you couldn't go out without permission." On slavery in Kerala more generally, see Saradamoni, *Emergence of a Slave Caste;* Radhakrishnan, "Peasant Struggles"; Nair, *Slavery in Kerala.*
18. The concept of "middle peasant" is hard to operationalize. Perhaps the closest equivalent is *kannakar,* etymologically related to sight, hence overseers in the traditional land tenure system. Kannakar were mistaken for "tenants" by colonial revenue authorities. Of relatively high social status and economic autonomy, kannakar met most criteria for middle peasant status. The best source for this period is Panikkar, "Peasant Revolts"; see also Oommen, *From Mobilization to Institutionalization;* Karat, "Agrarian Relations"; Karat, "Peasant Movement"; Karat, "Organized Struggles." On the Mappila rebellions, see Arnold, "Islam." For a very different perspective on the dislocation of the nineteenth century, see Dale, *Islamic Society.*

19. For a summary of this long and complex process, see Herring, "Stealing Congress's Thunder"; Kannan, *Of Rural Proletarian Struggles;* Panikkar, *Against Lord and State;* Nossiter, *Communism in Kerala.*
20. Herring, *Land to Tiller,* chaps. 6 and 7.
21. One older laborer corrected this formulation of the union secretary by saying that the land did not belong to Brahmins but "to god," with obvious sarcasm. Temple trusts were an important mechanism for land control.
22. The *torthu* is a thin wrap somewhere between a towel and a handkerchief, which serves both functions, worn loosely across the shoulder or wrapped around the head. The refusal to remove the torthu in the presence of social superiors is still a cause of anger among some farmers, as it connotes disrespect (and is meant that way).
23. This theme is central to one of the great novels in the powerful movement for "socialist realism" in literature in Kerala, *Two Measures of Rice* [*Rende Edangazhi*] by Takazhi Shivasankaran Pillai.
24. Polanyi, *The Great Transformation,* p. 46.
25. At the time of the first communist agrarian reforms in the mid-1950s, pure rentiers constituted 1 percent of agricultural households in Travancore, 2 percent in Cochin, and 2 percent in Malabar (Varghese, *Agrarian Change,* p. 201).
26. Herring, "Stealing Congress's Thunder."
27. This account is based on interviews with local officials of the Labour and Revenue Departments, union leaders, representatives to the IRC, local activists and leadership of the National Farmers' Association (DKS), and political operatives of various parties, agricultural workers, and farmers. Written sources include local press reports, memoranda from tehsildars to the Collector, memoranda from the Collector to department heads in the capital, police files, Labour Department records, transcriptions of telephone conversations, minutes of meetings of the IRC, Proceedings of the Legislative Assembly, and Reports of the Government of Kerala.
28. The DKS proposal is curious in the context of its complaints about the declining interest of laborers in cultivation. The harvest share system gives labor an incentive to maximize production. What the DKS feared was a ratchet effect in which laborers became essentially sharecroppers who would over time increase their share of the crop. Because many farmers had historically been tenants, they well understood the importance of their own precedent.
29. This is clear from the actual implementation of High Court orders by police officials, with the backing of the Collector, as indicated by police reports (for example, DI-36026/73-P, dated 31 August 1973). One of the conditions for the High Court's granting police protection to farmers was compliance with resolution of permanency claims as established by labor officers or tehsildars (memorandum from the government pleader to the Collector, O.P. No. 29 84-73). Police files are especially telling as evidence for the vector sum of intra-state conflicts; what the police adopt as policy is far more important than disembodied policy pronouncements from other branches of the local state.
30. G.O. No. 136/Ei/73/LRB, dated 29 June 1973. On the farmers' view of the legislation and its practical effect on agriculture, see Herring, "Dilemmas of Agrarian Communism."
31. The Collector made estimates of union strength as of the 1973 strike. The KSKTU dwarfed all other unions, with six times the number of workers of the CPI union, which

was closely followed by the Congress. The Socialist Party, Bharatiya Jana Sangh, and Muslim League also had local branches involved in the strike, but in very small numbers. A small number of organizations on strike were affiliated with no party. The aggregate number of strikers given by the Collector was 157,869, but if ever there were an example of pseudo-precision, this is it.

32. This advantage is declining; high-yielding varieties are spreading peaks in labor demand (as farmers employ different varieties), and many farmers are abandoning labor-intensive crops. For details, see Herring, "Dilemmas of Agrarian Communism."
33. Specifically prohibited were public displays of weapons, corpses, or effigies, processions or public assemblies (exempting marriages and agricultural operations), and public songs, slogans, or music (again exempting wedding processions). Order of District Magistrate, D3 26242/74 under the Kerala Police Act (V of 1961). One hundred and eighty volunteers from the DKS side were removed by police for violation of the magistrate's order the following day.
34. Paige, *Agrarian Revolution.* Data are from Government of Kerala, Bureau of Economics and Statistics; Government of Kerala, Department of Economics and Statistics.
35. Government of Kerala, Department of Economics and Statistics, p. 7.
36. For example, the dilemma of loss of jobs from pressures in the global economy was raised in *KLAG* (14 March 1985); the proposal for a regional minimum wage for all of South India to staunch the loss of jobs was initiated in *KLAG* (14 March 1989). The effect of Delhi's import liberalization laws on employment and profits in the rubber sector was debated in *KLAR* (26 March 1985); import policy and the coconut and rubber industries were addressed forcefully by the Kerala Congress in particular, reflecting its political base (*KLAR,* 25 June 1986). The connection between unprofitability of paddy agriculture and unemployment was made in the government's response to a member's question in the assembly in *KLAG* (30 April 1985 and 9 July 1986), relying heavily on enumeration of the various subsidies provided to paddy cultivators. In the debate on paddy profitability and unemployment in the second session of 1987 (*KLAR,* 31 July 1987), the government acknowledged the connection but pleaded poverty with a long list of projections of aggregate costs of extending subsidies.
37. Rudolph and Rudolph, *In Pursuit of Lakshmi,* p. 2.
38. See Kannan, *Of Rural Proletarian Struggles;* Kannan and Pushpangadan, "Agricultural Stagnation in Karala"; Government of Kerala, Bureau of Economics and Statistics; Government of Kerala, Department of Economics and Statistics; Mencher, "Lessons and Non-Lessons"; Mencher, "Agrarian Lessons." Although labor scarcity periodically occurs in some areas, the dilemmas stressed in the text remain.
39. Heller, *The Labor of Development.*
40. Tornquist and Tharakan, "Democratization."
41. Swardson, "Conference."
42. For a sophisticated framework for understanding reformist projects on the fine line between state-centric and overly society-driven models, see Jonathan Fox's treatment of Mexico (*The Politics of Food,* pp. 31–40).
43. Scott, *Weapons of the Weak.*
44. See Glade, *State Shrinking.*
45. Gulati, "Agricultural Workers' Pension."

46. Lindblom, "The Market as Prison."
47. Tharamangalam, "Perils."

References

Arnold, David. 1982. "Islam, the Mappilas and Peasant Revolt in Malabar." *Journal of Peasant Studies* 9, no. 4 (July): 255–65.

Block, Fred, and Margaret R. Somers. 1984. "Beyond the Economistic Fallacy: The Holistic Social Science of Karl Polanyi." In Theda Skocpol, ed., *Vision and Method in Historical Sociology.* Cambridge: Cambridge University Press.

Dale, Stephen. 1980. *Islamic Society on the South Asian Frontier: The Mappilas of Malabar, 1498–1922.* Oxford: Clarendon Press.

Dreze, Jean, and Amartya Sen. 1989. *Hunger and Public Action.* Oxford: Clarendon Press.

Fox, Jonathan. 1992. *The Politics of Food in Mexico: State Power and Social Mobilization.* Ithaca: Cornell University Press.

Glade, William P. 1986. *State Shrinking.* Austin, Tex.: Institute of Latin American Studies.

Government of Kerala. 1977a. *Report of the Committee Appointed by Government to Hold Enquiries and Advise Government in Respect of Fixation of Minimum Wages for Employment in Agricultural Operations in Kerala* (Malayalam) (Thiruvananthapuram).

———. 1977b. Labour Department. *Report of the Sub-committee under the Palghat Agricultural Area.* No. 1382/77/LD (Palghat).

———. 1985. Bureau of Economics and Statistics. *Report on the Survey on Socio-Economic Conditions of Agricultural and Other Rural Laborers in Kerala, 1983–84* (Thiruvananthapuram).

———. 1988. Department of Economics and Statistics. *Report on the Survey of Unemployment in Kerala, 1987* (Thiruvananthapuram).

Gulati, Leela. 1990. "Agricultural Workers' Pension in Kerala." *Economic and Political Weekly* 25, no. 6 (February): 339–43.

Hechter, Michael. 1981. "Karl Polanyi's Social Theory: A Critique." *Politics and Society* 10, no. 4: 399–430.

Heller, Patrick. 1999. *The Labor of Development: Workers and the Transformation of Capitalism in Kerala, India.* Ithaca: Cornell University Press.

Herring, Ronald J. 1980. "Abolition of Landlordism in Kerala: A Redistribution of Privilege." *Economic and Political Weekly* 15, no. 26 (28 June).

———. 1983. *Land to Tiller: The Political Economy of Agrarian Reform in South Asia.* New Haven: Yale University Press; Delhi: Oxford University Press.

———. 1988. "Stealing Congress's Thunder: The Rise to Power of a Communist Movement in South India." In Peter Merkl and Kay Lawson, eds., *When Parties Fail.* Princeton: Princeton University Press.

———. 1989. "Dilemmas of Agrarian Communism." *Third World Quarterly* 11, no. 1 (January): 89–115.

Jeffrey, Robin. 1993. *Politics, Women and Well Being: How Kerala Became "a Model."* Delhi: Oxford University Press.

Kannan, K. P. 1988. *Of Rural Proletarian Struggles.* Delhi: Oxford University Press.

Kannan, K. P., and K. Pushpangadan. 1988. "Agricultural Stagnation in Kerala." *Economic and Political Weekly* 23, no. 39 (24 September): A120–A128.

Karat, Prakash. 1973. "Agrarian Relations in Malabar, 1925–1948." *Social Scientist* 2, nos. 2–3 (September–October).

———. 1976. "Peasant Movement in Malabar, 1934–1940." *Social Scientist* 5, no. 2 (September).

———. 1977. "Organized Struggles of Malabar Peasantry, 1934–1940." *Social Scientist* 5, no. 8 (March).

Krishnaji, N. 1979. "Agrarian Relations and the Left Movement in Kerala." *Economic and Political Weekly* 15, no. 9 (March).

Lindblom, Charles E. 1978. *Politics and Markets.* New York: Basic Books.

———. 1982. "The Market as Prison." *Journal of Politics* 44, no. 2 (May): 324–36.

Logan, William. 1887/1981, *A Malabar Manual* (Thiruvananthapuram, India: Charithram).

Lowi, Theodore. 1969. *The End of Liberalism: Ideology, Policy and the Crisis of Public Authority.* New York: Norton.

Mencher, Joan P. 1978. "Agrarian Relations in Two Rice Regions of Kerala." *Economic and Political Weekly* 14, no. 9 (3 March).

———. 1980. "The Lessons and Non-Lessons of Kerala: Agricultural Labourers and Poverty." *Economic and Political Weekly* 15: 41–43.

Nair, Adoor K. K. Ramachandran. 1986. *Slavery in Kerala.* Delhi: Mittal Publications.

Nayar, P. K. B. 1979. "Agrarian Movements in Rural Development: A Case Study." Typescript. Kariavattom, India: University of Kerala.

North, Douglass C. 1977. "Markets and Other Allocative Systems in History: The Challenge of Karl Polanyi." *Journal of European Economic History* 6, no. 3 (Winter): 703–16.

Nossiter, T. J. 1981. *Communism in Kerala.* Berkeley: University of California Press.

———. 1988. *Marxist State Governments in India.* London: Pinter.

Oommen, T. K. 1985. *From Mobilization to Institutionalization: The Dynamics of Agrarian Movement in Twentieth Century Kerala.* Bombay: Popular Prakashan.

Paige, Jeffrey. 1975. *Agrarian Revolution: Social Movements and Export Agriculture in the Underdeveloped World.* New York: Free Press.

Panikkar, K. N. 1979. "Peasant Revolts in Malabar in the Nineteenth and Twentieth Centuries." In A. R. Desai, ed., *Peasant Struggles in India.* Bombay: Oxford University Press.

———. 1989. *Against Lord and State: Religion and Peasant Uprisings in Malabar, 1836–1921.* Delhi: Oxford University Press.

Parayil, Govindan. 1996. "The 'Kerala Model' of Development: Development and Sustainability in the Third World." *Third World Quarterly* 17, no. 5: 941–57.

Pillai, Thakazhi Shivasankaran. 1967. *Two Measures of Rice.* Bombay: Jaico.

Polanyi, Karl. 1957. *The Great Transformation.* Boston: Beacon Press.

———. 1977. *The Livelihood of Man.* New York: Academic Press.

Prosterman, Roy L., and Jeffrey M. Riedinger. 1987. *Land Reform and Democratic Development.* Baltimore: Johns Hopkins University Press.

Radhakrishnan, P. 1980. "Peasant Struggles and Land Reforms in Malabar." *Economic and Political Weekly* 15: 50.

Rudolph, Lloyd I., and Susanne Hoeber Rudolph. 1981. "Transformation of Congress Party: Why 1980 Was Not a Restoration." *Economic and Political Weekly* 16: 18.

———. 1987. *In Pursuit of Lakshmi: The Political Economy of the Indian State.* Chicago: University of Chicago Press.

Saradamoni, K. 1980. *Emergence of a Slave Caste: Pulayas of Kerala.* New Delhi: People's Publishing House.

Scott, James C. 1976. *The Moral Economy of the Peasant: Rebellion and Subsistence in Southeast Asia.* New Haven: Yale University Press.

———. 1985. *Weapons of the Weak: Everyday Forms of Peasant Resistance.* New Haven: Yale University Press.

Swardson, Anne. 1999. "Conference Considers Global Code of Conduct." *Washington Post* (1 February).

Tharamangalam, Joseph. 1998. "The Perils of Social Development without Economic Growth: The Development Debacle of Kerala, India." *Bulletin of Concerned Asian Scholars* 30, no. 1: 23–34.

Tornquist, Olle, and P. K. Michael Tharakan. 1996. "Democratization and Attempts to Renew the Radical Political Development Project: Case of Kerala." *Political and Economic Weekly* (13 July): 1847–58.

Varghese, T. C. 1970. *Agrarian Change and Economic Consequences: Land Tenure in Kerala, 1850–1960.* Bombay: Allied.

World Bank. 1997. *India: Achievements and Challenges in Reducing Poverty.* Washington, D.C.

CHAPTER NINE

Policies for Sustainable Development

HERMAN E. DALY

This chapter presents four interrelated policies for sustainable development. The policies are presented from the perspective of the United States but should, in principle, apply to any country. Before getting to the specific policies, I discuss a basic point of view within which the policies appear most sensible and urgent, even though I think they are also defensible to a degree within the standard neoclassical framework. The four policies are then presented in order of increasing radicalism. The first two are fairly conservative, fundamentally neoclassical, and should be relatively noncontroversial, although often they are not. The third will be hotly debated by many, and the fourth will be considered outrageous by most economists. It would be politic to omit the fourth, but I cannot because it is the complementary external policy that is logically required if the first three internal policies are not to be undercut by free trade and capital mobility. At the end of the chapter, I defend this controversial suggestion by countering the two most frequently raised objections to it.

Point of View

Much depends on which paradigm one accepts: the economy as subsystem or the economy as total system. For those who, understandably, have become allergic to the word *paradigm,* I suggest Joseph Schumpeter's earlier and more descriptive term, "preanalytic vision." Because I think that preanalytic visions are fundamental, I shall take the time to illustrate their importance for the issue at hand with a story about the evolution of the World Bank's 1992 World Development Report (WDR), *Development and the Environment.*

An early draft of the 1992 WDR had a diagram titled "the relationship between the economy and the environment." It consisted of a square labeled "economy," with an arrow coming in labeled "inputs" and an arrow going out labeled "outputs"—nothing more. I worked in the Environment Department of the World Bank at that time and was asked to review and comment on the draft. I remarked that the picture was a good idea but failed to show the environment, and that it would help to have a larger box, representing the environment, surrounding the one depicting the economy. Then the relation between the environment and the economy would be clear: specifically, the economy is a subsystem of the environment and depends on the environment both as a source of raw material inputs and as a sink for waste outputs. The text accompanying the diagram should explain that the environment physically sustains the economy by regenerating the low-entropy inputs that it requires and by absorbing the high-entropy wastes that it cannot avoid generating, as well as by supplying other systemic ecological services. Environmentally sustainable development could then be defined as development that does not destroy these natural support functions.

The second draft had the same diagram but with an unlabeled box drawn around the economy, like a picture frame, with no change in the text. I commented that the larger box had to be labeled "environment" to avoid being merely decorative, and that the text had to explain that the economy was related to the environment in the ways just described. The third draft omitted the diagram altogether. There was no further effort to draw a picture of the relation between the economy and the environment. I thought that was very odd.

By coincidence, a few months later the chief economist of the World Bank, Lawrence Summers, under whom the 1992 WDR was being written, happened to be on a review panel at the Smithsonian Institution discussing the book *Beyond the Limits* by Donella Meadows, which he considered worthless. In that book there was a diagram showing the relation of the economy to the ecosystem as subsystem to total system, identical to what I had suggested (Figure 9.1). During the question-and-answer period I asked the chief economist if, looking at that diagram, he felt that the issue of the physical size of the economic subsystem relative to the total ecosystem was important, and if he thought that economists should be asking the question "What is the optimal scale of the macro economy relative to the environment that supports it?" His reply was immediate and definite: "That's not the right way to look at it," he said.

Reflecting on these two experiences has strengthened my belief that the differ-

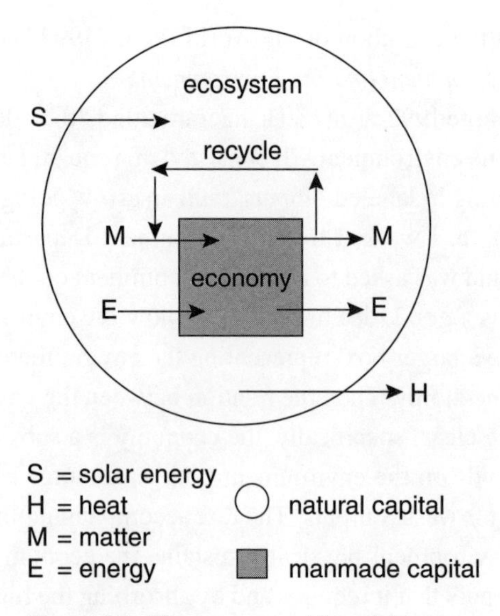

Figure 9.1. *The empty-world model.*

ence truly lies in our preanalytic visions. My preanalytic vision of the economy as subsystem leads immediately to the questions: How big is the subsystem relative to the total system? How big *can it be* without disrupting the functioning of the total system? How big *should it be,* what is its optimal scale, beyond which further growth in scale would be anti-economic—would increase environmental costs more than it increased production benefits? The chief economist had no intention of being sucked into these subversive questions: that is not the right way to look at it, and any questions arising from that way of looking at it are simply not the right questions.

That attitude sounds rather unreasonable and peremptory, but in a way that had also been my response to the diagram in the first draft of *Development and the Environment* showing the economy receiving raw material inputs from nowhere and exporting waste outputs to nowhere. That is not the right way to look at it, I basically said, and any questions arising from that picture—say, how to make the economy grow as fast as possible by speeding up throughput from an infinite source to an infinite sink—were not the right questions. Unless one has in mind the preanalytic vision of the economy as subsystem, the whole idea of sustainable development—of an economic subsystem being sustained by a larger ecosystem whose carrying capacity it must respect—makes no sense whatsoever. It was not surprising therefore that the WDR of 1992 was incoherent on the subject of sustainable development, placing it in solitary confinement in a half-page box where it was implicitly defined as nothing other than good development policy. It is the preanalytic vision of the economy as a box floating in infinite space that allows people to speak of "sustainable *growth*" (as opposed to *development*)—a clear oxymoron to those who see the economy as a subsystem of a finite and nongrowing ecosystem. The difference could not be more fundamental, more elementary, or more irreconcilable.

It is interesting that so much should be at stake in such a simple picture. The required tool of thought here is not a thousand-equation general equilibrium model on a Cray computer; all we really need is a wide-lined Big Chief Tablet with one crayola! Once you draw the boundary of the environment around the economy, you have implicitly admitted that the economy cannot expand forever. You have said that John Stuart Mill was right, that populations of human bodies and populations of capital goods cannot grow forever. At some point quantitative growth must give way to qualitative development as the path of progress, and we must come to terms with Mill's vision of the classical stationary state.

The World Bank, however, cannot say that—at least not yet and not publicly, because growth is the official solution to poverty. If growth is physically limited, or if it begins to cost more than it is worth at the margin and thereby becomes uneconomic, then how will we lift poor people out of poverty? We pretend there is no answer, but the answer is painfully obvious: by population control; by redistribution; and by improvements in resource productivity, both technical and managerial. But population control and redistribution are considered politically impossible. Increasing resource productivity is considered a good idea until it conflicts with capital and labor productivity—until we realize that historically we have bought high productivity and high incomes for capital and labor by using resources lavishly, thereby sacrificing resource productivity in exchange for a reduction in class conflict between capital and labor. Yet resources are the limiting factor in the long run—the very factor whose productivity, according to economic logic, should be maximized. When we draw that containing boundary of the environment around the economy, we move from "empty-world" economics to "full-world" economics (Figure 9.2). Economic logic stays the same, but the perceived pattern of scarcity changes radically, and policies must change radically if they are to remain economic. That is why there is such resistance to a simple picture. The fact that the picture is so simple and so obviously realistic is why it cannot be contemplated by the growth economists. That is why they react to it much as vampires react to crucifixes—"no, no, take it away, please!—that's not the right way to look at it!"

Let us persevere in looking at it that way, and turn now to consider some economic policies consistent with this "full-world" vision.

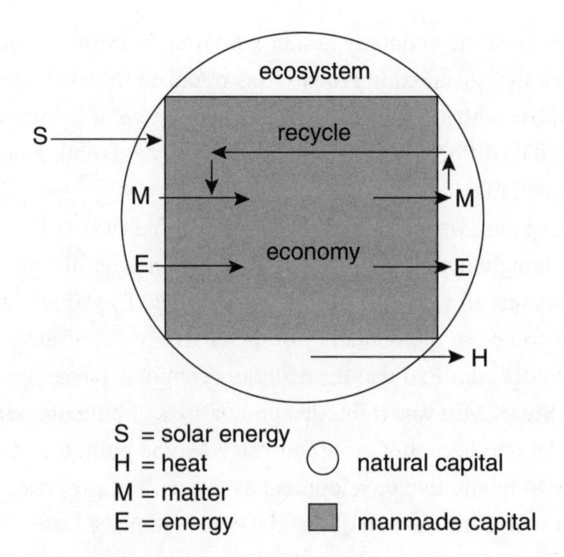

Figure 9.2. *The full-world model.*

Four Policy Suggestions

1. *Stop counting the consumption of natural capital as income.* Income is by definition the maximum amount that a society can consume this year and still be able to consume the same amount next year. That is, consumption this year, if it is to be called income, must leave intact the capacity to produce and consume the same amount next year. Thus sustainability is built into the very definition of income. But the productive capacity that must be maintained intact has traditionally been thought of as manmade capital only, excluding natural capital. We have habitually counted natural capital as a free good. This might have been justified in yesterday's empty world, but in today's full world it is anti-economic. The error of implicitly counting natural capital consumption as income is customary in three areas: (1) the System of National Accounts (SNA); (2) evaluation of projects that deplete natural capital; and (3) international balance of payments accounting.

The first (SNA) is well recognized, and efforts are under way to correct it—indeed, the World Bank played a pioneering role in this important initiative, and I hope it will regain its earlier interest in "greening the GNP.''

The second (project evaluation) is well recognized by standard economics, which has long taught the need to count "user cost" (depletion charges) as part of the opportunity cost of projects that deplete natural capital. World Bank *best* practice counts user costs, but World Bank *average* practice ignores it. Uncounted user costs show up in inflated net benefits and an overstated rate of return for depleting projects. This biases investment allocation toward projects that deplete natural capital and away from more sustainable projects. Correcting this bias is the logical first step toward a policy of sustainable development. User cost must be counted not only for depletion of nonrenewables but also for projects that divest renewable natural capital by exploiting it beyond sustainable yield. The sink or absorptive services of natural capital, as well as its source or regenerative services, can also be depleted if used beyond sustainable capacity. Therefore a user cost must be charged to projects that deplete sink capacity, such as the capacity of a river to carry off wastes or, most notably, the atmosphere's ability to absorb carbon dioxide. Measuring user cost is admittedly highly uncertain, but attempting to avoid the issue simply means that we assign to depleted natural capital the precise default value of zero, which is frequently not the best estimate.[1] Even when zero is the best estimate, it should be arrived at not by default but by reasoned calculation based on explicit assumptions about backstop technologies, discount rates, and reserve lifetimes.[2]

Third, in balance of payments accounting the export of depleted natural capital, whether petroleum or timber cut beyond sustainable yield, is entered in the current account and thus treated entirely as income. This is an accounting error. Some portion of those nonsustainable exports should be treated as the sale of a capital asset and entered on capital account. If this were properly done, some countries would see their apparent balance of trade surplus converted into a true deficit, one that is being financed by drawdown and transfer abroad of their stock of natural capital. Reclassifying transactions in a way that might convert a country's balance of trade from a surplus to a deficit would trigger an entirely different set of recommendations and actions by the International Monetary Fund (IMF). This reform of balance of payments accounting should be the initial focus of the IMF's new interest in environmentally sustainable development. The World Bank should warmly encourage its sister institution to get busy on this; such reform does not come naturally.

2. *Tax labor and income less, and tax resource throughput more.* In the past it has been customary for governments to subsidize the throughput of resources to stimulate growth.[3] Thus energy, water, fertilizer, and even deforestation are even now frequently subsidized. To its credit, the World Bank has generally opposed these subsidies. But it is necessary to go beyond removal of explicit financial subsidies to the removal of implicit environmental subsidies as well. By "implicit environmental subsidies" I mean external costs to the community that are not charged to the commodities whose production generates them.

Economists have long advocated internalizing external costs, either by calculating and charging Pigouvian taxes (taxes that when added to marginal private costs make them equal to marginal social costs) or by Coasian redefinition of property rights (such that values that used to be public property and not valued in markets become private property whose values are protected by their new owners). These solutions are elegant in theory but often quite difficult in practice. A blunter but much more operational instrument would be simply to shift our tax base away from labor and income and onto throughput. We have to raise public revenue somehow, and the present system is highly distortional in that by taxing labor and income in the face of high unemployment in nearly all countries, we are discouraging exactly what we want more of. The present signal to firms is to shed labor and substitute more capital and resource throughput, to the extent feasible. It would be better to economize on throughput because of the high external costs of its associated depletion and pollution, and at the same time to use more labor because of the high social benefits associated with reducing unemployment—a "double dividend" in terms of efficiency.

As a bumper sticker slogan the idea is "tax bads, not goods." In more theoretical terms the idea is to stop taxing value added and to start taxing that to which value is added, namely, the natural resource flow yielded by natural capital.[4] Because natural capital is the limiting factor in the long run (a point to be argued in the next section) and because its true opportunity cost is only poorly reflected in markets, it is reasonable to raise its effective price through taxation. Shifting the tax base to throughput induces greater resource efficiency and internalizes, in a gross, blunt manner, the externalities from depletion and pollution. It also avoids the distortions of taxing income. True, the exact external costs will not have been precisely calculated and attributed to exactly those activities that caused them, as with a Pigouvian tax that aims to equate marginal social costs and benefits for each activity. But those calculations and attributions are so difficult and uncertain that insisting on them in the interests of "crackpot rigor" would be equivalent to a full-employment act for econometricians.

Politically the shift toward ecological taxes could be sold under the banner of revenue neutrality. However, the income tax structure should be maintained so as to keep progressivity in the overall tax structure by taxing very high incomes and subsidizing very low incomes. The bulk of public revenue would be raised from taxes on throughput either at the depletion or pollution end, but especially the former. The goal of the vestigial income tax would be redistribution, not net public revenue. The shift could be carried out gradually by a pre-announced schedule to minimize disruption.[5] Ecological tax reform should be a key part of structural adjustment but should be pioneered in the North. Indeed, sustainable development itself must be achieved in the North first. It is absurd to expect much sacrifice for sustainability in the South if similar measures have not first been taken in the North.[6] The major weakness in the World Bank's ability to foster environmentally sustainable development is that it has leverage only over the South, not over the North. Some way must be found for the World Bank to serve as an honest broker, an agent for reflecting the legitimate demands of the South back to the North.

3. *Maximize the productivity of natural capital in the short run and invest in increasing its supply in the long run.* Economic logic requires that we behave in two ways toward the *limiting factor* of production: maximize its productivity today and invest in its increase tomorrow. Those principles are not in dispute. Disagreements do exist about whether natural capital is really the limiting factor. Some argue that manmade and natural capital are such good substitutes that the very idea of a limiting factor, which requires that the factors be complementary, is irrelevant.[7] It is true that without complementarity there is no limiting factor. So the question is, are manmade capital and natural capital basically complements or substitutes? Here again we can provide perpetual full employment for econometricians, and I would welcome more empirical work on this, even though I think it is sufficiently clear to common sense that natural and manmade capital are fundamentally complements and only marginally substitutable (see Appendix 9.1).[8]

In the past, natural capital has been treated as superabundant and priced at zero, so it did not really matter whether it was a complement or a substitute for manmade capital. Now remaining natural capital appears to be both scarce and complementary, and therefore limiting. For example, the fish catch is limited not by the number of fishing boats but by the remaining populations of fish in the sea. Cut timber is limited not by the number of sawmills but by the remaining standing forests. Pumped crude oil is limited not by manmade pumping capacity but by remaining stocks of petroleum in the ground. The natural capital of the atmosphere's capacity to serve as a sink for carbon dioxide is likely to be even more limiting to the rate at which petroleum can be burned than is the source limit of remaining oil in the ground.

In the short run, raising the price of natural capital by taxing throughput, as advocated above, will give the incentive to maximize natural capital productivity. Investing in natural capital over the long run is also needed. But how do we invest in something that, by definition, we cannot make? If we could make it, it would be manmade capital! For renewable resources we have the possibility of "fallowing investments," or more generally "waiting" in the Marshallian sense—allowing this year's growth increment to be added to next year's stock rather than consuming it.[9] For nonrenewables we do not have this option. We can only liquidate them. So the question is how fast do we liquidate, and how much of the proceeds can we count as income if we invest the rest in the best available renewable substitute? And how much of the correctly counted income do we then consume and how much do we invest?

One renewable substitute for natural capital is the mixture of natural and manmade capital represented by tree plantations, fish farms, and so on, which we may call "cultivated natural capital." But even within this important hybrid category we have a complementary combination of natural and manmade capital components—for example, a plantation forest may use manmade capital to plant trees, control pests, and choose the proper rotation, but the complementary natural capital services of rainfall, sunlight, and soil are still there and eventually still become limiting. Also, cultivated natural capital usually requires a reduction in biodiversity relative to natural capital proper.

For both renewable and nonrenewable resources, investments in enhancing throughput productivity are needed. Increasing resource productivity is indeed a good substitute for finding more of the resource. But the main point is that investment should be in the limiting factor, and to the extent that natural capital has replaced manmade capital as the limiting factor, the World Bank's investment focus should shift correspondingly. I do not believe that it has. In fact, the failure to charge user cost on natural capital depletion, noted earlier, surely biases investment away from replenishing projects.

The three policies suggested thus far all require the recognition and counting of costs heretofore not counted. It is difficult to imagine a global authority imposing a more complete and uniform cost-accounting regime on all nations. It is also difficult to imagine nations agreeing on an international treaty to that effect. What is easy to imagine is just what we observe—different national cost-counting standards leading to an international standards-lowering competition to reduce wages, environmental controls, social security standards, and so on. The best way to avoid such a situation is to give up the ideology of global economic integration by free trade and free capital mobility, and to accept the need for national tariffs to protect, not inefficient industries, but efficient national standards of cost accounting.

4. *Move away from the ideology of global economic integration by free trade, free capital mobility, and export-led growth— and toward a more nationalist orientation that seeks to develop domestic production for internal markets as the first option, having recourse to international trade only when clearly much more efficient.*[10] At the present time global interdependence is celebrated as a self-evident good. The royal road to development, peace, and harmony is thought to be the unrelenting conquest of each nation's market by all other nations. The word *globalist* has politically correct connotations, whereas the word *nationalist* has come

to be pejorative. This is so much the case that it is necessary to remind ourselves that the World Bank exists to serve the interests of its members, *which are nation-states, national communities*— not individuals, not corporations, not even non-governmental organizations. It has no charter to serve the one-world-without-borders cosmopolitan vision of global integration—of converting many relatively independent national economies, loosely dependent on international trade, into one tightly integrated world economic network upon which the weakened nations depend for even basic survival.

The model of international community upon which the Bretton Woods institutions rests is that of a "community of communities," an international federation of *national* communities cooperating to solve global problems under the principle of *subsidiarity*. The model is not the cosmopolitan one of direct global citizenship in a single integrated world community without intermediation by nation-states.

To globalize the economy by erasure of national economic boundaries through free trade, free capital mobility, and free, or at least uncontrolled, migration, is to wound fatally the major unit of community capable of carrying out any policies for the common good. That includes not only national policies for purely domestic ends but also international agreements required to deal with those environmental problems that are irreducibly global (buildup of carbon dioxide, or ozone depletion). International agreements presuppose the ability of national governments to carry out policies in their support. If nations have no control over their borders, they are in a poor position to enforce national laws designed to serve the common good, including those laws necessary to secure national compliance with international treaties.[11]

Cosmopolitan globalism weakens national boundaries and the power of national and subnational communities while strengthening the relative power of transnational corporations. Because there is no world government capable of regulating global capital in the global interest, and because the desirability and possibility of a world government are both highly doubtful, it will be necessary to make capital less global and more national. I know that is an unthinkable thought right now, but take it as a prediction: ten years from now the buzz words will be "renationalization of capital" and the "community rooting of capital for the development of national and local economies," not the current shibboleths of export-led growth stimulated by whatever adjustments are necessary to increase global competitiveness. "Global competitiveness" (frequently a thought-substituting slogan) usually reflects not so much a real increase in resource productivity as a standards-lowering competition to reduce wages, externalize environmental and social costs, and export natural capital at low prices while calling it income.[12]

The World Bank should reflect deeply on the forgotten words of one of its founders, John Maynard Keynes: "I sympathize therefore, with those who would minimize, rather than those who would maximize, economic entanglement between nations. Ideas, knowledge, art, hospitality, travel—these are the things which should of their nature be international. But let goods be homespun whenever it is reasonably and conveniently possible; and, above all, let finance be primarily national."[13]

Replies to the Two Most Frequent Objections to Abandoning Free Trade

1. *Growth will compensate.* Some globalists will admit that the problems just outlined are real, but whatever costs they entail are, in their view, more than compensated by the welfare increase from economic growth brought about by free trade and global integration. While it may be true that free trade increases economic growth, the other link in the chain of argument, that growth increases welfare, is shown below to be devoid of empirical support in the United States since 1947.

It is very likely that we have entered an era in which growth is increasing environmental and social costs faster than it is increasing production benefits. Growth that increases costs by more than it increases benefits is anti-economic growth and should be called that. But Gross National Product can never register anti-economic growth because nothing is ever subtracted. It is much too gross.

Although economists did not devise GNP to be a direct measure of welfare, nevertheless welfare is assumed to be highly correlated with GNP. Therefore if free trade promotes growth in GNP, it is assumed that it also promotes growth in welfare. But the link between GNP and welfare has become questionable, and with it the argument for deregulated international trade, and indeed for all other growth-promoting policies.

Evidence for doubting the correlation between GNP and welfare in the United States is taken from two sources. First, William Nordhaus and James Tobin, in 1972, asked, whether growth was obsolete as a measure of welfare and hence as a proper guiding objective of policy. To answer their question, they developed a direct index of welfare, called Measured Economic Welfare (MEW), and tested its correlation with GNP over the period 1929–65. They found that for the period as a whole, GNP and MEW were indeed positively correlated: for every six units of increase in GNP, there was, on average, a four-unit increase in MEW.[14] Economists breathed a sigh of relief, forgot about MEW, and concentrated on GNP.

Some twenty years later (in *For the Common Good,* 1989), John Cobb, Clifford Cobb, and I revisited the issue and began development of our Index of Sustainable Economic Welfare (ISEW) with a review of the Nordhaus and Tobin MEW. We discovered that if one takes only the second half of the period (the eighteen years from 1947 to 1965), the correlation between GNP and MEW falls dramatically. In this more recent period—surely the more relevant for projections into the future—a six-unit increase in GNP yielded on average only a one-unit increase in MEW. This suggests that GNP growth at this stage of U.S. history may be a quite inefficient way of improving economic welfare —certainly less efficient than in the past.

The ISEW was then developed to replace MEW, because MEW omitted any correction for environmental costs, did not correct for distributional changes, and included leisure, which dominated the MEW and introduced many arbitrary valuation decisions.[15] The ISEW, like the MEW though less dramatically, was correlated with GNP up to a point, beyond which the correlation turned slightly negative.[16]

Measures of welfare are difficult and subject to many arbitrary judgments, so sweeping conclusions should be resisted. However, it seems fair to say that for the United States since 1947, the empirical evidence that GNP growth has increased welfare is *very* weak. Consequently, any impact on welfare via free trade's contribution to GNP growth would also be very weak. In other words, the "great benefit," for which we are urged to sacrifice community standards and industrial peace, turns out on closer inspection not to exist.

2. *Comparative advantage proves that global integration is beneficial.* Because I am an economist and really do revere David Ricardo, the great champion of classical free trade, I think that it is important to point out that if he were alive now, he would *not* support a policy of free trade and global integration as it is understood today.

The reason is simple: Ricardo was careful to base his comparative advantage argument for free trade on the explicit assumption that capital was immobile between nations. Capital, as well as labor, stayed at home; only goods were traded internationally. It was the fact that capital could not, in this model, cross national boundaries that led directly to replacement of absolute advantage by comparative advantage. Capital follows absolute advantage as far as it can within national boundaries. But because by assumption it cannot pursue absolute advantage across national boundaries, it has recourse to the next best strategy, which is to reallocate itself within the nation according to the principle of comparative advantage.

If, for example, Portugal produces both wine and cloth cheaper than England does, then capital would love to leave England and follow absolute advantage to Portugal, where it would produce both wine and cloth. But by assumption it cannot. The next best thing is to specialize domestically in the production of cloth and trade it for Portuguese wine. This is because England's disadvantage relative to Portugal in cloth production is less than its disadvantage relative to Portugal in wine production. England has a comparative advantage in cloth, Portugal a comparative advantage in wine. Ricardo showed that each country would be better off specializing in the product in which it had a comparative advantage and trading for the other, regardless of absolute advantage. Free trade between the countries, and competition within each country, would lead to this mutually beneficial result.[17]

Economists have been giving Ricardo a standing ovation for this demonstration ever since 1817. So wild has been the enthusiasm for the conclusion that some economists forgot the assumption on which the argument leading to that conclusion was based, namely, internationally immobile capital. Whatever the case in Ricardo's time, in our day it would be hard to imagine anything more contrary to fact than the assumption that capital is immobile internationally. It is vastly more mobile than goods.

The argument for globalization based on comparative advantage is therefore embarrassed by a false premise. When starting from a false premise, one would have a better chance of hitting a correct conclusion if one's logic were also faulty! But Ricardo's logic is not faulty. Therefore I conclude that he would not be arguing for free trade—at least not on the basis of comparative advantage, which requires such a wildly counterfactual assumption.

Unlike some of today's economists and politicians, Ricardo would never argue that because free trade in goods is beneficial, adding free trade in capital must be even more beneficial! To use the conclusion of an argument that was premised on capital *immobility* in order to support an argument in favor of capital *mobility* is too illogical for words.[18]

In the classical vision of Smith and Ricardo the national community embraced both national labor and national capital, and these classes cooperated, albeit with conflict, to produce national goods, which then competed in international markets against the goods of other nations produced by their own national capital-labor teams.

Nowadays, in the globally integrated free-trade world it no longer makes sense to think of national teams of labor and capital. Global capitalists now communicate by mobile telephone with their former national workers in the following manner:

> Sorry to inform you of your dismissal, old Union Joe, but as everybody knows, we live in a global economy now—I can buy labor abroad at one-tenth the wage your union wants, and with lower environmental and social taxes, and still sell my product in this market or any other. Your severance check is in the mail. Good luck, Joe. —What's that? What do you mean "bonds of national community"? I just told you that we live in a global economy, and have abandoned all that nationalistic stuff that caused two World Wars. Haven't you heard of Smoot and Hawley? Factor mobility is

necessary for maximum efficiency, and without maximum efficiency we will lose out in global competition. —Yes, of course there will be a tendency to equalize wages worldwide, but profits will also equalize. —Well, yes, of course wages will be equalized downward and profits equalized upward. What else would you expect in a global economy that reflects world supply and demand? I can't change the law of supply and demand, Joe. Besides, don't you want the Chinese and Mexican workers to be as rich as you are? You're not a racist, are you, Joe? Furthermore, economists have proved that free trade benefits everyone. So be grateful, and now that you have some time, why not enroll in Econ 101 at your local community college—try to learn some economics, Joe. It will help you feel better.

At this stage in the dialogue there is not much community left. We have here the abrogation of a basic social agreement between labor and capital over how to divide up the value that they jointly add to raw materials. That agreement has been reached nationally, not internationally. It was not reached by economic theory, but through generations of national debate, elections, strikes, lockouts, court decisions, and violent conflicts. And now, that agreement, on which national community and industrial peace depend, is being repudiated in the interests of global integration. That is a poor trade, even if you call it "free trade."

Free trade, specialization, and global integration mean that nations are no longer free *not* to trade. Yet freedom not to trade is surely necessary if trade is to remain mutually beneficial. National production for the national market should be the dog, and international trade its tail. But the globalist free traders want to tie the dogs' tails together so tightly that the international knot will wag the national dogs in what they envision as a harmoniously choreographed canine ballet. But I foresee a multilateral dogfight. High-consuming countries, whether their high consumption results from many people or from high consumption per capita, will, in a finite and globally integrated full world, more and more be at each others' throats.

To avoid war, nations must both consume less and become more self-sufficient. But free traders say that we should become less self-sufficient and more globally integrated as part of the overriding quest to consume ever more. That is the worst advice I can think of.

Appendix 9.1

A point sure to be contested is the assertion that manmade and natural capital are complements. Many economists insist that they are substitutes. It is therefore necessary to argue the case for complementarity, which I shall do in three ways.

One way to make an argument is to assume the opposite of what you want to demonstrate and then show that it is absurd. If manmade capital were a near-perfect substitute for natural capital, then natural capital would also be a near-perfect substitute for manmade capital. But in the "empty" world we already had an abundance of natural capital, which, as a near-perfect substitute for manmade capital, would have made the effort to accumulate manmade capital absurd, given that we already possessed an abundant supply of a near-perfect substitute. Historically, however, we did accumulate manmade capital—precisely because it is complementary to natural capital.

Second, manmade capital is itself a physical transformation of natural resources, which are the flow yield from the stock of natural capital. Therefore, producing more of the alleged substitute (manmade capital) physically requires more of the very thing being substituted for (natural capital)—the defining condition of complementarity!

Finally, manmade capital (along with labor) is an agent of transformation of the resource flow from raw material inputs into product outputs. The natural resource flow (and the natural capital stock that generates it) are the *material cause* of production; the capital stock that transforms raw material inputs into product outputs is the *efficient cause* of production. One cannot substitute efficient cause for material cause—just as one cannot build the same wooden house with half the timber no matter how many saws and carpenters one tries to substitute. Also, to process more timber into more wooden houses in the same time period requires more saws, carpenters, and so on. Clearly the basic relation between manmade and natural capital is one of complementarity, not substitutability. Of course one could substitute bricks for timber, but that is the substitution of one resource input for another, not the substitution of capital for resources.[19] In making a brick house one would face the analogous inability of trowels and masons to substitute for bricks.

The complementarity of manmade and natural capital is made obvious at a concrete and commonsense level by asking, what good is a sawmill without a forest, a fishing boat without populations of fish, a refinery without petroleum deposits, an irrigated farm without an aquifer or river? We have long recognized the complementarity between public infrastructure and private capital—what good is a car or truck without roads to drive on? Following Alfred Lotka and Nicholas Georgescu-Roegen we can take the concept of natural capital even further and distinguish between endosomatic (within-skin) and exosomatic (outside-skin) natural capital. We can then ask, what good is the private endosomatic capital of our lungs and respiratory system without the public exosomatic capital of green plants that take up our carbon dioxide in the short run while in the long run replenishing the enormous atmospheric stock of oxygen and keeping the atmosphere at the proper mix of gases—that is, the mix to which our respiratory system is adapted and therefore complementary?

If natural and manmade capital are obviously complements, how is it that economists have overwhelmingly treated them as substitutes? First, not all economists have: Wassily Leontief's input-output economics, with its assumption of fixed-factor proportions, treats all factors as complements. Second, the formal, mathematical definitions of complementarity and substitutability are such that in the two-factor case the factors must be substitutes.[20] Because most textbooks are written on two-dimensional paper, this case receives most attention. Third, mathematical convenience continues to dominate reality in the general reliance on Cobb-Douglas and other constant elasticity of substitution production functions in which there is near-infinite substitutability of factors, in particular of capital for resources.[21]

Notes

1. Depletion of a nonrenewable resource has two costs: the opportunity cost of labor, capital, and other resources used to extract the resource in question, and the opportunity cost of not having the resource tomorrow because we used it up today. The second of these is referred to as "user cost" and is calculated by estimating the extra cost per unit of the best substitute that will have to be used when the resource in question is depleted, and then discounting that cost difference from the estimated date of depletion back to the present. That discounted amount is then added to the current cost of extraction to obtain the proper price that measures full opportunity cost.
2. See J. Kellenberg and H. Daly, "Counting User Costs in Evaluation of Projects that Deplete Natural Capital," working paper, ENVPE, World Bank, 1994.
3. The term *throughput* is an inelegant but highly useful derivative of the terms *input* and *output.* The matter-energy that goes into a system and eventually comes out is what goes through—the "throughput," as engineers have dubbed it. A biologist's synonym might be "the metabolic flow" by which an organism maintains itself. This physical flow connects the economy to the environment at both ends, and is of course subject to the physical laws of conservation and entropy.
4. See H. Daly, "Consumption and Welfare: Two Views of Value Added," *Review of Social Economy* 53, no. 4 (Winter 1995): 451–73.
5. See Ernst von Weizsacker, *Ecological Tax Reform* (London: Zed Books, 1992).
6. Even in its 1992 World Development Report (*Development and the Environment*) the World Bank has proved unable to face the most basic question: Is it better or worse for the South if the North continues to grow in its resource use? The standard view is that it is better, because growth in the North increases markets for Southern resource exports, as well as funds for aid and investment by the North in the South. The alternative view is that it makes things worse by preempting the remaining resources and ecological space needed to support Southern growth. Northern growth also increases income inequality and world political tensions. The alternative view urges continued *develop-*

ment in the North, but not *growth.* The two answers to the basic question cannot both be right. The absence of that fundamental question from World Bank policy research represents a failure of both nerve and intellect, as well as a continuing psychology of denial regarding limits to growth.

7. Both goods and factors of production can be either complements or substitutes. For consumer goods shoes and socks are complements (used together); shoes and boots are substitutes (one used instead of the other). In building a house bricks and wood are substitutes; bricks and masons are complements. If factors are good substitutes, the absence of one does not limit the usefulness of the other. For complements, the absence of one greatly reduces the usefulness of the other. The complementary factor in short supply is then the *limiting* factor.
8. Keep in mind that no one questions that some resources can be substituted for others, such as bricks for wood. But to substitute capital stock (saws and hammers) for wood is only marginally possible, if at all. Capital is the agent of transformation of the natural resource flow from raw material into finished product. Resources are the *material cause* of the finished product; capital is the *efficient cause.* One material cause may substitute for another (bricks for wood); one efficient cause may substitute for another (power saws for hand saws, or capital for labor), but efficient cause and material cause are related as complements rather than substitutes. If manmade capital is complementary with the natural resource flow, then it is also complementary with the natural capital stock that yields that flow.
9. Forgoing consumption today in exchange for greater consumption tomorrow is the essence of investment. Consumption is reduced by reducing either per capita consumption or population. Therefore investment in natural capital regeneration includes investment in population control, and in technical and social structures that demand less resource use per capita.
10. For an earlier analysis tending strongly in this direction, see W. Arthur Lewis, *The Evolution of the International Economic Order* (Princeton: Princeton University Press, 1978).
11. As a thought experiment, imagine first a world of free migration. What reason would there be in such a world for any country to try to reduce its birth rate? Now imagine that people do not migrate, but capital and goods, under free trade, migrate freely. With wages tending to equality worldwide, and cheap labor being a competitive advantage, what reason is there for any country to reduce its birth rate, especially that of its working-class majority? Does anyone think the United Nations will limit the global birth rate?
12. See H. E. Daly, "The Perils of Free Trade," *Scientific American* (November 1993).
13. J. M. Keynes, "National Self-Sufficiency," in *The Collected Writings of John Maynard Keynes,* vol. 21, ed. Donald Moggeridge (London: Macmillan and Cambridge University Press, 1933).
14. William Nordhaus and James Tobin, "Is Grown Obsolete?" In *Economic Growth,* National Bureau of Economic Research, General Series, no. 96E (New York: Columbia University Press, 1972).
15. For critical discussion and the latest revision of the ISEW, see Clifford W. Cobb and John B. Cobb, Jr., et al., *The Green National Product* (New York: University Press of America, 1994). For a presentation of the ISEW see the appendix of H. Daly and J. Cobb, *For the Common Good,* 2nd ed. (Boston: Beacon Press, 1994). See also Clifford W. Cobb et al., "If the GDP is Up, Why Is America Down?" *Atlantic Monthly* (October 1995).

16. Neither the MEW nor the ISEW considered the effect of an individual country's GNP growth on the *global* environment, and consequently on welfare at geographic levels other than the nation. Nor was there any deduction for harmful products, such as tobacco or alcohol. Nor did we try to correct for diminishing marginal utility of income. Such considerations, we suspect, would further weaken the correlation between GNP and welfare. Also, GNP, MEW, and ISEW all begin with Personal Consumption. Because all three measures have in common the largest single category, there is a significant autocorrelation bias, which makes the poor correlations with GNP all the more dramatic.
17. Absolute advantage is the rule for maximizing returns to capital when capital is mobile. Comparative advantage is the rule for maximizing returns to capital subject to the constraint that capital stays at home. The relevant comparison for comparative advantage is the difference between the internal cost ratios of the two countries—the unit cost ratio of cloth to wine in England compared to the unit cost ratio of cloth to wine in Portugal. Note that the units in which costs are measured may be different in each country without affecting the comparison of ratios; the units cancel out in each ratio. This means that absolute cost differences between countries do not matter for comparative advantage. It also means that in Ricardo's world, where capital stays at home, nations would unilaterally be able to adopt different standards of internalization of environmental and social costs without upsetting the principle of comparative advantage; that is, harmonization would not be necessary. It is only when capital is mobile, and absolute advantage reigns, that differing national cost-internalization practices would initiate an international standards-lowering competition to keep and attract capital.
18. In their July 1993 draft, "Trade and Environment: Does Environmental Diversity Detract from the Case for Free Trade?" Jagdish Bhagwati and T. N. Srinivasan reaffirm the view that free trade remains the optimal policy in spite of environmental issues. But their conclusion is based on what they call "a fairly general model . . . but one where resources, such as capital, do not move across countries" (p. 11). Of course, if capital is immobile between nations, comparative advantage arguments still hold. The point is that in today's world capital is highly mobile. Nor do they advocate restricting capital mobility to make the world conform to the assumptions of their argument—much the contrary! Their willingness to draw concrete policy conclusions from such an injudiciously abstracted model is a classic example of what A. N. Whitehead called the "fallacy of misplaced concreteness." At least they honestly pointed out their assumption and did not leave it under a pile of mathematics—but they did not note the triviality of their conclusion, given that assumption.
19. Regarding the house example I am frequently told that insulation (capital) is a substitute for resources (energy for space heating). If the house is considered the final product, then capital (agent of production, efficient cause) cannot end up as a part (material cause) of the house, whether as wood, brick, or insulating material. The insulating material is a resource like wood or brick, not capital. If the final product is not taken as the house but the service of the house in providing warmth, then the entire house, not only insulating material, is capital. In this case more or better capital (a well-insulated house) does reduce the waste of energy. Increasing the efficiency with which a resource is used is certainly a good substitute for more of the resource. But these kinds of waste-reducing efficiency measures (recycling prompt scrap, sweeping up sawdust and using

it for fuel or particle board, reducing heat loss from a house, and so on) are all marginal substitutions that soon reach their limit.

20. The usual definition of complementarity requires that for a given constant output a rise in the price of one factor would reduce the quantity of both factors. In the two-factor case both factors means all factors, and it is impossible to keep output constant while reducing the input of all factors. But complementarity might be defined back into existence in the two-factor case by avoiding the constant-output condition. For example, two factors could be considered complements if an increase in one alone would not increase output but an increase in the other would, and they could be considered perfect complements if an increase in neither factor alone would increase output but an increase in both would. It is not sufficient to treat complementarity as if it were nothing more than "limited substitutability." That means that we could get along with only one factor well enough, with only the other less well, but that we do not need both. Complementarity means that we need both and that the one in shortest supply is limiting. Imagine the L-shaped isoquants that depict the case of perfect complementarity. Now erase the right angle and replace it with a tiny ninety-degree arc. This seems to me the most realistic case—a very marginal range of substitution quickly giving way to a dominant relation of complementarity.
21. For a discussion, see Herman E. Daly, *Ecological Economics and the Ecology of Economics* (Cheltenham, England: Edward Elgar, 1999), especially part 3.

References

Cobb, Clifford W., et al. "If the GDP Is Up, Why is America Down?" *Atlantic Monthly* (October 1995).

Cobb, Clifford W., and John B. Cobb, Jr., et al. *The Green National Product.* New York: University Press of America, 1994.

Daly, H. "The Perils of Free Trade." *Scientific American* (November 1993).

———. "Consumption and Welfare: Two Views of Value Added." *Review of Social Economy* 53, no. 4 (Winter 1995): 451–73.

———. *Ecological Economics and the Ecology of Economics.* Cheltenham, England: Edward Elgar, 1999).

Daly, H., and J. Cobb. *For the Common Good.* 2nd ed. Boston: Beacon Press, 1994.

Kellenberg, J., and H. Daly. "Counting User Costs in Evaluation of Projects that Deplete Natural Capital." Working paper, ENVPE, World Bank, 1994.

Keynes, J. M. "National Self-Sufficiency." In *The Collected Writings of John Maynard Keynes,* vol. 21, ed. Donald Moggeridge. London: Macmillan and Cambridge University Press, 1933.

Nordhaus, William, and James Tobin. "Is Grown Obsolete?" In *Economic Growth,* National Bureau of Economic Research, General Series, no. 96E. New York: Columbia University Press, 1972.

Weizsacker, Ernst von. *Ecological Tax Reform.* London: Zed Books, 1992.

CHAPTER TEN

Weaving and Surviving in Laichingen, 1650–1900: Micro-History as History and as Research Experience

HANS MEDICK

Microstoria ne veut pas dire regarder des petites choses, mais regarder petit.
[To practice micro-history does not mean to look at little objects, but to regard things on a small scale.]
—Giovanni Levi

I shall take this remark on micro-history made by Italian historian Giovanni Levi at a 1990 discussion as my point of departure from which to consider how my own work has experimented with social, cultural, and economic perspectives on the study of history. While micro-historical perspectives originated in Italy in the late 1970s, simultaneous and independent work had begun elsewhere in Europe and in the United States as well. In Göttingen, for example, David Sabean, Peter Kriedte, Alf Lüdtke, Jürgen Schlumbohm, and I came together in the 1970s to develop micro-historical approaches that were different from, but in important respects parallel to, those developed in Italy.

I myself experimented with micro-historical approaches in my 1996 book *Weaving and Surviving at Laichingen, 1650–1900,* which was the culmination of fifteen years of work.[1] This book is a history of a small rural industrial community in a remote corner of the Swabian Alps in the former duchy of Württemberg. It is an example of what I call "history from the edge" (*entlegene Geschichte*) in that it is a local history, which at the same time tries to be more than just that. As the subtitle of the book, *Lokalgeschichte als Allgemeine Geschichte,* suggests, the book uses an intense micro-historical focus on local history as a possibility to confront problems of general history, or *Allgemeine Geschichte.*

In this chapter I shall discuss the methods, contents, and subject of micro-history as it applied to my work. But before then, I should like—by way of preface, so to speak—to make some observations on the history of this topic. As a scholarly term, "micro-history" was used for the first time in Europe by a historian whom one at first would not think of as a likely candidate to do so: Fernand Braudel. The Nestor of the Annales school in 1958 used the word in a specific way, however. His intention at this time was not to bring into play a strategic term meant to achieve a breakthrough for a new direction of social history. Rather, he used the word in an aside and in a critical fashion. "Microhistory" was for Braudel a synonym for a history of events of rather short duration, for those "appearances" on the surface of historical processes that have to be questioned by an investigation of their conjunctural and structural depths of *longue durée*.[2]

Paradoxically, it was not the usage of the term by a great opinion-making historian that led to its reception into professional language, but—as Carlo Ginzburg has pointed out[3]—its literarization in a work of fiction, namely, the popular novel by Raymond Queneau, *Les Fleurs bleues,* which appeared in 1965. It was only through this "literary turn"—and the negative-ironic accent it received thereby—that the word seems to have received attention from historians, above all in Italy, but this time with a thorough change of meaning in a positive direction. In a similar manner to Braudel, Queneau places macro- and micro-history in a hierarchical scheme, but does so in a changed and radicalized manner. In his novel, micro-history is the lowest and meanest variety of all possible kinds of history. It ranks not just beneath "world history" or "general history," but even below a mere "history of events." Queneau presents the following dialogue between the duke of Auge and his chaplain to illustrate the different kinds of histories:

> The Duke of Auge rubbed his hands and manifested all signs of the most lively satisfaction, then, suddenly, he began to look anxious.
>
> "And that world history I was asking you about, quite some time ago, now, I'm still waiting for your answer."
>
> "What do you want to know, exactly?"
>
> "What do you think of world history in general and of general history in particular? I'm listening."
>
> "I'm really tired," said the Chaplain.
>
> "You can have a rest later. Tell me, this Council of Basle, is that world history?"
>
> "Yes indeed. World history in general."
>
> "And my little cannons?"
>
> "General history in particular."

"And the marriage of my daughters?"

"Hardly even factual history. It's microhistory at the very most."

"It's what?" yells the Duke of Auge. "What the devil sort of language is that? Is today your Pentecost or what?"[4]

Since 1965, the term "micro-history" has witnessed a conceptual change in its meaning seen most evidently in the term's detachment from the negative and objectified image of a tiny, despicable, private reminder of history at large. But what is more important is that the subject matter itself has developed and constituted itself as a new historical research practice. Beginning with the ironic orientation toward a small reminder of history, a new perspective on historical research and knowledge emerged which, since the late 1970s, has opened up new territories for social and cultural history.

This new direction in historical research and writing has several salient characteristics. Whereas it is not easy at present to disentangle them to assess their relative importance, many historians since the 1970s cite "change of experience" and "change of historical awareness" as standing at the beginning of a new departure, but in the last resort a "change in method" seems to have been decisive. The two concepts "change of experience" and "change in method" have been introduced by Reinhart Koselleck to explain paradigmatic historiographic changes since antiquity,[5] but it is worth remarking that micro-history was not in his observational focus.

Whereas the micro-historical change of method may have derived its initial impulses from this change of experience, it cannot be entirely reduced to it. Further constituting aspects were the scholarly controversies and the development of a more encompassing intellectual discourse in the humanities and—within this—especially the challenge that came to social history from cultural anthropology. In one of the texts that stood at the beginning of systematic reflections on the problems of micro-history, Carlo Ginzburg and Carlo Poni pointed to this change of experience (*Erfahrungswandel,* as Koselleck called it). According to these authors, there were good reasons for claiming that the "great success of micro-historical reconstructions was connected to the emergence of doubts with relation to certain macro-historical processes."[6] The German historian Christian Meier further contextualizes the interest in micro-history by locating its origin in "certain social and political experiences of the more recent past and present," as a consequence of a weakening of "identifications with larger political units, be they a nation or a state, be they big parties or trade unions or even the movement of progress itself." Meier maintains that the "weakening of certain pathways into macro-history" has directly led to the interest in micro-history, which has subsequently seen the discovery of new historical objects and themes, as an interest taken in private, historical "micro-worlds," small *Lebenswelten,* those tiny domains of life and history that have the individual at their centers.[7]

Whether in the form of doubts regarding an identification with the assumptions of progress, the refusal of an evolutionist understanding of history, or the critique of ethnocentric global historical perspectives, these questionings of accustomed assumptions of historical philosophy and theory resulting from a contemporary "change of experience" were, and are, seen as connected with the origin of micro-history. Surely they have been important as initiating factors, but their specific effect may be to start a dynamic of research and methodological reflection going beyond the impetus exerted by an Erfahrungswandel.

Methodologically, micro-history moves beyond the concepts of macro-historical synthesis and the master narratives attached to them. It also tries to move beyond claims that see historical research, interpretation, and writing inescapably tied to a state of postmodern "fragmentation." Texts of individual historical fragments may offer a starting point for micro-analysis; from that point, it tries to reach the level of actions, behaviors, and experiences of historical subjects by exploring and reconstructing their manifold contexts, connections, and wider historical relations with the worlds they move in. It is precisely this departure that Giovanni Levi hinted at with his sentence of 1990, in which he—without irony—made the point that micro-history in the first instance does not intend to "look at little objects" but to "regard things on a small scale." Levi would not agree with a warning issued by Jürgen Kocka against a "tendency towards micro-historical Klein-Klein" in the history of everyday life.[8] Levi certainly does not move within the tradition of a German discourse on history, which, with its critique of micro-dimensions and perspectives, has maintained a strong position among Germany's professional historians since the nineteenth century. This position unites otherwise opposed critics like Heinrich von Treitschke, who in his *Deutsche Geschichte* and elsewhere made strong attacks on what he called *ideenlose Mikrologie in der Geschichte,* and Hans Ulrich Wehler, the Bielefeld historian, who repeatedly criticized the "theoretical incompetence of micro-history," an incompetence he attributes to its confinement to the *mikro-historische Besenkammer,* that is, to its concentration on small historical objects instead of "big" and encompassing structural and political issues.[9]

In my opinion, a critique like this is mistaken, because it confuses a methodological perspective with the smallness of an object of knowledge. In criticizing small objects, a small-scale historical perspective is—more subconsciously than consciously—done away with, explicitly for the reason that it has nothing to contribute to an investigation of large, encompassing, and "central" historical processes, such as industrialization, modernization, and the formation of states and nations. This hierarchization of historical objects and of perspectives of knowledge is questioned by a micro-historical approach. Such an approach defines itself—and this is what Levi meant in his sentence—not by the micro-dimensions and the smallness of its objects. It gains its heuristic potential, rather, from the microscopic view, as it is achieved by a radical reduction of the scale of observation. "Historians do not study villages, they study in villages," Levi has said in paraphrasing a sentence from Clifford Geertz, meaning that concentrating on limited fields of observation—whether villages, urban quarters, social groups, or one or several individuals—enables a qualitative enlargement of historical reconstructions and interpretations.[10]

A decisive gain for social history using micro-historical procedures consists in the fact that by manifold and precise investigations of historical specificities and details of the totality of individuals of a small society, the reciprocal relationships between cultural, economic, and political moments may be looked upon as a *lebensgeschichtlicher Zusammenhang*. What is needed to achieve this goal is a specific method of nominal record and data linkage, a method differing from that applied in traditional local and regional histories. Instead of preliminary categorization in the form of assumed macro-historical substances (*the* family, *the* individual, *the* state, *the* industrialization process), an experimental investigation of social networks and chains of action is undertaken, but never by a fixation on them alone, and always with a view to the social, economic, cultural, and political conditions and relationships that are expressed in and through them. Through this, new insights into the constitution of historical structures as well as into shorter long-term processes may be achieved.

My own research for *Weaving and Surviving in Laichingen,* concerning a small, proto-industrial rural society on the high plains of the Swabian Alps between the seventeenth and nineteenth centuries, pursued the what I have called a "detailed history of the whole,"[11] which, however, always tries to be aware of its partial and constructed character. My work in this respect does not—I should add—follow the synecdochal approach advocated, for instance, by Geertz, which involves the investigation of a significant social fact, for instance a cockfight ritual in Bali, to symbolize and represent conflicts, norms, and processes of a whole society. Instead, my work takes up the impulses of the French *histoire totale,* not in the direction of an anticipated *histoire globale* as initiated by Braudel. I concentrate on and limit myself to regional and local historical relations, in a manner similar to Le Roy Ladurie, who summarized his intentions in introducing his classic

Les Paysans du Languedoc in the following words: "With the best tools at my disposal, and within the limited framework of a single human group, I embarked upon the adventure of a 'total' history."[12]

My investigation, however, differs from the histoire totale approach in the reconstruction of its source and data basis on a micrological, nominative foundation. This provided me with the possibility both of pursuing a qualitative life-history approach as well as quantitative analyses encompassing the relations of all persons ever registered in Laichingen sources. What was at issue was to find a procedure providing methodological safeguards that the individual life to be found in the sources and to be reconstructed from them was not a priori drowned in statistical averages, but at the same time providing the advantage of statistical-serial analysis. Starting from the church registers for the village and using additional documentation from a lot of other sources, a nominative documentation of all individuals and families who appeared in the records between the middle of the seventeenth and the end of the nineteenth centuries was reconstructed. Through this enlarged type of family reconstitution, the elementary dates for each individual life and for each family could be linked with other texts and documents concerning these individuals and families. These other documents included, for example, tax lists, with inventories at death and at marriage that inform about the immovable and movable property of a household, from land, credits, and debts to household items like clothes, religious books, plates, spoons, knives, and forks (if there were any). Village court records are another important source I have linked with my nominative documentation.

To take the persons and names of the individual men, women, and children of Laichingen as the methodological point of departure for the investigation by no means implied that the individual was taken as the a priori of the history of Laichingen. On the contrary, the reconstruction of the data by nominal record linkage from sources and texts was undertaken to make sure that the significance of individual acts and life courses, but also of the details of everyday life, could be investigated and interpreted within the manifold contexts of local and supra-local social, economic, and cultural relationships and conditions. Whereas individual cases marked the beginning of the investigation, typical and significant cases emerged in the end. The aim was an interpretation and presentation of acts, lives, and details of the normal and exceptional history of everyday life, which would offer insights not only into the contexts but also into the constitution, the "making," of the encompassing relations of life, work, and domination in this local society and beyond.

In light of classical social and economic history, the material results of my investigations could be described as paradoxical and nonrepresentative. I originally chose to investigate the *Flecken* of Laichingen in southwest Germany because from very early on, this rural society owed its survival to the combined practices of artisanal linen weaving and agriculture on small plots. Since the late middle ages, Laichingen had practiced commercial linen weaving. Between the seventeenth and nineteenth centuries, Laichingen underwent a classical case of "proto-industrialization." By the 1740s, for example, more than half the households of Laichingen were engaged in commercial linen weaving. Their numbers grew more than threefold during the century, whereas the population of Laichingen between 1720 and 1820 almost remained stable. This is not the only paradox of the history of Laichingen to be explained. What is even more striking is the longue durée of its proto-industrial modes and relations of production. Handloom weaving on the basis of household or small workshop industries remained the dominant form of industrial activity until the beginning of the twentieth century. Quality production, "flexible specialization," skillful and unconventional forms of marketing by local *weber-marchands* and itinerant traders and later by *Muster-* and *Handlungsreisende,* but also the support of the Württemberg state during the nineteenth century in the form of trade schools, technological counseling, and subsidies for new and innovative forms of hand technology helped the producers of Laichingen overcome severe crises and maintain a successful position in niches of growing national and international markets.

Other centers of handloom linen production in Württemberg and southwest Germany underwent deindustrialization beginning at the end of the eighteenth century under the double impact of increased competition from such linen-producing regions as Silesia, Flanders, and Northern Ireland and of "substitutive competition" from the cotton industry. This Flecken, by contrast, situated in a remote and climatically adverse corner of "Swabian Siberia"—as this part of the Swabian Alps has been called since the eighteenth century—maintained its craft industrial character. This was not solely because of the craft and commercial skills of the Laichingen weavers but also because the survival of weaving was closely linked to the agricultural industry for small peasant and subpeasant landed property.

Since the start of the eighteenth century, the subpeasant group of *Gewerbesöldner* had become and remained the most dynamic social backbone of rural industrial production. Usually, the size of their landed properties was below the threshold of small peasant property but distinctly above the level of cottager landholdings, who just owned a house and, at the most, a patch of garden land. This critical size of their landed possessions enabled *Söldner* households to add on a substantial agrarian *Zubrot,* or "bread supplement," to their craft earnings by applying part of their family labor, especially that of the women, to agricultural activities. This substantial zubrot earning from an agrarian base distinguished the Söldner from the cottager class as well as from the weaver peasants. Söldners did not have to invest a disproportionate share of their work capacity in agriculture as the weaver peasants had to do, thereby permitting an intensification of their craft and commercial activities, which became more and more crucial as the margins of craft earnings grew smaller after the end of the eighteenth century. The possession of land and the supplementary livelihood reaped from it remained an essential precondition for industrial production and commercial activity. Paradoxically, it was only the full-time specialization, increased professionalism, and adoption of new hand technologies brought on during the second half of the nineteenth century that enabled landless weaver households to bring gains instead of losses and debts to the family income.

The history of Laichingen is not typical in that it does not follow concepts of proto-industrialization that linked the development of rural industries to increasing proletarianization and a corresponding loss of land on the part of rural industrial producers.[13] Its importance also cannot be grasped from notions of representativeness, which link what is thought to be relevant and typical to criteria of statistical representativeness. Laichingen's proto-industrial history could rather be called an "exceptionally normal case," to use an intriguing oxymoron that Edoardo Grendi has brought into circulation.[14] This history is an extreme case of the historically possible, but one that well illuminates the peculiarities of a society like that of Württemberg, which found its own way to industrial mass production rather late.

Before the nineteenth century, the expansion of rural industries in central Europe typically occurred in environments open to competition and free from guild restrictions. Laichingen did not fit this pattern. Here, the linen industry survived because of its experience as a craft trade guild. From early on, Laichingen's competitive advantage derived from its standards of craftsmanlike quality production and from distant markets. During the ancien régime, "free trade for the guild," or *Freihandel für die Zunft,* was the strange political slogan with which Laichingen weavers publicly registered their demands and protests against the interests of big commercial capital in the neighboring city of Urach, which acted in close cooperation with the state. In doing so, they successfully defended their interests and became the most important linen-producing center in the country. It was not least this sticking to standards of quality weaving that secured the survival of the Laichingen linen trades over centuries. In Laichingen, it might be said with a pointed formulation, "the long eighteenth century" lasted from the time of the Thirty Years' War to the beginning of the twentieth century.

Mentioning these issues and themes by no means exhausts the area of my investigation. What began as a search for the origins, the dynamic, and the specificity of a local and regional variety of proto-industrial capitalism terminated in an investigation of the peculiarities of Pietism in Laichingen and Württemberg. This did not happen by chance. Weaving and surviving in the society of Laichingen during the long eighteenth century did not rest only on an economic base. Rural industrial structures owed their longue durée to the "culture of survival," which in specific ways was formed by the mentality of Württemberg Lutheran Pietism. A characteristic feature of this mentality, which constituted the spirit of Pietism in Laichingen but perhaps also in Württemberg at large, consisted in a world of religious ideas from which emerged a specific form of "Protestant ethic." Essential to this Protestant ethic was a personal and "awakened" "succession of Christ" to be emulated on the narrow and toilsome path through life. Its primary orientation was not toward economic success and worldly reward but toward the "enduring" (*durchhalten*) of toil and misery, of labor and exertion, of illness and dying. Even when work and calling were sanctified as an essential part of this succession of Christ, they led less to an ethic of pragmatic and successful "getting through" (*durchkommen*), described by David Sabean as typical for the men and women in Württemberg peasant and craft households, than to one of persevering and "continuing steadfastly in the expectation of good," "holding out till the end"—as the influential writer of devotional treatises and pastor Immanuel Gottlob Brastberger has formulated.[15] This spirit and the ethic emanating from it were appropriate to the realities of Laichingen and Württemberg societies, which from the seventeenth until the twentieth century stood under the imperatives of partible inheritance and small property holding. For Laichingen weavers, craftsmen, and small peasants, "sanctification" of their lives by "perseverance" in their work made more sense and was at the same time more necessary than was upholding the abstract goal of successful *Gnadenwahl* by ascetic entrepreneurial activity, which Max Weber pointed out as the religious foundation of that ideal type of Protestant ethic. The mentality of Laichingen Lutheran Pietism, it might be said in distinction, amounted to a Protestant ethic that was lacking in the spirit of capitalism.

The consequences of this local formation of a Protestant ethic led to industry and even proto-industry. But they produced neither a capitalist spirit nor even proto-capitalist structures. It becomes clear that the relations of small and minute property in this society, and above all the process of permanent partition of property among kin and children, determined to a large extent the local dynamic of the Protestant ethic and its secularized expression, the Swabian mentality of *Schaffe, schaffe, Häusle baue* (work, work, build a little house). But it also becomes obvious how the religious attitudes of Württemberg Pietism for their part "sanctified"

a work ethic standing in close connection to the social and economic structures of small property. Pietist belief in holiness, Swabian proficiency, and the clinging to small property here entered into a characteristic synthesis, which thoroughly suffused a local way of life.

Toil and the misery of labor dictated attitudes toward surviving and dying in this society, which included the deaths of many infants and children. To a much greater extent than in other central European regions and settlements, these children were the casualties of their parents' conditions of life, work, and mental attitudes. Close microanalytic scrutiny does not indicate that women and men developed strategies that directly led to the death of their children, but it does show that adults did not and could not prevent a large proportion of them from dying in infancy or at a very early age. The infant death rate stood between 35 percent and 50 percent in all decades between the beginning of the eighteenth and the second half of the nineteenth centuries. It becomes clear that under conditions of extremely high fertility, which since the seventeenth century consistently surpassed that of the Hutterites, the model population for assessing high fertility, it was primarily women's workloads as tied to conditions of property that were the decisive factors contributing to the exceptional Laichingen death rates for infants and children. But this effect was manifested in paradoxical ways. Day-laborer families without significant landed property of their own had the lowest infant death rates of all social groups. In contrast, the highest infant death rates occurred among the richest group of proprietors active in nourishment occupations as innkeepers, bakers, or butchers. Among this group of the village aristocracy, it was clearly the resistance to breastfeeding by women, and not their involvement in arduous work tasks, that contributed to the extreme infant death rates, for in the other social groups different patterns obtained. For them, it seems to have been work imperatives related to the family economy's survival, especially under conditions of small and dissipated landed property with their demands of greater or lesser shares of women's labor, which were both a necessary and effective cause for the neglect of infants and children, with a consequently high risk of death. Infant and child mortality in Laichingen was highest among those households of small and medium landed proprietors that could not afford servants or day laborers, in which dual occupations in agriculture and weaving prevailed and in which women's work contributions were so crucial for the survival of the household that there was a fatal lack of care and time available for infants.

In short, the demographic micro-analysis of the "regime of death" in the society of Laichingen from the seventeenth to the middle of the nineteenth centuries brings to light once more the extreme degree to which weaving and working for survival conditioned the behavior patterns especially of the small and minute landed proprietors in this society, who were constantly living on the edge of their own reproduction. The Pietist advocacy of following the narrow path on the way to heaven and taking arduous work, toil, suffering, pain, and death rather than entrepreneurial success as signs of sanctification made sense under these conditions, too. The demographic micro-analysis clearly shows why the Laichingen variant of a Protestant ethic, which was part of its total social, cultural, and economic environment, did not lead to a "rationale *Lebensführung,*" which for Max Weber was the essence of a spirit of capitalism.[16] The high-pressure demographic equilibrium of extreme fertility and extreme mortality amounted to a system of wastage of human life whose main rationale was survival on the basis of small property and industry in a local society possibly exhibiting the highest density of craft occupations of any rural community in early modern Württemberg.

To draw some more general methodological conclusions from my investigations, I emphasize that studying history through the details of everyday life and work of a small and remote local society does not preclude a wider historical perspective or the discussion of more general historical problems. On the contrary, it is a provocation to do this in a new way. Even general historical interpretations—yes, even the categories and questions—change within a micro-historic field of observation. In this respect, micro-historical perspectives should be seen as fundamental but by no means *allumfassend* (all-encompassing). In 1966 Sigfried Kracauer, an early theoretician of micro-history, wrote:

> Not all of historical reality can be broken down into microscopic elements. The whole of history also comprises events and the developments which occur above the micro-dimension. For this reason histories at higher levels of generality are as much of the essence as studies of detail. But they suffer from incompleteness; and if the historian does not want to fill the gaps in them "out of his own wit and conjecture," he must explore the world of small events as well. Macro-history cannot become history in the ideal sense unless it involves micro-history. . . .
>
> The higher the level of generality at which a historian operates, the more historical reality thins out. What he retains of the past when he looks at it from a great distance are wholesale situations, long-term developments, ideological trends, etc.—big chunks of events whose volume wanes or waxes in direct ratio to distance. They are scattered over time; they leave many gaps to be filled. We do not learn enough about the past if we concentrate on the macrounits . . . with increasing distance the historian will find it increasingly difficult to lay hands on historical phenomena which are sufficiently specific and unquestionably real.[17]

Kracauer formulated these far-sighted remarks in his posthumous work *History: The Last Things before the Last.* Here he is pleading for the multi-perspectivity of historical investigations and texts. To both historical perspectives mentioned and also to those in between, Kracauer attributed a specific cognitive power and a justification to contribute to the reconstruction of the "whole of history." But he also stressed the dependence of these perspectives on their corresponding scales of observation and, for this reason, on their respective relations of sharpness or fuzziness (*Schärfe-Unschärfe-Relationen*). Regarding professional history's claim for synthesis or representation of the whole, he limited it to partial and conditional, provisional, but by no means arbitrary insights into reality and experience. In doing so, he insisted on the inevitability of a micro-historical perspective. This demand was new in the 1960s, and at the beginning of the twenty-first century it has by no means become obsolete. Kracauer has pointed out a direction of micro-historical considerations, which are open for wider historical scope and for theoretical enrichment, but whose historical close-ups at the same time, through a precise eye for details, change the view of the whole.

For micro-history at present, more is at stake than a mere change of view or perspective. Following Kracauer, micro-history is more concerned with developing methodologies that, through "construction in the material of history" (his words),[18] bring to light hitherto invisible changes and undetected permanences of historical processes and of human history.

Acknowledgments

This chapter is a rewritten—and, from the author's point of view, improved—version of a paper first presented at the Agrarian Studies Colloquium at Yale University on 12 November 1993. I wish to thank all colleagues and friends who contributed to the discussion of this material on this and other occasions, especially David Sabean, Robert Brenner, Ramachandra Guha, Salma Jayyusi, and James Scott.

Notes

1. Hans Medick, *Weben und Überleben in Laichingen, 1650–1900: Lokalgeschichte als Allgemeine Geschichte* (Göttingen, 1996). A second, slightly revised edition appeared in 1997.
2. Fernand Braudel, "Histoire et sociologie," in *Traité de sociologie,* ed. George Gurvitch, 2 vols. (Paris, 1958–60), vol. 1, p. 86.
3. Carlo Ginzburg, "Microhistory: Two or Three Things That I Know about It," *Critical Inquiry* 20 (1996): 13 ff.
4. Translated by the author from the German edition: Raymond Queneau, *Die blauen Blumen* (Frankfurt, 1985), p. 73.
5. Reinhart Koselleck, "Erfahrungswandel und Methodenwechsel: Eine historisch-

anthropologische Skizze," in *Zeitschichten: Studien zur Historik* (Frankfurt, 2000), pp. 27–77.

6. Carlo Ginzburg and Carlo Poni, "Was ist Mikrogeschichte?" *Geschichtswerkstatt* 6 (1985): 48.
7. Christian Meier, "Notizen zum Verhältnis von Makro- und Mikrogeschichte," in *Teil und Ganzes: Zum Verhältnis von Einzel- und Gesamtanalyse in Geschichts- und Sozialwissenschaften,* ed. Karl Acham and Winfried Schulze (Munich, 1990), pp. 120 and 122.
8. Jürgen Kocka, "Sozialgeschichte zwischen Struktur und Erfahrung: Die Herausforderung der Alltagsgeschichte," in *Geschichte und Aufklärung* (Göttingen, 1989), p. 43.
9. Heinrich von Treitschke, *Deutsche Geschichte im 19. Jahrhundert,* part 3, 5th ed. (Leipzig, 1903), p. 721; Hans-Ulrich Wehler, "Alltagsgeschichte: Königsweg zu neuen Ufern oder Irrgarten der Illusionen?" in *Aus der Geschichte lernen? Essays* (Munich, 1988), p. 144.
10. Giovanni Levi, "On Microhistory," in *New Perspectives on Historical Writing,* ed. Peter Burke (Oxford, 1991), p. 93.
11. Medick, *Weben und Überleben in Laichingen,* p. 24.
12. Emmanuel Le Roy Ladurie, *Les Paysans de Languedoc,* 2 vols. (Paris, 1966), vol. 1, p. 11.
13. For a recent assessment of the international debate on proto-industrialization, see Peter Kriedte, Hans Medick, and Jürgen Schlumbohm, *Eine Forschungslandschaft in Bewegung: Die Proto-Industrialisierung am Ende des 20. Jahrhunderts,* in *Jahrbuch für Wirtschaftsgeschichte,* no. 2 (1998), Special Issue on Proto-Industrialization, pp. 9–21.
14. Edoardo Grendi, "Micro-analisi e storia sociale," *Quaderni Storici* 35 (1977): 512.
15. David Sabean, "Intensivierung der Arbeit und Alltagserfahrung auf dem Lande—ein Beispiel aus Württemberg," *Sozialwissenschaftliche Informationen für Unterricht und Studium* 6 (1977): 151. On Immanuel Gottlob Brastberger and other influential writers of pietist devotional treatises in eighteenth-century Württemberg, see chapter 6 of my book *Weben und Überleben in Laichingen,* especially pp. 532 ff.
16. Max Weber, *Gesammelte Aufsätze zur Religionssoziologie,* vol. 1 (Tübingen, 1920).
17. Siegfried Kracauer, *Geschichte—vor den letzten Dingen* (Frankfurt, 1971), pp. 115–16.
18. Kracauer uses this concept critically in a letter to Theodor W. Adorno written on March 3, 1964. It is directed against what Kracauer thought was an overspeculative historical-philosophical emphasis in Adorno's cultural criticism. The letter is quoted in Dagmar Barnow, "An den Rand geschriebene Träume: Kracauer über Zeit und Geschichte," in *Siegfried Kracauer: Neue Interpretationen,* ed. Michael Kessler and Thomas Levin (Tübingen, 1990), p. 6.

References

Barnow, Dagmar. "An den Rand geschriebene Träume: Kracauer über Zeit und Geschichte." In *Siegfried Kracauer: Neue Interpretationen,* ed. Michael Kessler and Thomas Levin. Tübingen, 1990.

Braudel, Fernand. "Histoire et sociologie." In *Traité de sociologie,* ed. George Gurvitch. 2 vols. Paris, 1958–60.

Ginzburg, Carlo. "Microhistory: Two or Three Things That I Know about It." *Critical Inquiry* 20 (1996): 10–36.

Ginzburg, Carlo, and Carlo Poni. "Was ist Mikrogeschichte?" *Geschichtswerkstatt* 6 (1985): 48–52.

Grendi, Edoardo. "Micro-analisi e storia sociale." *Quaderni Storici* 35 (1977): 506–20.

Kocka, Jürgen. "Sozialgeschichte zwischen Struktur und Erfahrung: Die Herausforderung der Alltagsgeschichte." In *Geschichte und Aufklärung.* Göttingen, 1989.

Koselleck, Reinhart. "Erfahrungswandel und Methodenwechsel: Eine historisch-anthropologische Skizze." In *Zeitschichten: Studien zur Historik.* Frankfurt, 2000.

Kracauer, Siegfried. *Geschichte—vor den letzten Dingen.* Frankfurt, 1971.

Kreidte, Peter, Hans Medick, and Jürgen Schlumbohm. *Eine Forschungslandschaft in Bewegung: Die Proto-Industrialisierung am Ende des 20. Jahrhunderts.* In Special Issue on Proto-Industrialization, *Jahrbuch für Wirtschaftsgeschichte,* no. 2 (1998): 9–21.

Le Roy Ladurie, Emmanuel. *Les Paysans de Languedoc.* 2 vols. Paris, 1966.

Levi, Giovanni. "On Microhistory." In *New Perspectives on Historical Writing,* ed. Peter Burke. Oxford, 1991.

Medick, Hans. *Weben und Überleben in Laichingen, 1650–1900: Lokalgeschichte als Allgemeine Geschichte.* Göttingen, 1996.

Meier, Christian. "Notizen zum Verhältnis von Makro- und Mikrogeschichte." In *Teil und Ganzes: Zum Verhältnis von Einzel- und Gesamtanalyse in Geschichts- und Sozialwissenschaften,* ed. Karl Acham and Winfried Schulze. Munich, 1990.

Queneau, Raymond. *Die blauen Blumen.* Frankfurt, 1985.

Sabean, David. "Intensivierung der Arbeit und Alltagserfahrung auf dem Lande—ein Beispiel aus Württemberg." *Sozialwissenschaftliche Informationen für Unterricht und Studium* 6 (1977): 148–52.

Treitschke, Heinrich von. *Deutsche Geschichte im 19. Jahrhundert.* Part 3. 5th ed. Leipzig, 1903.

Weber, Max. *Gesammelte Aufsätze zur Religionssoziologie.* Vol. 1. Tübingen, 1920.

Wehler, Hans-Ulrich. "Alltagsgeschichte: Königsweg zu neuen Ufern oder Irrgarten der Illusionen?" In *Aus der Geschichte lernen? Essays.* Munich, 1988.

Contributors

DAVID ARNOLD is professor of South Asian history at the School of Oriental and African Studies in London. His work, which has ranged widely over the political and social history of modern India, includes *Science, Technology, and Medicine in Colonial India* (Cambridge University Press, 2000) and *Colonizing the Body: State Medicine and Epidemic Disease in Nineteenth-Century India* (University of California Press, 1993).

NINA BHATT has worked as a rural sociologist for the World Bank in Nepal and Washington, D.C., and as a micro-enterprise consultant for the Ford Foundation in India. She is completing a doctorate in cultural anthropology at Yale University.

HERMAN E. DALY is professor at the University of Maryland School of Public Affairs. From 1988 to 1994 he was Senior Economist at the World Bank. His books include *Steady-State Economics* (1977; reprint, Island Press, 1991), *Valuing the Earth* (MIT Press, 1992), *Beyond Growth* (Beacon Press, 1996), and *Ecological Economics and the Ecology of Economics* (Edward Elgar, 1999). He is co-author with theologian John B. Cobb, Jr., of *For the Common Good* (1989; reprint, Beacon Press, 1994).

PAUL GREENOUGH is professor of modern Indian history at the University of Iowa. His research interests include Indian social responses to famine, the history of Indian immunization, and the history of Indian environmentalism. Greenough is the author of *Prosperity and Misery in Modern Bengal* (Oxford University Press, 1982) and the co-editor (with Anna L. Tsing) of the forthcoming book *Imagination and Distress in Southern Asian Projects.*

RONALD J. HERRING is the John S. Knight Professor of International Relations and professor of government at Cornell University, where he is director of the Mario Einaudi Center for International Studies. His recent work centers on issues of authority in nature, environmental policy, and economic liberalization in South Asia. He is the author of *Land to the Tiller: The Political Economy of Agrarian Reform in South Asia* (Yale University Press and Oxford University Press, 1983), and he, along with Milton Esman, is the editor of *Carrots, Sticks, and Ethnic Conflict: Rethinking Development Assistance* (University of Michigan Press).

PETER JONES is professor of French history at the University of Birmingham, England. He has published a number of works on the agrarian history of France in the eighteenth and nineteenth centuries, including *Politics and Rural Society: The Southern Massif Central, 1750–1880* (Cambridge University Press, 1985) and *The Peasantry in the French Revolution* (Cambridge University Press, 1988). His more recent research explores agrarian issues in the context of state formation, and he is completing a comparative study of how six French villages responded to the challenge of reform and revolution in the period 1760–1820.

DAVID LUDDEN is professor of history at the University of Pennsylvania, where he teaches graduate and undergraduate courses in South Asian history, world history, and comparative social and economic history. His most recent book is *An Agrarian History of South Asia* (Cambridge, 1999), and he is the editor of *Reading Subaltern Studies: Critical Histories, Contested Meaning, and the Globalization of South Asia* (Permanent Black, forthcoming). His new research concerns the formation of modern agrarian economies in southern India.

HANS MEDICK is professor of modern history and historical anthropology at the University of Erfurt and fellow at the Max-Planck-Institut für Geschichte, Göttingen. He is co-founder and co-editor of the journal *Historische Anthropologie: Kultur–Gesellschaft–Alltag.* He is the co-editor with Benigna von Krusenstjern of *Zwischen Alltag und Katastrophe: Der Dreißigjährige Krieg aus der Nähe* (Göttingen, 1999), to which he contributed "Historisches Ereignis und zeitgenössische Erfahrung—die Eroberung und Zerstörung Magdeburgs 1631" and the introductory essay (with Benigna von Krusenstjern) "Die Nähe und Ferne des Dreißigjährigen Krieges"; and the co-editor with Anne-Charlott Trepp of *Geschlechtergeschichte und Allgemeine Geschichte: Herausforderungen und Perspektiven* (Göttingen, 1998).

HERMANN REBEL is associate professor of history at the University of Arizona, where he teaches European social and cultural history and interdisciplinary studies (history, economics, and anthropology). His publications include *Peasant*

Classes: The Bureaucratization of Property and Family Relations under Early Habsburg Absolutism, 1511–1636 (Princeton, 1983). He is currently working on a monograph that applies concepts of franchise bidding and labor hoarding to peasant household inventories and other sources from Austria in the period 1650–1850 in order to better understand the operations of peasant family firms.

JAMES C. SCOTT is director of the Program in Agrarian Studies and professor of political science and anthropology at Yale University. He is author of *The Moral Economy of the Peasant* (1976); *Weapons of the Weak: Everyday Forms of Peasant Resistance* (1985); *Domination and the Arts of Resistance: Hidden Transcripts* (1990); and *Seeing like a State* (1997), all published by Yale University Press.

PETER TAYLOR is associate professor of history at Dominican University, River Forest, Illinois, and the author of *Indentured to Liberty: Peasant Life and the Hessian Military State, 1688–1815* (Cornell University Press, 1994), an extension of the article published in this volume. He is currently working on a study of the death registers of the same parish that provided the main casework for the earlier volume.

LIANA VARDI is associate professor of history at the State University of New York at Buffalo. She is the author of a study of proto-industrialization, *The Land and the Loom: Peasants and Profit in Northern France, 1680–1800* (Duke University Press, 1993) and is completing a study on the Physiocrats.

Index